W9-ADT-868

Nonlinear Semigroups

Recent Titles in This Series

(*Continued in the back of this publication*)

Translations of

MATHEMATICAL
MONOGRAPHS

Volume 109

Nonlinear Semigroups

Isao Miyadera

Translated by
Choong Yun Cho

American Mathematical Society
Providence, Rhode Island

非線形半群

HISENKEI HANGUN (Nonlinear Semigroups)
by Isao Miyadera
Copyright © 1977 by Isao Miyadera
Originally published in Japanese by Kinokuniya Company Ltd., Publishers,
Tokyo in 1977

Translated from the Japanese by Choong Yun Cho

1991 *Mathematics Subject Classification.* Primary 47H20, 47H06; Secondary 35F25, 35G25.

ABSTRACT. This book provides a systematic exposition of the theory of nonlinear semigroups in Banach spaces and an introduction to nonlinear evolution equations. Chapters 2 and 3 present the basic properties of dissipative operators and nonlinear contraction semigroups in Banach spaces respectively. Chapter 4 is devoted to the generation of nonlinear contraction semigroups—the Crandall-Liggett theorem and the Komura theorems, etc. Chapter 5 develops the convergence of difference approximations of Cauchy problems for ω-dissipative operators and then the Kobayashi generation theorem of nonlinear semigroups. Among the topics covered are perturbations of nonlinear semigroups and their application to nonlinear evolution equations; the convergence and approximation of nonlinear semigroups and its application to first order quasilinear equations.

Library of Congress Cataloging-in-Publication Data

Miyadera, Isao, 1925–
  [Hisenkei hangun. English]
  Nonlinear semigroups/Isao Miyadera; translated by Choong Yun Cho.
  p. cm.—(Translations of mathematical monographs, ISSN 0065-9282; v. 109)
  Translation of: Hisenkei hangun.
  Includes bibliographical references and index.
  ISBN 0-8218-4565-9
  1. Semigroups of operators. 2. Nonlinear operators. I. Title. II. Series.
QA329.8.M5813   1992   92-11318
515′.7248—dc20   CIP

Copyright © 1992 by the American Mathematical Society. All rights reserved.
Translation authorized by the
Kinokuniya Company Ltd.
The American Mathematical Society retains all rights
except those granted to the United States Government.
Printed in the United States of America

Information on Copying and Reprinting can be found at the back of this volume.

The paper used in this book is acid-free and falls within the guidelines established to ensure permanence and durability. ∞

This publication was typeset using $\mathcal{AMS}$-TEX,
the American Mathematical Society's TEX macro system.

10 9 8 7 6 5 4 3 2 1   97 96 95 94 93 92

Cstack
JLIMON
12-2-92

Contents

Preface

The theory of generation of semigroups of linear contractions, which is the basis of the evolution equation, was developed by Hille and Yosida in 1948. Through this theory the existence and uniqueness for a solution of Cauchy's problem for $(d/dt)u(t) = Au(t)$ with $u(0) = x \in D(A)$ were proved for an m-dissipative operator A that has a dense domain in a Banach space. The theory of linear semigroups was further deepened by results of Phillips and many other mathematicians, and the theory of linear semigroups has now secured its position as an important area in the field of analysis. In 1953 Kato extended the theory of Hille-Yosida to the case where A depends on time t. Afterward, the theory of evolution equations of parabolic type was brought forth by Kato and Tanabe, and the theory of linear evolution equations has made marked progress. Moreover, during the first half of the 1960s the semilinear evolution equation with a nonlinear perturbation term $(d/dt)u(t) = A(t) + f(t, u(t))$ was studied by Segal, Browder and Kato, who obtained excellent results.

Under this historical background, Komura attracted much attention in 1967 when he announced the theory of generation of nonlinear semigroups in a Hilbert space. The theory was immediately extended by Kato to the case of a Banach space with a uniformly convex conjugate space. Afterward, in 1971, Crandall-Liggett obtained the splendid result that "in a general Banach space, an arbitrary m-dissipative operator always generates a semigroup of (nonlinear) contractions." This, together with the work of Komura, has become the basis for the study of nonlinear evolution equations. In addition to the problem of generation, the study of convergence and perturbation of semigroups, and concrete applications of the theory to nonlinear partial differential equations have been carried out vigorously by many mathematicians within and without the country. Developments in the field have led to remarkable progress during the last decade.

This book is concerned mainly with the theory of nonlinear semigroups in a Banach space. In Chapter 1, we summarize the basic results of functional analysis that are necessary for later chapters; Chapter 2 describes the dissipative operators that are closely related to semigroups of contractions;

Chapter 3 presents several properties of the semigroup of contractions. In Chapter 4, we discuss the theory of generation of semigroups of Crandall-Liggett (§§4.1–2), its applications (§4.4), and also the theory of generation of Komura and its extension (§4.3). In Chapter 5 we examine the convergence of the difference approximation to Cauchy's problem for the nonlinear evolution equation $(d/dt)u(t) \in Au(t)$, and we describe an extension of the generation theorem of Crandall-Liggett. In first half of Chapter 6 we describe the theory of convergence and approximation of semigroups. The second half contains the perturbation theory of semigroups and its application to the evolution equation $(d/dt)u(t) = A(t)u(t)$. As an application of the theorems of convergence and approximation of the semigroups given in Chapter 6, we introduce in Chapter 7 the work of Oharu and Takahashi on Cauchy's problem for quasilinear partial differential equations of first order.

As a whole, I have emphasized the explanation of the theoretical part and attempted to prove theorems with care so that the reader with a knowledge of functional analysis at the level of a college junior is able to fully understand the material. (By doing so, I am afraid that my writing might have become wordy.) Because of this and the limitation on the size of the book and also my inability, I have not described many examples. From the standpoint of keeping the discussion focused on Banach spaces, I have not touched on the subdifferential evolution equation at all. The reader may refer to the literature quoted in the postscript of this book.

Finally, I wish to express my profound gratitude to Professors A. Ichida and S. Izumi for their guidance over the years since my school days, and my sincere thanks to Professor S. Itô of the University of Tokyo, who recommended that I write this book. I am deeply grateful also to my friends: Mr. S. Oharu, who has given me much valuable advice, and those who conducted seminars on topics related to evolution equations every Thursday at Waseda University and studied together. I also wish to thank Mr. K. Yokota of the publication division of the Kinokuniya Publishing Company for his continuous assistance in the publication of this book.

Isao Miyadera
January 1977

CHAPTER 1

Basic Results of Functional Analysis

In this chapter we describe basic results of functional analysis that are needed later. Many of the theorems are given without proof. For the details, refer, for example, to Dunford-Schwartz [45], Hille-Phillips [49], and Yosida [132] (also Miyadera [97] for §3).

§1. Banach spaces

DEFINITION 1.1. Let Φ be the field of complex or real numbers. A set X satisfying the following properties (I) through (III) is called a Banach space.

(I) X is a linear space with Φ as the field of scalars.

(II) To each element x in X there corresponds a real number $\|x\|$, satisfying the following three conditions:

$$\|x\| \geq 0; \qquad \|x\| = 0 \rightleftarrows x = 0, \tag{1.1}$$

$$\|\alpha x\| = |\alpha| \, \|x\| \qquad (\alpha \in \Phi, \, x \in X), \tag{1.2}$$

$$\|x + y\| \leq \|x\| + \|y\| \qquad (x, y \in X). \tag{1.3}$$

$\|x\|$ is called the norm of x.

(III) X is complete; that is, for any sequence $\{x_n\}$ in X that satisfies $\lim_{m,n\to\infty} \|x_m - x_n\| = 0$ (we call such a sequence a Cauchy sequence), there exists an $x \in X$ such that $\lim_{n\to\infty} \|x_n - x\| = 0$. (We say the sequence $\{x_n\}$ converges to x, and write $\lim_{n\to\infty} x_n = x$, or $x_n \to x$ as $n \to \infty$, and call x the limit of the sequence $\{x_n\}$.)

In the above definition, when Φ is the field of complex numbers, we call X a complex Banach space; and, when Φ is the field of real numbers, we call X a real Banach space.

Unless specified otherwise, X denotes a Banach space (either complex or real).

DEFINITION 1.2. (i) $X_0 \subset X$ is called a linear set if $\alpha x + \beta y \in X_0$ for all $\alpha, \beta \in \Phi$ and $x, y \in X_0$.

(ii) $\alpha_1 x_1 + \alpha_2 x_2 + \cdots + \alpha_n x_n$ with $x_i \in X$ $(i = 1, 2, \ldots, n)$ $(\alpha_i \in \Phi)$ is called a linear combination of $x_1, x_2, \ldots, x_n$. For $S \subset X$, the set consisting of all linear combinations of elements of S is obviously a linear set. We say this linear set is generated by S (or spanned by S).

1

DEFINITION 1.3. (i) $X_0 \subset X$ is called a convex set if $\alpha x + (1 - \alpha)y \in X_0$ for all α with $0 < \alpha < 1$ and all elements $x, y \in X_0$.

(ii) $\alpha_1 x_1 + \alpha_2 x_2 + \cdots + \alpha_n x_n$ with $x_i \in X (i = 1, 2, \ldots, n)$, $\alpha_i \geq 0$, and $\alpha_1 + \alpha_2 + \cdots + \alpha_n = 1$, is called a convex combination of $x_1, x_2, \ldots, x_n$. For $S \subset X$, the set consisting of all convex combinations of elements of S is obviously a convex set. We denote this convex set by $\operatorname{co} S$, and we say that the convex set is generated by S (or spanned by S). The closure $\overline{\operatorname{co} S}$ of $\operatorname{co} S$ also forms a convex set. We denote this closed convex set by $\overline{\operatorname{co}} S$.

DEFINITION 1.4. Let X and Y be Banach spaces, $X_0 \subset X$, and $Y_0 \subset Y$. For an operator T that has X_0 as its domain and takes values in Y_0, we write $T: X_0 \to Y_0$. (Also, we denote the domain of the operator T and its range by $D(T)$ and $R(T)$, respectively. For the case above we have $D(T) = X_0$, $R(T) \subset Y_0$.)

(i) An operator $T: X_0 \to Y_0$ such that

$$\|Tx - Ty\| \leq \|x - y\| \qquad (x, y \in X_0) \tag{1.4}$$

is called a contraction operator.

(ii) An operator $T: X_0 \to Y_0$ is called a strict contraction if there exists a constant α with $0 \leq \alpha < 1$ such that

$$\|Tx - Ty\| \leq \alpha\|x - y\| \qquad (x, y \in X_0), \tag{1.5}$$

where the norm on the left-hand sides of (1.4) and (1.5) is the norm on the space Y.

We have the following fixed point theorem.

THEOREM 1.1. *Let* X_0 *be a closed subset of* X. *If* $T: X_0 \to X_0$ *is a strict contraction operator, then* T *has a unique fixed point; that is, there exists a unique* $x \in X_0$ *such that* $Tx = x$.

PROOF. There exists $0 \leq \alpha < 1$ such that

$$\|Tx - Ty\| \leq \alpha\|x - y\| \qquad (x, y \in X_0).$$

Now, choose $x_0 \in X_0$ arbitrarily and set $x_n = T^n x_0$ $(n = 1, 2, \ldots)$. Then, from

$$\|x_{n+p} - x_n\| \leq \sum_{k=n}^{n+p-1} \|x_{k+1} - x_k\| \leq \left(\sum_{k=n}^{n+p-1} \alpha^k \right) \|Tx_0 - x_0\| \qquad (n, p \geq 1)$$

it follows that $\{x_n\}$ $(\subset X_0)$ is a Cauchy sequence. Therefore $\{x_n\}$ converges. Set $\lim_{n \to \infty} x_n = x$; then $x \in X_0$ is a unique fixed point of T. □

DEFINITION 1.5. Let X be a Banach space.

(i) We say that X is strictly convex if $\|x + y\| \neq \|x\| + \|y\|$ for all linearly independent elements x and y of X.

(ii) If $\|x_n\| = \|y_n\| = 1$ and $\lim_{n \to \infty} \|x_n + y_n\| = 2$, where $x_n, y_n \in X$, imply that $\lim_{n \to \infty} \|x_n - y_n\| = 0$, then we say that X is uniformly convex.

As we can easily see, a necessary and sufficient condition for X to be strictly convex is that $x \neq y$, $\|x\| = \|y\| = 1$, imply that $\|\alpha x + (1-\alpha)y\| < 1$ for all α with $0 < \alpha < 1$. Therefore, X being strictly convex means that the points on the segment connecting two distinct points on the surface of the unit sphere in X are contained in the interior of the unit sphere $\{x \in X; \|x\| \leq 1\}$ except for the end points. A uniformly convex Banach space X is strictly convex. In fact, let X be uniformly convex and suppose $\|x + y\| = \|x\| + \|y\|$, where $x \neq 0$ and $y \neq 0$. Without loss of generality we can assume $\|x\| \leq \|y\|$. Then, from

$$\|x/\|x\| + y/\|y\| \| = (\|x\| \|y\|)^{-1} \| \|y\|(x + y) - (\|y\| - \|x\|)y\|$$

$$\geq (\|x\| \|y\|)^{-1} [\|y\|(\|x\| + \|y\|) - (\|y\| - \|x\|)\|y\|] = 2,$$

we have $\|x/\|x\| + y/\|y\| \| = 2$; and $\|x/\|x\| - y/\|y\| \| = 0$, since X is uniformly convex. Hence, we obtain $\|y\|x = \|x\|y$, and x and y are not linearly independent. Therefore, X is strictly convex.

DEFINITION 1.6. Let Φ be the field of complex numbers (the field of real numbers). If a set H satisfies I through III below, then we call H a Hilbert space (a real Hilbert space).

(I) H is a linear space with Φ as the field of scalars.

(II) To each pair $\{x, y\}$ of elements of H there corresponds a number $(x, y) \in \Phi$ satisfying the following:

$$(x, x) \geq 0 \quad \text{and} \quad (x, x) = 0 \rightleftarrows x = 0, \tag{1.6}$$

$$(x, y) = \overline{(y, x)}, \tag{1.7}$$

(where $\overline{(y, x)}$ denotes the complex conjugate of (y, x)),

$$(x + z, y) = (x, y) + (z, y) \qquad (x, y, z \in H), \tag{1.8}$$

$$(\alpha x, y) = \alpha(x, y) \qquad (x, y \in H, \alpha \in \Phi). \tag{1.9}$$

(x, y) is called the inner product of x and y. If we set $\|x\| = \sqrt{(x, x)}$ $(x \in H)$, then $|(x, y)| \leq \|x\| \|y\|$ (the Schwarz inequality) holds, and we see that the norm $\| \cdot \|$ satisfies the conditions (1.1) through (1.3).

(III) H is complete with respect to the norm $\|x\| = \sqrt{(x, x)}$ introduced above.

As we see from the definition, a Hilbert space is a Banach space. Also, in any Hilbert space H the parallelogram law

$$\|x + y\|^2 + \|x - y\|^2 = 2\|x\|^2 + 2\|y\|^2 \qquad (x, y \in H) \tag{1.10}$$

holds. From this, we see that a Hilbert space is a uniformly convex Banach space.

§2. Conjugate spaces and the weak topology

An operator that takes values in the field of real or complex numbers is called a functional. Let X be a Banach space with Φ as the field of scalars.

If a functional $f: X \to \Phi$ is continuous (that is, if $\lim_{n \to \infty} x_n = x$ implies $\lim_{n \to \infty} f(x_n) = f(x)$) and satisfies

$$f(\alpha x + \beta y) = \alpha f(x) + \beta f(y) \qquad (\alpha, \beta \in \Phi; x, y \in X), \qquad (1.11)$$

then f is called a continuous linear functional. As is well known, if f is a continuous linear functional, then

$$\sup_{x \neq 0} |f(x)|/\|x\| < \infty, \qquad (1.12)$$

and

$$\sup_{x \neq 0} |f(x)|/\|x\| = \sup_{\|x\| \leq 1} |f(x)| = \sup_{\|x\| = 1} |f(x)|.$$

Conversely, it is easy to see that a functional $f: X \to \Phi$, which satisfies (1.11) and (1.12), is a continuous linear functional. Because of this, we also call a continuous linear functional a bounded linear functional.

For a bounded linear functional f, the norm $\|f\|$ of f is defined by

$$\|f\| = \sup_{x \neq 0} |f(x)|/\|x\|.$$

The set of all bounded linear functionals (defined on X) forms a Banach space with norm $\|f\|$ introduced above. We denote this Banach space by X^* and call it the conjugate space of X. We sometimes use $x^*, y^*, \ldots$, instead of f, to denote a bounded linear functional. We denote the conjugate space $(X^*)^*$ of X^* by X^{**}, and its elements by $x^{**}, y^{**}, \ldots$.

For $x \in X$, a functional $x_x^{**}: X^* \to \Phi$ is defined by

$$x_x^{**}(x^*) = x^*(x) \qquad (x^* \in X^*).$$

Then x_x^{**} is a bounded linear functional. Therefore, $x_x^{**} \in X^{**}$. Furthermore, $x_{\alpha x + \beta y}^{**} = \alpha x_x^{**} + \beta x_y^{**}$ $(\alpha, \beta \in \Phi; x, y \in X)$ and

$$\|x_x^{**}\| \left(= \sup_{x^* \neq 0} |x_x^{**}(x^*)|/\|x^*\| = \sup_{x^* \neq 0} |x^*(x)|/\|x^*\| \right) = \|x\| \qquad (x \in X)$$

holds. From this, we can identify x_x^{**} with x, and consider X as a closed linear subset of X^{**}.

DEFINITION 1.7. If $X = X^{**}$, then X is said to be reflexive. In other words, a Banach space X is reflexive, if for every $x^{**} \in X^{**}$ there exists an $x \in X$ such that

$$x^{**}(x^*) = x^*(x) \qquad (x^* \in X^*).$$

We have the following

THEOREM 1.2. *A necessary and sufficient condition for a Banach space X to be reflexive is that its conjugate space X^* is reflexive.*

THEOREM 1.3. *A uniformly convex Banach space is reflexive.*

Next, we describe the weak topology and the weak* topology. Let x_0 be an arbitrary element of a Banach space X, and let

$$U(x_0; x_1^*, x_2^*, \ldots, x_n^*, \varepsilon) = \{x \in X; |x_i^*(x - x_0)| < \varepsilon \ (i = 1, 2, \ldots, n)\}$$

for $x_i^* \in X^*$ $(i = 1, 2, \ldots, n)$, $n = 1, 2, \ldots$, and $\varepsilon > 0$. Then, since the class of all sets obtained from the above set by arbitrarily changing $x_1^*, x_2^*, \ldots, x_n^*$, n and $\varepsilon > 0$ satisfies the axioms for a neighborhood system, we can introduce a topology on X with these sets as the neighborhoods of x_0. We call this topology the weak topology on X. The weak topology satisfies the Hausdorff separation axiom but does not satisfy the first axiom of countability. The topology on X introduced by the norm is called the strong topology (in order to distinguish it from the weak topology). The weak topology is weaker than the strong topology. When a sequence $\{x_n\}$ in X converges to x with respect to the weak topology, we say that the sequence $\{x_n\}$ converges weakly to x, and express by $\text{w-}\lim_{n\to\infty} x_n = x$. We see easily that a necessary and sufficient condition for $\{x_n\}$ to converge weakly to x is that $\lim_{n\to\infty} x^*(x_n) = x^*(x)$ holds for arbitrary $x^* \in X^*$. If $\{x_n\}$ converges weakly to x, then $\{\|x_n\|\}$ is a bounded sequence of numbers and $\|x\| \leq \liminf_{n\to\infty} \|x_n\|$ holds. In contrast to weak convergence, when a sequence $\{x_n\}$ converges to x with respect to the strong topology, that is, when $\lim_{n\to\infty} \|x_n - x\| = 0$, then we say the sequence $\{x_n\}$ converges strongly to x.

We have the following

THEOREM 1.4. *A necessary and sufficient condition for a Banach space X to be reflexive is that the unit sphere in X be compact with respect to the weak topology.*

If every sequence $\{x_n\}$ in X_0 $(\subset X)$ has a subsequence that converges weakly to some point of X, then X_0 is said to be weakly sequentially compact. We have

THEOREM 1.5. *Let X be a reflexive Banach space. A necessary and sufficient condition for a subset X_0 of X to be weakly sequentially compact is that X_0 be a bounded set (that is, $\sup\{\|x\|; x \in X_0\} < \infty$).*

If $\lim_{n\to\infty} x_n = x$, then $\text{w-}\lim_{n\to\infty} x_n = x$, and $\lim_{n\to\infty} \|x_n\| = \|x\|$. Conversely,

THEOREM 1.6. *Let X be a uniformly convex Banach space, and let x, $x_n \in X$ $(n = 1, 2, \ldots)$. If $\text{w-}\lim_{n\to\infty} x_n = x$, and $\lim_{n\to\infty} \|x_n\| = \|x\|$, then $\lim_{n\to\infty} x_n = x$.*

PROOF. Assume $\text{w-}\lim_{n\to\infty} x_n = x$ and $\lim_{n\to\infty} \|x_n\| = \|x\|$. Let $x = 0$. Then $\lim_{n\to\infty} x_n = 0$. Let $x \neq 0$. Then $\text{w-}\lim_{n\to\infty} (x_n/\|x_n\| + x/\|x\|) = 2x/\|x\|$. Therefore,

$$2 = \|2x/\|x\| \| \leq \liminf_{n\to\infty} \|x_n/\|x_n\| + x/\|x\| \|$$

$$\leq \limsup_{n\to\infty} \|x_n/\|x_n\| + x/\|x\| \| \leq 2,$$

that is, $\|x_n/\|x_n\| + x/\|x\| \| \to 2$ as $n \to \infty$. Then from the uniform convexity of X, we have $\|x_n/\|x_n\| - x/\|x\| \| \to 0$ as $n \to \infty$. Also,

$\|x_n/\|x_n\| - x_n/\|x\| \| = \|x\|^{-1} |\|x_n\| - \|x\| | \to 0$ as $n \to \infty$. From this, we have

$$\|x\|^{-1}\|x_n - x\| \le \|x_n/\|x\| - x_n/\|x_n\| \| + \|x_n/\|x_n\| - x/\|x\| \| \to 0 \qquad (n \to \infty)$$

and hence $\lim_{n\to\infty} \|x_n - x\| = 0$. $\square$

REMARK. In Theorem 1.6 above, we can replace the assumption that $\lim_{n\to\infty} \|x_n\| = \|x\|$ by $\limsup_{n\to\infty} \|x_n\| \le \|x\|$, since we can obtain $\|x\| \le \liminf_{n\to\infty} \|x_n\|$ from w-$\lim_{n\to\infty} x_n = x$.

If a subset X_0 of X is closed with respect to the weak topology, then X_0 is said to be a weakly closed set. On the other hand, the term *closed set* is reserved for sets that are closed with respect to the strong topology. A weakly closed set is always a closed set. Conversely,

THEOREM 1.7. *A closed convex subset of a Banach space* X *is weakly closed.*

COROLLARY 1.8. *If a sequence* $\{x_n\}$ *in* X *converges weakly to* x, *then there is a sequence in* $\mathrm{co}\{x_1, x_2, \dots, x_n, \dots\}$ *which converges (strongly) to* x.

We let x_0^* be an arbitrary element of X^* (the conjugate space of X), and set

$$U(x_0^*; x_1, x_2, \dots, x_n, \varepsilon)$$
$$= \{x^* \in X^*; |x^*(x_i) - x_0^*(x_i)| < \varepsilon (i = 1, 2, \dots, n)\}$$

for $x_i \in X$ $(i = 1, 2, \dots, n)$, $n = 1, 2, \dots$, and $\varepsilon > 0$. The class of all sets obtained by arbitrarily changing $x_1, x_2, \dots, x_n$, n, and ε in the above set satisfies the axioms for a neighborhood system. Therefore, we can introduce a topology on X^* with these sets as the neighborhoods of x_0^*. We call this topology the weak* topology of X^*. The weak* topology satisfies the separation axiom of Hausdorff but does not satisfy the first axiom of countability. For X^*, we can think of the weak topology and the weak* topology mentioned above, but the weak* topology is weaker than the weak topology. If X is reflexive, then the two coincide. When a sequence $\{x_n^*\}$ in X^* converges to x^* with respect to the weak* topology, then $\{x_n^*\}$ is said to be weakly* convergent to x^*, and we write w*-$\lim_{n\to\infty} x_n^* = x^*$. A necessary and sufficient condition for $\{x_n^*\}$ to be weakly* convergent to x^* is that $\lim_{n\to\infty} x_n^*(x) = x^*(x)$ holds for every $x \in X$. If w*-$\lim_{n\to\infty} x_n^* = x^*$, then $\{\|x_n^*\|\}$ is a bounded sequence of numbers and $\|x^*\| \le \liminf_{n\to\infty} \|x_n^*\|$.

We have the following theorem.

THEOREM 1.9. *The unit sphere* $\{x^* \in X^*; \|x^*\| \le 1\}$ *in the conjugate space* X^* *of a Banach space* X *is compact with respect to the weak* topology.*

§3. Functions with values in a Banach space

Let Ω be an interval, (either finite or infinite) in $R^1 = (-\infty, \infty)$. We describe the integral and differential for functions which take their values in a Banach space X and have domain Ω.

We start with the definition of the Bochner integral. Let $\mathfrak{M}$ be the class of all Lebesgue measurable sets contained in Ω, and denote the Lebesgue measure of $A \in \mathfrak{M}$ by $m(A)$. $x(s) : \Omega \to X$ is called a simple function if there exist a countable number of $A_n \in \mathfrak{M}$ $(n = 1, 2, \ldots)$ which are mutually disjoint (no common points among them), such that $\Omega = \bigcup_{n=1}^{\infty} A_n$ and $x(s)$ is constant on each A_n.

DEFINITION 1.8. Given $x(s) : \Omega \to X$:

(i) If there exists a sequence of simple functions $\{x_n(s)\}$ such that $x(s) = \lim_{n \to \infty} x_n(s)$ for almost every s (a.e. s), then we say $x(s)$ is strongly measurable. Hence, a simple function is strongly measurable.

(ii) If $x^*(x(s))$ is a real- or complex-valued measurable function for all $x^* \in X^*$, then $x(s)$ is said to be weakly measurable.

We see from the definition that if $x(s)$ is strongly measurable, then $x(s)$ is weakly measurable and $\|x(s)\|$ is a (real-valued) measurable function. If the range $x(\Omega)$ of a function $x(s) : \Omega \to X$ is separable, then $x(s)$ is said to be separably valued; in addition, if there is a set A with measure 0 such that $x(\Omega \setminus A)$ is separable, then we say that $x(s)$ is almost separably valued.

THEOREM 1.10. *A necessary and sufficient condition for $x(s) : \Omega \to X$ to be strongly measurable is that $x(s)$ be weakly measurable and almost separably valued.*

From this we obtain the following corollary:

COROLLARY 1.11. *Let X be a separable Banach space. Then, a necessary and sufficient condition for $x(s) : \Omega \to X$ to be strongly measurable is that $x(s)$ be weakly measurable.*

Functions that are representable by a linear combination of two strongly measurable functions, a product of a real-valued (or complex-valued) measurable function that is finite-valued at a.e. $s \in \Omega$ and a strongly measurable function, and the weak limit of a sequence of strongly measurable functions are all strongly measurable functions.

DEFINITION 1.9. (i) Let $x(s) : \Omega \to X$ be a simple function. Then by the definition there is a sequence $\{x_n\} \subset X$ and a sequence of mutually disjoint measurable sets $\{A_n\}$ such that $x(s) = x_n$ $(s \in A_n)$ and $\Omega = \bigcup_{n=1}^{\infty} A_n$.

If $\|x(s)\|$ is Lebesgue integrable on Ω, we say that $x(s)$ is Bochner integrable on Ω and define the Bochner integral by

$$\int_{\Omega} x(s)\, dm = \sum_{n=1}^{\infty} x_n m(A_n) \qquad \left(= \lim_{k \to \infty} \sum_{n=1}^{k} m(A_n) x_n \right).$$

(From $\sum_{n=1}^{\infty} \|x_n\| m(A_n) = \int_\Omega \|x(s)\| \, dm < \infty$, it follows that $\sum_{n=1}^{k} m(A_n) x_n$ converges to an element of X as $k \to \infty$, and the limit does not depend on the representation of $x(s)$. Hence, this definition has meaning.)

(ii) Let $x(s): \Omega \to X$. If there is a sequence $\{x_n(s)\}$ of Bochner integrable simple functions on Ω such that

$$\lim_{n\to\infty} x_n(s) = x(s) \qquad (\text{a.e. } s),$$

and

$$\lim_{n\to\infty} \int_\Omega \|x(s) - x_n(s)\| \, dm = 0,$$

then we say that $x(s)$ is Bochner integrable on Ω. We define the Bochner integral of $x(s)$ by

$$\int_\Omega x(s) \, dm = \lim_{n\to\infty} \int_\Omega x_n(s) \, dm. \qquad (1.13)$$

(Since we can show that the limit on the right-hand side of the above equation exists, and the limit does not depend on the choice of a sequence of Bochner integrable simple functions that approximate $x(s)$, Definition (1.13) has meaning.) Furthermore, if $\Omega = (a, b)$ or $\Omega = [a, b]$, we will express $\int_\Omega x(s) \, dm$ by $\int_a^b x(s) \, ds$.

We know the following:

THEOREM 1.12. *A necessary and sufficient condition for $x(s): \Omega \to X$ to be Bochner integrable on Ω is that $x(s)$ is strongly measurable and also $\|x(s)\|$ is Lebesgue integrable on Ω.*

COROLLARY 1.13. *If $x(s): \Omega \to X$ is Bochner integrable on Ω, then for every $\varepsilon > 0$ there exists a sequence $\{A_n\}$ of measurable sets which are mutually disjoint such that $\Omega = \bigcup_{n=1}^{\infty} A_n$, and the following (1.14) holds:*

> *Let s_n be an arbitrary element of A_n and define $x_\varepsilon(s) = x(s_n)$ $(s \in A_n; n = 1, 2, \ldots)$. Then, $x_\varepsilon(s)$ is a Bochner integrable* $\qquad$ (1.14) *simple function on Ω and $\int_\Omega \|x(s) - x_\varepsilon(s)\| \, dm \leq \varepsilon$.*

We denote the set of all Bochner integrable functions on Ω by $L^1(\Omega; X)$. The following hold in the same way as in the case of the Lebesgue integral:

(i)
$$\left\| \int_\Omega x(s) \, dm \right\| \leq \int_\Omega \|x(s)\| \, dm \qquad (x(\cdot) \in L^1(\Omega; X)).$$

(ii) *If $x_i(\cdot) \in L^1(\Omega; X)$ and $\alpha_i \in \Phi$ $(i = 1, 2)$, then $\alpha_1 x_1(\cdot) + \alpha_2 x_2(\cdot) \in L^2(\Omega; X)$ and*

$$\int_\Omega [\alpha_1 x_1(s) + \alpha_2 x_2(s)] \, dm = \alpha_1 \int_\Omega x_1(s) \, dm + \alpha_2 \int_\Omega x_2(s) \, dm.$$

(iii) (Convergence theorem). *Let $x_n(\cdot) \in L^1(\Omega; X)$ $(n = 1, 2, \ldots)$ and $\lim_{n\to\infty} x_n(s) = x(s)$ for a.e. s. If there is a Lebesgue integrable function $g(s)$*

with $\|x_n(s)\| \le g(s)$ *(a.e. s; $n = 1, 2, \ldots$), then $x(\cdot) \in L^1(\Omega; X)$ and*

$$\lim_{n \to \infty} \int_\Omega x_n(s)\, dm = \int_\Omega x(s)\, dm.$$

(iv) *We define* $\|x(\cdot)\| = \int_\Omega \|x(s)\|\, dm$ *for* $x(\cdot) \in L^1(\Omega; X)$. *Then,* $L^1(\Omega; X)$
is a Banach space with the norm $\|x(\cdot)\|$.

(v) *Let* $x(\cdot) \in L^1(\Omega; X)$; *then for a.e.* $s \in \Omega$,

$$\lim_{h \to 0} h^{-1} \int_s^{s+h} \|x(t) - x(s)\|\, dt = 0$$

and hence

$$\lim_{h \to 0} h^{-1} \int_s^{s+h} x(t)\, dt = x(s). \; (^1)$$

Also, for the function of two variables $x(s, t): \Omega \times \Omega \to X$, *the theorem of
interchange of the order of integration holds just as for the Lebesgue integral.*

Next, we describe the concepts of continuity and differentiability of the
function $x(s): \Omega \to X$.

DEFINITION 1.10. Let Ω be an open interval, $x(s): \Omega \to X$, and $s_0 \in \Omega$.

(i) If w-$\lim_{s \to s_0} x(s) = x(s_0)$, i.e., $\lim_{s \to s_0} x^*[x(s)] = x^*[x(s_0)]$ holds for
every $x^* \in X^*$, then we say that $x(s)$ is weakly continuous at s_0.

(ii) If $\lim_{s \to s_0} x(s) = x(s_0)$, i.e., $\lim_{s \to s_0} \|x(s) - x(s_0)\| = 0$ holds, then
we say $x(s)$ is continuous at s_0. (We sometimes say strongly continuous in
contrast to weakly continuous.)

If $x(s)$ is weakly continuous (continuous) at each point of an open interval
Ω, then $x(s)$ is said to be weakly continuous (continuous) on Ω.

Also, when Ω is a closed interval (for example, $\Omega = [a, b]$), if $x(s)$
is continuous on the open interval (a, b), and $\lim_{s \to a+0} x(s) = x(a)(^2)$ and
$\lim_{s \to b-0} x(s) = x(b)(^3)$, then $x(s)$ is called continuous on the closed interval
$\Omega = [a, b]$. Similarly, we can define weak continuity of a function on a
closed interval. Obviously, if $x(s)$ is strongly continuous, then it is weakly
continuous.

THEOREM 1.14. (i) *If $x(s): \Omega \to X$ is weakly continuous on Ω, then $x(s)$
is strongly measurable.*

(ii) *Let $[a, b]$ be a bounded closed interval. If $x(s): [a, b] \to X$ is weakly
continuous on $[a, b]$, then $x(s)$ is Bochner integrable on $[a, b]$.*

REMARK. If $x(s)$ is continuous on a bounded closed interval $[a, b]$, then
from the above theorem, $x(s)$ is Bochner integrable on $[a, b]$. In this case,
we can define the integral of $x(s)$ on $[a, b]$ by the method of Riemann,

---

$(^1)$We set $\int_a^b x(s)\, ds = -\int_b^a x(s)\, ds$ if $a > b$.

$(^2), (^3)$ In this case, $x(s)$ is said to be right continuous at a, and left continuous at b.

and the integral defined by the method of Riemann can be shown to coincide with the Bochner integral.

DEFINITION 1.11. Let $x(s): (a, b) \to X$ and $s_0 \in (a, b)$:

(i) If w-$\lim_{h \to 0} h^{-1}[x(s_0 + h) - x(s_0)] = x_0$, then we say $x(s)$ is weakly differentiable at s_0 and x_0 is called the weak derivative of $x(s)$ at s_0.

(ii) If $\lim_{h \to 0} h^{-1}[x(s_0 + h) - x(s_0)] = x_0$, then we say $x(s)$ is strongly differentiable at s_0 and x_0 is called the strong derivative of $x(s)$ at s_0.

If $x(s)$ is strongly (weakly) differentiable at each point of (a, b), then we say $x(s)$ is strongly (weakly) differentiable on (a, b). We denote the strong derivative of $x(s)$ by $(d/ds)x(s)$ or $x'(s)$ or $Dx(s)$, and the weak derivative of $x(s)$ by (w-$d/ds)x(s)$ (or by $x'(s)$ or $Dx(s)$).

Moreover, if $\lim_{h \to 0+} h^{-1}[x(s_0 + h) - x(s_0)] = x_0$, then we say x_0 is the strong right derivative of $x(s)$ at s_0. We define the strong left derivative similarly. We denote the strong right (left) derivative of $x(s)$ by $D^+ x(s)$ $(D^- x(s))$. If $x(s): [a, b] \to X$ is strongly differentiable on (a, b), and has strong right derivative and strong left derivative at a and b, respectively, then we say $x(s)$ is strongly differentiable on $[a, b]$. We can introduce the concepts of the weak right (left) derivative and weakly differentiable on a closed interval similarly.

THEOREM 1.15. *Let $x(s): (a, b) \to X$ be continuous on (a, b). If there is the strong right derivative $D^+ x(s)$ at each $s \in (a, b)$, and $D^+ x(s)$ is continuous on (a, b), then $x(s)$ is strongly differentiable on (a, b).*

Finally, we describe theorems of the Radon-Nikodym type.

DEFINITION 1.12. Let $[a, b]$ be a bounded closed interval and let $x(s): [a, b] \to X$. $x(s)$ is called (strongly) absolutely continuous on $[a, b]$ if the following holds:

> For every $\varepsilon > 0$ there exists a $\delta > 0$ such that if $[a_i, b_i) \subset [a, b]$, $[a_i, b_i)$ $(i = 1, 2, \ldots, n)$ are mutually disjoint and $\sum_{i=1}^{n}(b_i - a_i) < \delta$, then $\sum_{i=1}^{n} \|x(b_i) - x(a_i)\| < \varepsilon$.

THEOREM 1.16. *Let $[a, b]$ be a bounded closed interval.*
(i) *Let $x(s): [a, b] \to X$ be Bochner integrable on $[a, b]$, and set*

$$y(s) = \int_a^s x(t)\,dt \qquad (a \leq s \leq b).$$

Then, $y(s)$ is (strongly) absolutely continuous on $[a, b]$, strongly differentiable at a.e. s, and the strong derivative $y'(s) = x(s)$ (a.e. s).

(ii) *If $y(s): [a, b] \to X$ is (strongly) absolutely continuous and weakly differentiable at a.e. s, then the weak derivative $x(s)$ of $y(s)$ is Bochner integrable on $[a, b]$ and*

$$y(s) = y(a) + \int_a^s x(t)\,dt \qquad (a \leq s \leq b)$$

holds. Therefore, $y(s)$ is strongly differentiable at a.e. s and $y'(s) = x(s)$ (a.e. s).

THEOREM 1.17. *Let X be a reflexive Banach space. A necessary and sufficient condition for $y(s)\colon [a, b] \to X$ to be (strongly) absolutely continuous on $[a, b]$ is that there exists a Bochner integrable function $x(s)$ on $[a, b]$ such that*

$$y(s) = y(a) + \int_a^s x(t)\,dt \qquad (a \le s \le b).$$

In this case, $y(s)$ is strongly differentiable at a.e. s and $y'(s) = x(s)$ (a.e. s).

REMARK. In a general Banach space, there are functions that are (strongly) absolutely continuous but not strongly differentiable everywhere.

CHAPTER 2

Dissipative Operators

Let X be a real or complex Banach space and let X^* be its conjugate space. We denote the elements of X by $x, y, \ldots$, and the elements of X^* by $f, g, \ldots$ or $x^*, y^*, \ldots$. We denote the value $f(x)$ of f $(\in X^*)$ at a point x $(\in X)$ by (x, f) and also, for convenience, we define the product of a scalar α and $f \in X^*$ by

$$(x, \alpha f) = \overline{\alpha}(x, f) \qquad (x \in X),$$

where $\overline{\alpha}$ is the complex conjugate of α. Therefore, in the case where X is a Hilbert space, $(\cdot, \cdot)$ is the ordinary inner product of the Hilbert space. The real part of (x, f) is denoted by $\operatorname{Re}(x, f)$.

§1. The duality mapping

DEFINITION 2.1. For each $x \in X$, we set

$$F(x) = \{f \in X^* : (x, f) = \|x\|^2 = \|f\|^2\}. \tag{2.1}$$

Then, by the Hahn-Banach theorem, we see that $F(x) \neq \varnothing$. Hence, F can be viewed as a multivalued mapping $X \to X^*$. We call this (multivalued) mapping $F: X \to X^*$ the dual mapping.

From this definition, we easily obtain the following lemma.

LEMMA 2.1.
(i) *The image $F(x)$ of x $(\in X)$ is a closed convex set.*
(ii) $F(\alpha x) = \alpha F(x)$ $(x \in X; \alpha$ *a scalar).*
(iii) *(Monotonicity)* $\operatorname{Re}(x - y, f - g) \geq 0$ $(x, y \in X, f \in F(x), g \in F(y))$.
(iv) *If X is reflexive, then F is a mapping from X onto X^*, that is, X^* is the range $F(X)$ $(= \bigcup_{x \in X} F(x)) = X^*$ of F.*

LEMMA 2.2. *If X^* is a strictly convex Banach space, then F is a single-valued mapping.*

PROOF. From the definition, $F(0) = \{0\}$. Next, we let $x \neq 0$, and let $f, g \in F(x)$. Then, from

$$(x, f) = \|f\|^2 = \|x\|^2 = \|g\|^2 = (x, g) \, (\neq 0)$$

13

we obtain $\|f + g\| \|x\| \geq (x, f + g) = 2\|x\|^2$. That is,

$$\|f + g\| \geq 2\|x\| = \|f\| + \|g\|.$$

Therefore, $\|f + g\| = \|f\| + \|g\|$.

Since X^* is strictly convex, there is a scalar α such that $g = \alpha f$. From $(x, f) = (x, g) = \overline{\alpha}(x, f)$ and $(x, f) \neq 0$, we have $\overline{\alpha} = 1$, i.e., $\alpha = 1$. Hence, $g = f$. □

THEOREM 2.3. *If X^* is a uniformly convex Banach space, then the duality mapping F is a single-valued mapping and also uniformly continuous on every bounded set of X. That is, if B is a bounded set of X, then given any $\varepsilon > 0$ there is a positive number δ such that if $x, y \in B$ and $\|x - y\| < \delta$, then $\|F(x) - F(y)\| < \varepsilon$.*

PROOF. Since a uniformly convex Banach space is strictly convex, from Lemma 2.2 we have that F is a single-valued mapping. Next, we let B be a bounded subset of X and show that F is uniformly continuous on B. Suppose that F is not uniformly continuous on B. Then there exist an $\varepsilon_0 > 0$ and sequences $\{x_n\}$ and $\{y_n\}$ in B such that

$$\lim_{n \to \infty} \|x_n - y_n\| = 0 \quad \text{and} \quad \|F(x_n) - F(y_n)\| \geq \varepsilon_0 \quad (n = 1, 2, \dots).$$

Then, since $\{\|x_n\|\}$ does not converge to 0, we can select $\alpha_0 > 0$ and a subsequence $\{x_{n_i}\}$ of $\{x_n\}$ such that

$$\|x_{n_i}\| \geq \alpha_0 \quad (i = 1, 2, \dots).$$

Since $\|x_{n_i} - y_{n_i}\| \to 0$ as $i \to \infty$, there is a natural number i_0 such that $\|y_{n_i}\| \geq \alpha_0/2$ $(i \geq i_0)$. Now, if we set $u_{n_i} = x_{n_i}/\|x_{n_i}\|$ and $v_{n_i} = y_{n_i}/\|y_{n_i}\|$ $(i \geq i_0)$, then $\|u_{n_i}\| = \|v_{n_i}\| = 1$ and also

$$\|u_{n_i} - v_{n_i}\| = \left\| \frac{x_{n_i} - y_{n_i}}{\|x_{n_i}\|} + \frac{(\|y_{n_i}\| - \|x_{n_i}\|)y_{n_i}}{\|x_{n_i}\| \|y_{n_i}\|} \right\|$$

$$\leq 2\|x_{n_i}\|^{-1} \|x_{n_i} - y_{n_i}\|$$

$$\leq (2/\alpha_0)\|x_{n_i} - y_{n_i}\| \to 0 \quad (i \to \infty).$$

From $\|F(u_{n_i})\| = \|u_{n_i}\| = 1$ and $\|F(v_{n_i})\| = \|v_{n_i}\| = 1$, we have

$$\limsup_{i \to \infty} \|F(u_{n_i}) + F(v_{n_i})\| \leq 2.$$

Furthermore, from

$$\operatorname{Re}(u_{n_i}, F(u_{n_i}) + F(v_{n_i}))$$

$$= \operatorname{Re}(u_{n_i}, F(u_{n_i})) + \operatorname{Re}(v_{n_i}, F(v_{n_i})) + \operatorname{Re}(u_{n_i} - v_{n_i}, F(v_{n_i}))$$

$$\geq 2 - \|u_{n_i} - v_{n_i}\| \to 2 \quad (i \to \infty)$$

we have $\liminf_{i\to\infty} \|F(u_{n_i}) + F(v_{n_i})\| \geq \liminf_{i\to\infty} \mathrm{Re}(u_{n_i}, F(u_{n_i}) + F(v_{n_i})) \geq 2$. Hence,

$$\lim_{i\to\infty} \|F(u_{n_i}) + F(v_{n_i})\| = 2.$$

Since X^* is uniformly convex, we obtain from the above equation

$$\lim_{i\to\infty} \|F(u_{n_i}) - F(v_{n_i})\| = 0.$$

Since $F(x_{n_i}) = F(\|x_{n_i}\|u_{n_i}) = \|x_{n_i}\|F(u_{n_i})$ and $F(y_{n_i}) = \|y_{n_i}\|F(v_{n_i})$ ((iii) of Lemma 2.1), we have

$$F(x_{n_i}) - F(y_{n_i}) = \|x_{n_i}\|(F(u_{n_i}) - F(v_{n_i})) + (\|x_{n_i}\| - \|y_{n_i}\|)F(v_{n_i}).$$

Therefore,

$$\|F(x_{n_i}) - F(y_{n_i})\| \leq M\|F(u_{n_i}) - F(v_{n_i})\| + \|x_{n_i} - y_{n_i}\| \to 0 \qquad (i \to \infty),$$

where $M = \sup\{\|x\| : x \in B\}$ (since B is a bounded set, M is finite). This contradicts the inequality $\|F(x_n) - F(y_n)\| \geq \varepsilon_0$ $(n = 1, 2, \ldots)$. $\quad\square$

LEMMA 2.4. *If X is a Hilbert space, then the duality mapping F coincides with the identity mapping I on X. That is, $F = I$.*

PROOF. Since $X^* = X$, F is a mapping of X into itself. Let $x \in X$ and $f \in F(x)$, then from $(x, f) = \|x\|^2 = \|f\|^2$, we have

$$\|x - f\|^2 = \|x\|^2 - 2\,\mathrm{Re}(x, f) + \|f\|^2 = 0.$$

Thus we obtain $f = x$, and hence $F = I$. $\quad\square$

Next, we consider the tangent functional of the unit sphere in a Banach space X: $\lim_{\alpha\to 0+} \alpha^{-1}(\|x + \alpha y\| - \|x\|)$ $(x, y \in X)$. If we choose and fix $x, y \in X$ arbitrarily and consider the function of $\alpha > 0$,

$$\alpha^{-1}(\|x + \alpha y\| - \|x\|),$$

then this is an increasing function. In fact, when we let $\beta > \alpha > 0$, from

$$(\beta - \alpha)\|x\| = \|(\beta x + \alpha\beta y) - (\alpha x + \alpha\beta y)\| \geq \beta\|x + \alpha y\| - \alpha\|x + \beta y\|$$

we have

$$\beta^{-1}(\|x + \beta y\| - \|x\|) \geq \alpha^{-1}(\|x + \alpha y\| - \|x\|).$$

Moreover, from $\alpha^{-1}(\|x + \alpha y\| - \|x\|) \geq -\|y\|$ $(\alpha > 0)$, this function is bounded from below. Thus,

$$\lim_{\alpha\to 0+} \alpha^{-1}(\|x + \alpha y\| - \|x\|) \qquad (= \inf_{\alpha>0} \alpha^{-1}(\|x + \alpha y\| - \|x\|))$$

exists and is finite valued. Now,

DEFINITION 2.2. We define functions $\tau_+ : X \times X \to (-\infty, \infty)$ and $\tau_- : X \times X \to (-\infty, \infty)$ for arbitrary $x, y \in X$, by the following equations:

$$\tau_+(y, x) = \lim_{\alpha\to 0+} \alpha^{-1}(\|x + \alpha y\| - \|x\|) \qquad (= \inf_{\alpha>0} \alpha^{-1}(\|x + \alpha y\| - \|x\|)), \tag{2.2}$$

$$\tau_-(y, x) = -\tau_+(-y, x). \tag{2.3}$$

From these equations, we have

$$\tau_-(y, x) = \lim_{\alpha \to 0-} \alpha^{-1}(\|x + \alpha y\| - \|x\|) = \sup_{\alpha < 0} \alpha^{-1}(\|x + \alpha y\| - \|x\|). \quad (2.4)$$

Also, from $2\|x\| \le \|x + \alpha y\| + \|x - \alpha y\|$, we have

$$\alpha^{-1}(\|x\| - \|x - \alpha y\|) \le \alpha^{-1}(\|x + \alpha y\| - \|x\|) \quad (\alpha > 0).$$

Thus,

$$\tau_-(y, x) \le \tau_+(y, x) \quad (x, y \in X). \quad (2.5)$$

THEOREM 2.5. *Let* $x, y \in X$.
(i) *There exists an element* $f_+ \in F(X)$ *such that*

$$\|x\|\tau_+(y, x) = \sup\{\operatorname{Re}(y, f) : f \in F(x)\} = \operatorname{Re}(y, f_+). \quad (2.6)$$

(ii) *There exists an element* $f_- \in F(x)$ *such that*

$$\|x\|\tau_-(y, x) = \inf\{\operatorname{Re}(y, f) : f \in F(x)\} = \operatorname{Re}(y, f_-). \quad (2.7)$$

PROOF. (i) Since $F(x) = \{0\}$ if $x = 0$, (2.6) is self-evident. Now, let $x \ne 0$ and set

$$c = \sup\{\operatorname{Re}(y, f) : f \in F(x)\} \quad \text{and} \quad c' = \|x\|\tau_+(y, x).$$

For $f \in F(x)$ and $\alpha > 0$, we have

$$\|x + \alpha y\| \|x\| \ge \operatorname{Re}(x + \alpha y, f) = \|x\|^2 + \alpha \operatorname{Re}(y, f).$$

From this,

$$\alpha^{-1}(\|x + \alpha y\| - \|x\|)\|x\| \ge \operatorname{Re}(y, f).$$

Here we let $\alpha \to 0$; then $c' \ge \operatorname{Re}(y, f)$ $(f \in F(x))$ and we obtain $c' \ge c$. Next, to prove that $c' \le c$ we set $c'' = \tau_+(y, x)$. From the definition of $\tau_+(y, x)$, we have

$$\|x + \alpha y\| \ge \|x\| + \alpha c'' \quad (\alpha > 0).$$

Since $\|x\| > 0$, $\|x + \alpha y\| > 0$ for sufficiently small $\alpha > 0$. That is, there is an $\alpha_0 > 0$ such that $\|x + \alpha y\| > 0$ $(0 < \alpha \le \alpha_0)$. Now, we choose $f_\alpha \in F(x + \alpha y)$ arbitrarily, and set $g_\alpha = f_\alpha / \|f_\alpha\|$ $(0 < \alpha \le \alpha_0)$. (Note that $\|f_\alpha\| = \|x + \alpha y\| > 0$.) Then $\|g_\alpha\| = 1$ and

$$\|x\| + \alpha c'' \le \|x + \alpha y\| = (x + \alpha y, g_\alpha) = \operatorname{Re}(x + \alpha y, g_\alpha)$$
$$= \operatorname{Re}(x, g_\alpha) + \alpha \operatorname{Re}(y, g_\alpha) \le \|x\| + \alpha \operatorname{Re}(y, g_\alpha).$$

Thus, we obtain

$$c'' \le \operatorname{Re}(y, g_\alpha), \quad (2.8)$$

$$\|x\| \le \liminf_{\alpha \to 0+} \operatorname{Re}(x, g_\alpha). \quad (2.9)$$

Since the unit sphere $\{f \in X^* : \|f\| \leq 1\}$ in X^* is compact with respect to the weak* topology (see Theorem 1.9), the net $\{g_\alpha : 0 < \alpha \leq \alpha_0\}$([1]) has a cluster point $g_+ \in X^*$ with $\|g_+\| \leq 1$ (with respect to the weak* topology). ([2]) Then, from (2.9) we have

$$\mathrm{Re}(x, g_+) \geq \liminf_{\alpha \to 0+} \mathrm{Re}(x, g_\alpha) \geq \|x\|.$$

Hence, $\|g_+\| \geq 1$. Thus, $\|g_+\| = 1$ and

$$\|x\| \geq |(x, g_+)| \geq \mathrm{Re}(x, g_+) \geq \|x\|.$$

Therefore we obtain

$$\|g_+\| = 1 \quad \text{and} \quad (x, g_+) = \|x\|. \tag{2.10}$$

Also, from (2.8)

$$c'' \leq \mathrm{Re}(y, g_+). \tag{2.11}$$

Now, we set $f_+ = \|x\| g_+$. Then from (2.10) and (2.11), we have $(x, f_+) = \|x\|^2 = \|f_+\|^2$, that is, $f_+ \in F(x)$. Also, we obtain

$$\mathrm{Re}(y, f_+) = \|x\| \mathrm{Re}(y, g_+) \geq \|x\| c'' = c'.$$

Thus, $c \geq \mathrm{Re}(y, f_+) \geq c'$. Together with $c' \geq c$ shown above, (2.6) is now proved.

(ii) If we consider $-y$ instead of y, then there exists an element $f_- \in F(x)$ from (i) and

$$\|x\| \tau_+(-y, x) = \sup\{\mathrm{Re}(-y, f) : f \in F(x)\} = \mathrm{Re}(-y, f_-).$$

From this and $\tau_-(y, x) = -\tau_+(-y, x)$, and

$$-\sup\{\mathrm{Re}(-y, f) : f \in F(x)\} = \inf\{\mathrm{Re}(y, f) : f \in F(x)\},$$

we obtain (2.7). □

Here, we introduce the following notation.

DEFINITION 2.3. We define functions $\langle \cdot, \cdot \rangle_s : X \times X \to (-\infty, \infty)$ and $\langle \cdot, \cdot \rangle_i : X \times X \to (-\infty, \infty)$ by the following: for every $x, y \in X$

$$\langle y, x \rangle_s = \sup\{\mathrm{Re}(y, f) : f \in F(x)\}, \tag{2.12}$$

$$\langle y, x \rangle_i = \inf\{\mathrm{Re}(y, f) : f \in F(x)\} \quad (= -\langle -y, x \rangle_s).([3]) \tag{2.13}$$

([1]), ([2]) We define $\beta \preceq \alpha$ if $\alpha \leq \beta$; then the set of real numbers $\{\alpha : 0 < \alpha \leq \alpha_0\}$ forms a directed set through "$\preceq$". In general, for $\mathfrak{A}$ a directed set and X a topological space, we call $\{x_\alpha \in X : \alpha \in \mathfrak{A}\}$ a net (in the topological space X). Also, a point x $(\in X)$ is said to be a cluster point of the net $\{x_\alpha \in X : \alpha \in \mathfrak{A}\}$ if for each $\alpha \in \mathfrak{A}$ and arbitrary neighborhood U of x, there exists $\beta \succeq \alpha$ such that $x_\beta \in U$. See for example Lemma 9 on page 29 of [45] in the bibliography at the end of this book.

([3]) Since F is a single-valued mapping if X^* is a strictly convex Banach space, $\langle y, x \rangle_s = \langle y, x \rangle_i = \mathrm{Re}(y, F(x))$ $(x, y \in X)$.

Then, from Theorem 2.5, for arbitrary $x, y \in X$,

$$\langle y, x \rangle_s = \|x\| \tau_+(y, x) \quad \text{and} \quad \langle y, x \rangle_i = \|x\| \tau_-(y, x). \tag{2.14}$$

COROLLARY 2.6. *Let* $x, y, z \in X$.
(i) $\langle \alpha x + y, x \rangle_j = (\mathrm{Re}\,\alpha)\|x\|^2 + \langle y, x \rangle_j$ (α *a scalar;* $j = i$ *or* s),
(ii) $\langle \alpha y, \beta x \rangle_j = \alpha \overline{\beta} \langle y, x \rangle_j$ ($\alpha \overline{\beta} \geq 0$; $j = i$ *or* s),
(iii) $\langle z + y, x \rangle_j \leq \|z\| \|x\| + \langle y, x \rangle_j$ ($j = i$ *or* s),
(iv) $\langle \cdot, \cdot \rangle_s \colon X \times X \to (-\infty, \infty)$, *is upper semicontinuous*,
(v) $\langle \cdot, \cdot \rangle_i \colon X \times X \to (-\infty, \infty)$ *is lower semicontinuous*.

PROOF. (i) and (iii) are self-evident. And, from $F(\beta x) = \beta F(x)$ and $(\alpha y, \beta f) = \alpha \overline{\beta}(y, f)$ ($f \in F(x)$), we obtain (ii). Next, we show (iv). From (2.14) and the definition of τ_+, we have

$$\langle y, x \rangle_s = \inf_{\alpha > 0} \|x\| \alpha^{-1}(\|x + \alpha y\| - \|x\|).$$

For each $\alpha > 0$, since $\|x\| \alpha^{-1}(\|x + \alpha y\| - \|x\|)$ is a continuous function $X \times X \to (-\infty, \infty)$, it follows that $\langle \cdot, \cdot \rangle_s \colon X \times X \to (-\infty, \infty)$ is upper semicontinuous. Since $\langle y, x \rangle_i = -\langle -y, x \rangle_s$, we obtain (v) from (iv). $\square$

COROLLARY 2.7. *Let* $u, v \in X$.
(i) *The following four propositions are equivalent*:

(a$_1$) $\mathrm{Re}(v, f) \leq 0$ *for a certain* $f \in F(u)$,
(a$_2$) $\|u - \alpha v\| \geq \|u\|$ *for all* $\alpha > 0$,
(a$_3$) $\tau_-(v, u) \leq 0$,
(a$_4$) $\langle v, u \rangle_i \leq 0$.

(ii) *When* $u \neq 0$, *the following three propositions are mutually equivalent*:

(b$_1$) $\mathrm{Re}(v, f) \leq 0$ *for all* $f \in F(u)$,
(b$_2$) $\tau_+(v, u) \leq 0$,
(b$_3$) $\langle v, u \rangle_s \leq 0$.

PROOF. (i) Since the result is self-evident when $u = 0$, we let $u \neq 0$. From (2.7) and (2.14) (a$_1$), (a$_3$), and (a$_4$) are equivalent. If we assume (a$_2$), then we have $\alpha^{-1}(\|u + \alpha(-v)\| - \|u\|) \geq 0$ ($\alpha > 0$). Hence, $\tau_+(-v, u) \geq 0$ and $\tau_-(v, u) = -\tau_+(-v, u) \leq 0$, that is, we obtain (a$_3$). Conversely, assume (a$_3$). Then, since $\tau_+(-v, u) \geq 0$ and

$$\tau_+(-v, u) = \inf_{\alpha > 0} \alpha^{-1}(\|u + \alpha(-v)\| - \|u\|),$$

we have $\alpha^{-1}(\|u - \alpha v\| - \|u\|) \geq 0$ for all $\alpha > 0$. That is, we obtain (a$_2$).
(ii) is self-evident from (2.6) and (2.14). $\square$
Finally, we mention a lemma which we will use often.

LEMMA 2.8. *If $u(t): (a, b) \to X$ has weak derivative $u'(t_0)$ $(\in X)$ at $t_0 \in (a, b)$ and also $\|u(t)\|$ is differentiable at $t = t_0$, then*

$$[(d/dt)\|u(t)\|^2]_{t=t_0} = 2\|u(t_0)\|[(d/dt)\|u(t)\|]_{t=t_0}$$
$$= 2\langle u'(t_0), u(t_0)\rangle_s = 2\langle u'(t_0), u(t_0)\rangle_i.$$

PROOF. Suppose that $f \in F(u(t_0))$ is given arbitrarily. Since

$$\text{Re}(u(t) - u(t_0), f) = \text{Re}(u(t), f) - \|u(t_0)\|^2 \le (\|u(t)\| - \|u(t_0)\|)\|u(t_0)\|,$$

if $t > t_0$ we have

$$\text{Re}\left(\frac{u(t) - u(t_0)}{t - t_0}, f\right) \le \frac{\|u(t)\| - \|u(t_0)\|}{t - t_0}\|u(t_0)\|,$$

and if $t < t_0$ we have

$$\text{Re}\left(\frac{u(t) - u(t_0)}{t - t_0}, f\right) \ge \frac{\|u(t)\| - \|u(t_0)\|}{t - t_0}\|u(t_0)\|.$$

Now we let $t \to t_0 + 0$ and $t \to t_0 - 0$; then we have

$$\text{Re}(u'(t_0), f) \le \|u(t_0)\|[(d/dt)\|u(t)\|]_{t=t_0},$$

and

$$\text{Re}(u'(t_0), f) \ge \|u(t_0)\|[(d/dt)\|u(t)\|]_{t=t_0},$$

respectively. That is, we obtain

$$\|u(t_0)\|[(d/dt)\|u(t)\|]_{t=t_0} = \text{Re}(u'(t_0), f). \quad \square$$

§2. Dissipative operators

In this section we consider an operator which has both its domain and range contained in X but is not necessarily single valued. When A is such an operator, we denote the domain by $D(A)$, the range by $R(A)$, and the graph by $\mathfrak{G}(A)$. In the same way for the case of the ordinary (single-valued) operator, we denote by Ax the image of a point $x \in D(A)$ under the operator A. Then Ax is a nonempty subset of X (if A is single valued, this set consists of a single point), and

$$R(A) = \bigcup_{x \in D(A)} Ax, \quad \text{and} \quad \mathfrak{G}(A) = \bigcup_{x \in D(A)} \{[x, y]: y \in Ax\}.$$

Here $[x, y]$ represents an element of $X \times X$. Furthermore, the product λA of a scalar λ and an operator A, and the inverse operator A^{-1} of A are defined by the following:

$$D(\lambda A) = D(A), \quad (\lambda A)x = \lambda Ax = \{\lambda y: y \in Ax\} \quad (x \in D(A));$$
$$(2.15)$$

$$D(A^{-1}) = R(A), \quad A^{-1}y = \{x \in D(A): y \in Ax\} \quad (y \in D(A^{-1})).$$
$$(2.16)$$

Also, we define the sum $(A_1 + A_2)$ of two operators A_1 and A_2 by

$$D(A_1 + A_2) = D(A_1) \cap D(A_2),$$
$$(A_1 + A_2)x = A_1 x + A_2 x \qquad\qquad\qquad (2.17)$$
$$= \{y + z : y \in A_1 x, \ z \in A_2 x\} \quad \text{for } x \in D(A_1 + A_2).$$

DEFINITION 2.4. Let A be an operator whose domain and range are contained in X, and which is not necessarily single valued. If the following condition (D_1) is satisfied, then A is called a dissipative operator:

> For each $x, y \in D(A)$, $x' \in Ax$ and $y' \in Ay$, there exists a
> proper $f \in F(x - y)$ such that $\mathrm{Re}(x' - y', f) \le 0$. (D_1)

If $-A$ is a dissipative operator, then A is called an accretive operator.

REMARK. Since $F = I$ when X is a Hilbert space, the above condition (D_1) becomes as follows: For each $x, y \in D(A)$, $x' \in Ax$, and $y' \in Ay$, we have $\mathrm{Re}(x' - y', x - y) \le 0$.

THEOREM 2.9. *The following (a_1) and (a_2) are equivalent:*

(a_1) *A is a dissipative operator.*

(a_2) *For each $x, y \in D(A)$, $x' \in Ax$, and $y' \in Ay$, any one of the following (D_2), (D_3), and (D_4) holds:*

(D_2) $\|x - y - \lambda(x' - y')\| \ge \|x - y\|$ *for all $\lambda > 0$;*

(D_3) $\tau_-(x' - y', x - y) \le 0$;

(D_4) $\langle x' - y', x - y \rangle_i \le 0$.

PROOF. Apply Corollary 2.7 with $u = x - y$ and $v = x' - y'$. $\square$

COROLLARY 2.10. *Let A be a dissipative operator and let I be the identity operator on X. Set*

$$J_\lambda = (I - \lambda A)^{-1} \qquad (\lambda > 0);$$

then for each $\lambda > 0$, J_λ is a single-valued operator such that $D(J_\lambda) = R(I - \lambda A)$ and $R(J_\lambda) = D(A)$, and J_λ satisfies the following inequality:

$$\|J_\lambda x - J_\lambda y\| \le \|x - y\| \qquad (x, y \in D(J_\lambda)). \qquad (2.18)$$

PROOF. Since A is a dissipative operator, (D_2) holds. Therefore, if $z_1, z_2 \in D(A)$ and $z_1 \ne z_2$, then

$$(I - \lambda A)z_1 \cap (I - \lambda A)z_2 = \varnothing$$

for all $\lambda > 0$. Hence we obtain

$$J_\lambda x = (I - \lambda A)^{-1} x = z \qquad (x \in (I - \lambda A)z)$$

for each $\lambda > 0$, and J_λ is single-valued. Next, fix $\lambda > 0$ and let $x, y \in D(J_\lambda) = R(I - \lambda A)$. Then, we can write $x = z - \lambda z'$, $y = w - \lambda w'$, where $z, w \in D(A)$, $z' \in Az$, and $w' \in Aw$. Since $J_\lambda x = z$ and $J_\lambda y = w$, we obtain

$$\|J_\lambda x - J_\lambda y\| = \|z - w\| \le \|z - w - \lambda(z' - w')\| = \|x - y\|$$

by applying (D_2) again. $\square$

For an operator A with $D(A) \subset X$ and $R(A) \subset X$ which is not necessarily single valued we define $\|\|Ax\|\|$ by $\|\|Ax\|\| = \inf\{\|x'\| : x' \in Ax\}$ for $x \in D(A)$. Thus, if A is single valued, we have

$$\|\|Ax\|\| = \|Ax\| \qquad (x \in D(A)).$$

LEMMA 2.11. *Let A be a dissipative operator and let $\lambda > 0$.*
(i) *If we set $A_\lambda x = \lambda^{-1}(j_\lambda - I)x$, then $A_\lambda x \in AJ_\lambda x$ $(x \in D(J_\lambda))$;*
(ii) $\|A_\lambda x\| \leq \|\|Ax\|\|$ $(x \in D(J_\lambda) \cap D(A))$;
(iii) *if $x \in D(J_\lambda)$ and $\mu > 0$, then*

$$\frac{\mu}{\lambda} x + \frac{\lambda - \mu}{\lambda} J_\lambda x \in D(J_\mu),$$

and

$$J_\lambda x = J_\mu \left(\frac{\mu}{\lambda} x + \frac{\lambda - \mu}{\lambda} J_\lambda x \right);$$

(iv) *if $\lambda \geq \mu > 0$ and $x \in D(J_\lambda) \cap D(J_\mu)$, then $\|A_\lambda x\| \leq \|A_\mu x\|$.*

PROOF. (i) Let $x \in D(J_\lambda)$ and set $J_\lambda x = y$. Then $x \in (I - \lambda A)y$, that is, $\lambda^{-1}(y - x) \in Ay$. Thus $A_\lambda x = \lambda^{-1}(y - x) \in Ay = AJ_\lambda x$.

(ii) Let $x \in D(J_\lambda) \cap D(A)$ and choose $x' \in Ax$ arbitrarily. Then we obtain $J_\lambda(x - \lambda x') = x$ from $x - \lambda x' \in (I - \lambda A)x$. Therefore,

$$\|A_\lambda x\| = \lambda^{-1}\|J_\lambda x - x\| = \lambda^{-1}\|J_\lambda x - J_\lambda(x - \lambda x')\|$$
$$\leq \lambda^{-1}\|x - (x - \lambda x')\| = \|x'\| \quad \text{(from (2.18))}.$$

Since $x' \in Ax$ is arbitrary, we have $\|A_\lambda x\| \leq \|\|Ax\|\|$.

(iii) Let $x \in D(J_\lambda) = R(I - \lambda A)$. Then we can write $x = y - \lambda y'$, where $y \in D(A)$ and $y' \in Ay$, and thus $J_\lambda x = y$. Hence,

$$\frac{\mu}{\lambda} x + \frac{\lambda - \mu}{\lambda} J_\lambda x = \frac{\mu}{\lambda}(y - \lambda y') + \frac{\lambda - \mu}{\lambda} y$$
$$= y - \mu y' \in R(I - \mu A) = D(J_\mu),$$

and

$$J_\mu \left(\frac{\mu}{\lambda} x + \frac{\lambda - \mu}{\lambda} J_\lambda x \right) = J_\mu(y - \mu y') = y = J_\lambda x.$$

(iv) By applying (iii) and (2.18) we have

$$\lambda\|A_\lambda x\| = \|J_\lambda x - x\| \leq \|J_\lambda x - J_\mu x\| + \|J_\mu x - x\|$$
$$= \left\| J_\mu \left(\frac{\mu}{\lambda} x + \frac{\lambda - \mu}{\lambda} J_\lambda x \right) - J_\mu x \right\| + \|J_\mu x - x\|$$
$$\leq \left\| \frac{\mu}{\lambda} x + \frac{\lambda - \mu}{\lambda} J_\lambda x - x \right\| + \|J_\mu x - x\|$$
$$= (\lambda - \mu)\|A_\lambda x\| + \mu\|A_\mu x\|.$$

Therefore, $\|A_\lambda x\| \leq \|A_\mu x\|$. $\square$

Consider two operators A_1 and A_2, whose domains and ranges are contained in X, and that are not necessarily single valued. If $A_1 x \subset A_2 x$ holds for each $x \in D(A_1)$ with $D(A_1) \subset D(A_2)$, then A_2 is called an extension of A_1, or A_1 is called a restriction of A_2, and is denoted by $A_2 \supset A_1$ or $A_1 \subset A_2$. Also, if A_1 and A_2 are dissipative operators and $A_1 \subset A_2$, then A_2 is called a dissipative extension of A_1.

DEFINITION 2.5. (i) Let S be a subset of X. A dissipative operator A is called a maximal dissipative operator on S if all dissipative extensions of A coincide with A on S; that is, for any dissipative extension B of A we have $D(B) \cap S = D(A) \cap S$ and $Bx = Ax$ $(x \in D(A) \cap S)$. In particular, a maximal dissipative operator on the whole space X is called simply a maximal dissipative operator.

(ii) A dissipative operator A with

$$R(I - \lambda A) = X \tag{2.19}$$

for all $\lambda > 0$ is called an m-dissipative operator.

LEMMA 2.12. (i) *Let* $S_1 \subset S_2$ $(\subset X)$. *If* A *is a maximal dissipative operator on* S_2, *then* A *is also a maximal dissipative operator on* S_1.

(ii) *If* A *is a dissipative operator, then* A *has a maximal dissipative extension, that is, there exists a maximal dissipative operator* $\widetilde{A}$ *such that* $A \subset \widetilde{A}$.

(iii) *An* m-*dissipative operator is a maximal dissipative operator.*

PROOF. (i) is self-evident.

(ii) Let $\mathfrak{A}$ be the set of all dissipative extensions of A. With the relation "$\subset$" which expresses the extension of operators, $\mathfrak{A}$ forms a partially ordered set. Then, by Zorn's lemma, $\mathfrak{A}$ has a maximal element $\widetilde{A}$. Obviously, $\widetilde{A}$ is a maximal dissipative extension of A.

(iii) Let A be an m-dissipative operator and $\widetilde{A}$ a dissipative extension of A. We need to show that $\widetilde{A} \subset A$. Let $x \in D(\widetilde{A})$ and $x' \in \widetilde{A}x$. Since $x - \lambda x' \in X = R(I - \lambda A)$, where $\lambda > 0$ is fixed, there exist a $y \in D(A)$ and $y' \in Ay$ such that $x - \lambda x' = y - \lambda y'$, i.e., $x - y - \lambda(x' - y') = 0$. From $\widetilde{A} \supset A$, $y \in D(\widetilde{A})$, and $y' \in Ay$, and since $\widetilde{A}$ is a dissipative operator

$$\|x - y\| \leq \|x - y - \lambda(x' - y')\| = 0$$

by (D_2). Hence, $x = y$ and $x' = y'$. Therefore, $x \in D(A)$ and $x' \in Ax$, and $D(\widetilde{A}) \subset D(A)$ and $\widetilde{A}x \subset Ax$ $(x \in D(\widetilde{A}))$; that is, we have $\widetilde{A} \subset A$. $\square$

LEMMA 2.13. *A necessary and sufficient condition for* A *to be an* m-*dissipative operator is that* A *is a dissipative operator and also*

$$R(I - \lambda_0 A) = X \tag{2.20}$$

holds for some $\lambda_0 > 0$.

PROOF. It is self-evident that the condition is necessary. We show that it is sufficient. From conditions (2.20) and (2.18), we have $J_{\lambda_0} = (I - \lambda_0 A)^{-1}$

is a contraction operator defined on all of X; that is, $\|J_{\lambda_0}x - J_{\lambda_0}y\| \leq \|x - y\|$ $(x, y \in X)$. We must note that

$$I - \lambda A = \frac{\lambda}{\lambda_0}\left[I - \left(1 - \frac{\lambda_0}{\lambda}\right)J_{\lambda_0}\right](I - \lambda_0 A) \qquad (\lambda > 0). \tag{2.21}$$

Now, we choose and fix $x \in X$ arbitrarily and define an operator $T: X \to X$ by

$$Ty = x + (1 - \lambda_0/\lambda)J_{\lambda_0}y \qquad (y \in X).$$

Since

$$\|Ty - Tz\| \leq |1 - \lambda_0/\lambda| \, \|y - z\| \qquad (y, z \in X),$$

we have $|1 - \lambda_0/\lambda| < 1$, that is ([4]), when $\lambda_0/2 < \lambda$ $T: X \to X$ becomes a strict contraction operator and hence has a fixed point z $(\in X)$. Therefore $Tz = z$; that is, $x = [I - (1 - \lambda_0/\lambda)J_{\lambda_0}]z$. From this and (2.20), we have

$$x \in [I - (1 - \lambda_0/\lambda)J_{\lambda_0}](I - \lambda_0 A)[D(A)];$$

that is, $X = [I - (1 - \lambda_0/\lambda)J_{\lambda_0}](1 - \lambda_0 A)[D(A)]$. From this and (2.21), we obtain

$$R(I - \lambda A) = X \quad \text{for } \lambda > \lambda_0/2.$$

Again we apply the above argument to obtain

$$R(I - \lambda A) = X \quad \text{for } \lambda > (1/2)(\lambda_0/2) = \lambda_0/2^2.$$

We repeat this process and obtain (2.19). $\quad\square$

Next we give some examples of dissipative operators.

EXAMPLE 2.1. Let $R^1 = (-\infty, \infty)$ and $g: R^1 \to R^1$ be a monotone decreasing function. Then, g is a single valued dissipative operator defined on R^1 (one-dimensional Euclidean space).

Now, we set

$$g_+(x) = \lim_{h \to 0+} g(x + h), \quad g_-(x) = \lim_{h \to 0-} g(x + h) \qquad (x \in R^1),$$

and define a function g (which is not necessarily single valued) by

$$\tilde{g}(x) = \{z : g_+(x) \leq z \leq g_-(x)\} \qquad (x \in R^1).$$

Hence, $\tilde{g}(x) = g(x)$ at points x where g is continuous. At points x where g is discontinuous, $\tilde{g}$ is a (multivalued) operator which associates to x the closed interval $[g_+(x), g_-(x)]$. Obviously, $\tilde{g}$ is an m-dissipative operator.

EXAMPLE 2.2. Let H be a Hilbert space, and let φ be a (single-valued) functional $H \to (-\infty, \infty]$ with $\varphi \not\equiv \infty$. (That is, $\{x \in H : \varphi(x) < \infty\} \neq \varnothing$.) We define the subdifferential $\partial\varphi$ of φ as follows:

$$(\partial\varphi)x = \{y \in H : \varphi(u) \geq \varphi(x) + \text{Re}(y, u - x), \forall u \in H\} \qquad (x \in D(\partial\varphi));$$

([4]) We fix such a λ and define T by the above equation.

here $x \in D(\partial\varphi)$ means that the set on the right-hand side of the above equation is not empty. Note that if $x \in D(\partial\varphi)$, then $\varphi(x) \neq \infty$, that is, $\varphi(x)$ has finite value.

Then, $-\partial\varphi$ is a dissipative operator. In fact, let

$$x_i \in D(-\partial\varphi) = D(\partial\varphi) \qquad \text{and} \qquad y_i \in (-\partial\varphi)x_i \qquad (i = 1, 2);$$

then $\varphi(x_i)$ is a finite value, and

$$\varphi(x_2) \geq \varphi(x_1) - \text{Re}(y_1, x_2 - x_1), \qquad \varphi(x_1) \geq \varphi(x_2) + \text{Re}(y_2, x_2 - x_1).$$

Therefore, $\text{Re}(y_2 - y_1, x_2 - x_1) \leq 0$.

Now, assume that the functional φ is convex; that is,

$$\varphi(tx + (1 - t)y) \leq t\varphi(x) + (1 - t)\varphi(y) \qquad (x, y \in H, \ 0 < t < 1),$$

and is also lower semicontinuous. Then,

$$-\partial\varphi \text{ is an m-dissipative operator}. \tag{2.22}$$

We prove this below. From Lemma 2.13, it suffices to show that

$$R(I - (-\partial\varphi)) = R(I + \partial\varphi) = H.$$

To do this, let $y \in H$ and set

$$\psi(x) = \varphi(x) + \|x - y\|^2/2 \qquad (x \in H).$$

Then, ψ is a functional $H \to (-\infty, \infty]$, $\psi \not\equiv \infty$, and ψ is convex, and also lower semicontinuous. Furthermore, we have that

$$\text{if } \|x\| \to \infty, \quad \text{then } \psi(x) \to \infty. \tag{2.23}$$

PROOF OF (2.23). Since $\varphi \not\equiv \infty$, we can choose $x_0 \in H$ such that $\varphi(x_0)$ is finite. Now set

$$\varphi_1(x) = \varphi(x + x_0) - \varphi(x_0) \qquad (x \in H),$$

then φ_1 is a functional $H \to (-\infty, \infty]$, and $\varphi_1(0) = 0$, φ_1 is convex, and φ_1 is lower semicontinuous. From the convexity of φ_1 and from $\varphi_1(0) = 0$, we obtain

$$\varphi_1(tx) \leq t\varphi_1(x) \qquad (x \in H, \ 0 < t < 1). \tag{2.24}$$

Furthermore,

$$\liminf_{\|x\| \to \infty} \varphi_1(x)/\|x\| \neq -\infty. \tag{2.25}$$

In fact, if we let $\liminf_{\|x\| \to \infty} \varphi_1(x)/\|x\| = -\infty$, then we can choose a sequence $\{x_n\}$ in H such that $\|x_n\| \to \infty$ and $\varphi_1(x_n) \leq -n\|x_n\|$ ($n = 1, 2, \ldots$). Then, we have from (2.24)

$$\varphi_1\left(\frac{1}{\sqrt{n}\|x_n\|}x_n\right) \leq \frac{1}{\sqrt{n}\|x_n\|}\varphi_1(x_n) < -\sqrt{n} \qquad (n = 1, 2, \ldots).$$

On the other hand, because $x_n/(\sqrt{n}\|x_n\|) \to 0$ as $n \to \infty$ and from the lower semicontinuity of φ_1, we have

$$-1 = \varphi_1(0) - 1 < \varphi_1\left(\frac{1}{\sqrt{n}\|x_n\|}x_n\right)$$

for sufficiently large n. This is a contradiction.

Since (2.25) holds, we can choose a proper constant c such that $\varphi_1(x) \geq c\|x\|$ for x which makes $\|x\|$ sufficiently large; that is, we can make $\varphi(x + x_0) \geq \varphi(x_0) + c\|x\|$ hold. If we write x for $x + x_0$, then we have

$$\psi(x) = \varphi(x) + \|x - y\|^2/2 \geq \varphi(x_0) + c\|x - x_0\| + \|x - y\|^2/2 \to \infty$$
$$(\|x\| \to \infty).$$

(2.23) is proved. □

Next, we set $\alpha = \inf\{\psi(x) : x \in H\}$; then α has a finite value. In fact, since $\psi \not\equiv \infty$, $\alpha \neq \infty$. If we let $\alpha = -\infty$, then there is a sequence $\{x_n\}$ in H such that $\psi(x_n) < -n$ $(n = 1, 2, \ldots)$. Then from (2.23), $\{x_n\}$ must be bounded. Since a bounded set in a Hilbert space is weakly sequentially compact, we can choose a weakly convergent subsequence $\{x_{n_k}\}$ of $\{x_n\}$. Now, let w-$\lim_{k\to\infty} x_{n_k} = x'$ $(\in H)$. If we set $X_i = \overline{\mathrm{co}}\{x_{n_k} : k \geq i\}$ for $i = 1, 2, \ldots$, then since the X_i are weakly closed sets, we obtain $x' \in X_i$ $(i = 1, 2, \ldots)$. Therefore, for each i $(= 1, 2, \ldots)$ there is a z_i such that

$$\|z_i - x'\| < 1/i, \qquad z_i \in \mathrm{co}\{x_{n_k} : k \geq i\}.$$

Then, $\psi(x_{n_k}) < -n_k \leq -n_i$ $(k \geq i)$, and since z_i can be expressed as a convex combination of a finite number of x_{n_k} $(k \geq i)$ and ψ is a convex functional, we have

$$\psi(z_i) < -n_i \qquad (i = 1, 2, \ldots).$$

Since $\lim_{i\to\infty} \|z_i - x'\| = 0$, the lower semicontinuity of ψ implies that

$$\psi(x') \leq \liminf_{i\to\infty} \psi(z_i) = -\infty,$$

that is, $\psi(x') = -\infty$. This disagrees with $\psi: H \to (-\infty, \infty]$. Thus, we obtain $\alpha \neq -\infty$ and we have shown that α has finite value.

Next, we show that

$$\text{there exists a point } x_0 \in H \text{ such that } \psi(x_0) = \alpha. \tag{2.26}$$

That is, ψ has a minimum value.

PROOF OF (2.26). Since α has finite value, for every natural number n we can choose a point $x_n \in H$ that satisfies $\psi(x_n) < \alpha + 1/n$ $(\leq \alpha + 1)$. Then from (2.23), $\{x_n\}$ is a bounded sequence. Therefore, $\{x_n\}$ contains a weakly convergent subsequence $\{x_{n_k}\}$. We let w-$\lim_{k\to\infty} x_{n_k} = x_0$ $(\in H)$ and set $X_i = \overline{\mathrm{co}}\{x_{n_k} : k \geq i\}$ $(i = 1, 2, \ldots)$. Then by the same method as applied to the above case, we have $x_0 \in X_i$ $(i = 1, 2, \dots)$ and we can select $z_i \in \mathrm{co}\{x_{n_k} : k \geq i\}$ such that

$$\|z_i - x_0\| < 1/i, \quad \psi(z_i) < \alpha + 1/n_i \quad (i = 1, 2, \dots).$$

Hence, from the lower semicontinuity of ψ, we have

$$\psi(x_0) \leq \liminf_{i \to \infty} \psi(z_i) \leq \alpha.$$

On the other hand, from the definition of α, $\alpha \leq \psi(x_0)$, and we have $\psi(x_0) = \alpha$. $\square$

Then, from the next lemma, we obtain

$$x_0 \in D(\partial\varphi), \quad y \in x_0 + (\partial\varphi)x_0 \subset R(I + \partial\varphi).$$

Since $y \in H$ is arbitrary, we have $R(I + \partial\varphi) = H$. Hence we have shown that $-\partial\varphi$ is an m-dissipative operator.

LEMMA 2.14. *Let φ be a (single-valued) convex functional $H \to (-\infty, \infty]$ and $\varphi \not\equiv \infty$. Let $\alpha \geq 0$, $y \in H$, and set*

$$\psi(x) = \varphi(x) + \frac{\alpha}{2}\|x - y\|^2 \quad (x \in H).$$

Then, a necessary and sufficient condition for ψ to take a minimum value at x_0 $(\in H)$ is that

$$x_0 \in D(\partial\varphi) \quad \text{and also} \quad \alpha(y - x_0) \in (\partial\varphi)x_0.$$

PROOF. (Sufficiency) Note that $\varphi(x_0)$ has finite value, because $x_0 \in D(\partial\varphi)$. Since $\alpha(y - x_0) \in (\partial\varphi)x_0$,

$$\varphi(u) \geq \varphi(x_0) + \mathrm{Re}(\alpha(y - x_0), u - x_0) \quad (u \in H)$$

from the definition of $\partial\varphi$. Because

$$\begin{aligned}
\mathrm{Re}(y - x_0, u - x_0) &= \mathrm{Re}(y - x_0, u - y + (y - x_0)) \\
&= \|y - x_0\|^2 + \mathrm{Re}(y - x_0, u - y) \\
&\geq (\|y - x_0\|^2 - \|u - y\|^2)/2,
\end{aligned}$$

$$\varphi(u) + \frac{\alpha}{2}\|u - y\|^2 \geq \varphi(x_0) + \frac{\alpha}{2}\|y - x_0\|^2 \quad (u \in H).$$

That is, $\psi(u) \geq \psi(x_0)$ for all $u \in H$, and ψ has a minimum value at x_0.

(Necessity) Suppose that ψ has a minimum value at x_0. Then

$$\psi(v) = \varphi(v) + \frac{\alpha}{2}\|v - y\|^2 \geq \psi(x_0) = \varphi(x_0) + \frac{\alpha}{2}\|x_0 - y\|^2;$$

that is,

$$\varphi(v) - \varphi(x_0) \geq \frac{\alpha}{2}(\|x_0 - y\|^2 - \|v - y\|^2) \quad (v \in H).$$

In particular, if we think of $v = (1-t)x_0 + tu$ $(0 < t < 1; \; u \in H)$, then $\varphi(v) \leq (1-t)\varphi(x_0) + t\varphi(u)$. That is, because $\varphi(v) - \varphi(x_0) \leq t(\varphi(u) - \varphi(x_0))$, we obtain

$$
\begin{aligned}
t(\varphi(u) - \varphi(x_0)) &\geq \frac{\alpha}{2}(\|x_0 - y\|^2 - \|(1-t)x_0 + tu - y\|^2) \\
&= \frac{\alpha}{2}[-t^2\|x_0 - u\|^2 + 2t\,\mathrm{Re}(y - x_0, u - x_0)] \quad (u \in H, \; 0 < t < 1).
\end{aligned}
$$

When we divide both sides by t and let $t \to 0$, we have

$$
\varphi(u) \geq \varphi(x_0) + \mathrm{Re}(\alpha(y - x_0), u - x_0) \quad (u \in H).
$$

Hence $x_0 \in D(\partial\varphi)$ and $\alpha(y - x_0) \in (\partial\varphi)x_0$. $\square$

EXAMPLE 2.3. We consider the following initial boundary value problem for a system of nonlinear partial differential equations of hyperbolic type.

Given nonnegative functions u_0 and v_0 defined on $(R^+)^2 = \{(x, y) : x \geq 0, \; y \geq 0\}$, the problem is to find nonnegative functions $u(t, x, y)$ and $v(t, x, y)$ (where $t \geq 0$, $(x, y) \in (R^+)^2$) that satisfy (2.27) through (2.29):

$$
\begin{cases}
u_t + u_x + u^2 - v^2 = 0, \\
v_t + v_y + v^2 - u^2 = 0,
\end{cases} \tag{2.27}
$$

$$
u(0, x, y) = u_0(x, y), \qquad v(0, x, y) = v_0(x, y), \tag{2.28}
$$

$$
u(t, 0, y) = v(t, x, 0) = 0. \tag{2.29}
$$

Later we will consider this problem from the viewpoint of nonlinear semigroups (see Example 4.10).

Here, we investigate the dissipativity of the differential operator A which arises in connection with this problem. We set $D_M(A) = \{w = [u, v] : u, v, u_x, v_y$ are continuous in $(R^+)^2$, and $u(0, y) = v(x, 0) = 0$ $(x, y \geq 0)$, $0 \leq u(x, y), v(x, y) \leq M$ $(x, y \geq 0)\}$ for $M > 0$, and define the operator A as follows:

$$
\begin{aligned}
D(A) = \left(\bigcup_{M>0} D_M(A)\right) \\
\cap \{w = [u, v] : [u, v] \in L^1((R^+)^2) \times L^1((R^+)^2), \\
[u_x + u^2 - v^2, v_y + v^2 - u^2] \in L^1((R^+)^2) \times L^1((R^+)^2)\},
\end{aligned}
$$

$$
Aw = A([u, v]) = [-u_x - u^2 + v^2, -v_y - v^2 + u^2] \quad (w = [u, v] \in D(A)).
$$

Then, the (single-valued) operator A is a dissipative operator such that $D(A) \subset L^1((R^+)^2) \times L^1((R^+)^2)$ and $R(A) \subset L^1((R^+)^2) \times L^1((R^+)^2)$. We need to show that for $w_1 = [u_1, v_1]$, $w_2 = [u_2, v_2] \in D(A)$ and $\lambda > 0$,

$$
\|(w_1 - \lambda Aw_1) - (w_2 - \lambda Aw_2)\| \geq \|w_1 - w_2\| \tag{2.30}
$$

holds (see Theorem 2.9). Here $\|\cdot\|$ denotes the norm on the space $L^1((R^+)^2) \times L^1((R^+)^2)$. That is, for $w = [u, v] \in L^1((R^+)^2) \times L^1((R^+)^2)$,

$$\|w\| = \int_0^\infty \int_0^\infty (|u(x, y)| + |v(x, y)|)\, dx\, dy.$$

To show (2.30), we put $\tilde{u} = u_1 - u_2$, $\tilde{v} = v_1 - v_2$, and consider

$$\int_0^T \int_0^T |\tilde{u} + \lambda(\tilde{u}_x + u_1^2 - u_2^2 + v_2^2 - v_1^2)|\, dx\, dy,$$

$$\int_0^T \int_0^T |\tilde{v} + \lambda(\tilde{v}_y + v_1^2 - v_2^2 + u_2^2 - u_1^2)|\, dx\, dy \qquad (T > 0).$$

We set $I_1 = \{x : \tilde{u}(x, y) > 0\} \cap [0, T]$ and $I_2 = \{x : \tilde{u}(x, y) < 0\} \cap [0, T]$ (with $y \geq 0$ fixed); then

$$\int_0^T |\tilde{u} + \lambda(\tilde{u}_x + u_1^2 - u_2^2 + v_2^2 - v_1^2)|\, dx$$

$$\geq \int_{I_1} (\tilde{u} + \lambda(\tilde{u}_x + u_1^2 - u_2^2 + v_2^2 - v_1^2))\, dx$$

$$+ \int_{I_2} (-1)(\tilde{u} + \lambda(\tilde{u}_x + u_1^2 - u_2^2 + v_2^2 - v_1^2))\, dx$$

$$\geq \int_0^T (|\tilde{u}| + \lambda(|u_1^2 - u_2^2| - |v_2^2 - v_1^2|))\, dx.$$

(Here we use the fact that $\int_{I_1} \tilde{u}_x\, dx \geq 0$, $\int_{I_2} \tilde{u}_x\, dx \leq 0$, and $u_1^2 - u_2^2$ have the same sign as $\tilde{u}$.) Integrating with respect to y on $[0, T]$ we have

$$\int_0^T \int_0^T |\tilde{u} + \lambda(\tilde{u}_x + u_1^2 - u_2^2 + v_2^2 - v_1^2)|\, dx\, dy$$

$$\geq \int_0^T \int_0^T (|\tilde{u}| + \lambda(|u_1^2 - u_2^2| - |v_2^2 - v_1^2|))\, dx\, dy.$$

Similarly,

$$\int_0^T \int_0^T |\tilde{v} + \lambda(\tilde{v}_y + v_1^2 - v_2^2 + u_2^2 - u_1^2)|\, dx\, dy$$

$$\geq \int_0^T \int_0^T (|\tilde{v}| + \lambda(|v_2^2 - v_1^2| - |u_2^2 - u_1^2|))\, dx\, dy.$$

When we add the sides of these two inequalities, respectively, and let $T \to \infty$, we obtain (2.30).

EXAMPLE 2.4. Let T be a contraction operator with $D(T) \subset X$ and $R(T) \subset X$ (X is a Banach space); that is, $\|Tx - Ty\| \leq \|x - y\|$ ($x, y \in D(T)$). Then T_h is a dissipative operator, where $T_h = h^{-1}(T - I)$ ($h > 0$). This is because if $x, y \in D(T_h) = D(T)$, then for every $f \in F(x - y)$ we have

$$\mathrm{Re}(T_h x - T_h y, f) = h^{-1}[\mathrm{Re}(Tx - Ty, f) - \|x - y\|^2]$$
$$\leq h^{-1}[\|Tx - Ty\| \|f\| - \|x - y\|^2]$$
$$\leq h^{-1}[\|x - y\|^2 - \|x - y\|^2] = 0.$$

Also, let X_0 be a closed convex set of X, and let T be a contraction operator $X_0 \to X_0$. Then $T_h = h^{-1}(T - I)$ $(h > 0)$ is a dissipative operator, and

$$R(I - \lambda T_h) \supset X_0 = D(T_h) \quad \text{for all } \lambda > 0. \tag{2.31}$$

In particular if $X_0 = X$, that is, T is a contraction operator $X \to X$, then T_h is an m-dissipative operator.

Now, to show (2.31) holds, let $\lambda > 0$ and $x \in X_0$, and define an operator U by

$$Uz = \frac{\lambda}{\lambda + h} Tz + \frac{h}{\lambda + h} x \quad (z \in X_0).$$

Then U is an operator $X_0 \to X_0$ such that

$$\|Uz - Uy\| \leq \frac{\lambda}{\lambda + h} \|z - y\| \quad (y, z \in X_0).$$

Since $\lambda/(\lambda + h) < 1$, $U : X_0 \to X_0$ becomes a strict contraction operator, and from the fixed point theorem, we have $Uz = z$; that is, there exists $z \in X_0$ such that $(I - \lambda T_h)z = x$. Since $\lambda > 0$ and $x \in X_0$ are arbitrary, (2.31) is proved.

REMARK. For the (single-valued) operator T_h of Example 2.4, $\mathrm{Re}(T_h x - T_h y, f) \leq 0$ holds for all $f \in F(x - y)$ with $x, y \in D(T_h)$. In general, if an operator A (not necessarily single-valued), with $D(A) \subset X$ and $R(A) \subset X$, satisfies the following conditions, then A is called a strictly dissipative operator:

> If $x, y \in D(A)$, $x' \in Ax$, and $y' \in Ay$, then we have
> $\mathrm{Re}(x' - y', f) \leq 0$ for all $f \in F(x - y)$. $\qquad (*)$

From the definition of $\langle \cdot, \cdot \rangle_s$, it follows that condition j $(*)$ is equivalent to $\langle x' - y', x - y \rangle_s \leq 0$. Moreover, if A is single-valued, then $(*)$ is equivalent to $\tau_+(Ax - Ay, x - y) \leq 0$ (see (ii) of Corollary 2.7). A strictly dissipative operator is a dissipative operator, but in general, the converse does not hold as we see in the next example. If X^* is a strictly convex Banach space, since the duality mapping F is single-valued, these two concepts coincide.

EXAMPLE 2.5. The set that consists of all ordered pairs (a, b) of real numbers forms a Banach space with the ordinary addition and scalar multiplication and the norm $\|(a, b)\| = |a| + |b|$. We denote this space by $l^1(2)$. Now, we define an operator A by $D(A) = \{(a, 0); -\infty < a < \infty\}$, $A(a, 0) = (a, -a)$ $((a, 0) \in D(A))$; then $A : D(A)$ $(\subset l^1(2)) \to l^1(2)$ is a continuous operator, and

$$\tau_-(A(a, 0) - A(b, 0), (a, 0) - (b, 0)) = 0$$

and

$$\tau_+(A(a, 0) - A(b, 0), (a, 0) - (b, 0)) = 2|a - b|$$

for every $(a, 0)$ and $(b, 0) \in D(A)$. Therefore, A is a dissipative operator but it is not a strictly dissipative operator.

§3. Properties of maximal dissipative and m-dissipative operators

In this section we examine the various properties of maximal dissipative and m-dissipative operators.

LEMMA 2.15. *Let X^* be a strictly convex Banach space. If A is a maximal dissipative operator on S $(\subset X)$, then Ax is a closed convex set for each $x \in D(A) \cap S$.*

PROOF. Note that the duality mapping F is single-valued (see Lemma 2.2). Let $x \in D(A) \cap S$. We show first that Ax is a convex set. Let $x_1', x_2' \in Ax$, and set $x' = tx_1' + (1-t)x_2'$ $(0 < t < 1)$. For each $y \in D(A)$ and $y' \in Ay$, we have

$$\begin{aligned} \operatorname{Re}(x' &- y', F(x - y)) \\ &= t\operatorname{Re}(x_1' - y', F(x - y)) + (1 - t)\operatorname{Re}(x_2' - y', F(x - y)) \leq 0. \end{aligned}$$

Thus, if we set

$$\tilde{A}z = \begin{cases} Az & (z \in D(A) \backslash \{x\}), \\ Ax \cup \{x'\} & (z = x), \end{cases}$$

then $\tilde{A}$ is a dissipative extension of A and $D(\tilde{A}) = D(A)$. Since A is a maximal dissipative operator on S, $\tilde{A}z = Az$ $(z \in D(A) \cap S)$; in particular $\tilde{A}x = Ax$. Hence, $x' \in Ax$ and Ax is a convex set. Next, to show that Ax is a closed set, we let $x_n' \in Ax$ and $\lim_{n \to \infty} x_n' = x'$. Since $\operatorname{Re}(x_n' - y', F(x - y)) \leq 0$ $(n = 1, 2, \dots)$ for each $y \in D(A)$ and $y' \in Ay$, we obtain $\operatorname{Re}(x' - y', F(x - y)) \leq 0$ as $n \to \infty$. If we define $\tilde{A}$ the same way as above, then we have $\tilde{A}x = Ax$ and obtain $x' \in Ax$. Therefore, Ax is a closed set. □

DEFINITION 2.6. Let A be an operator, not necessarily single-valued, with $D(A) \subset X$ and $R(A) \subset X$. A is called a closed operator if $x_n \in D(A)$, $x_n' \in Ax_n$, $\lim_{n \to \infty} x_n = x$, and $\lim_{n \to \infty} x_n' = x'$ imply that $x \in D(A)$ and $x' \in Ax$, that is, the graph $\mathfrak{G}(A)$ of A is a closed set in $X \times X$. An operator A is called demiclosed if $x_n \in D(A)$, $x_n' \in Ax_n$, $\lim_{n \to \infty} x_n = x$, and w-$\lim_{n \to \infty} x_n' = x'$ imply that $x \in D(A)$ and $x' \in Ax$. In the conclusion of the definition of demiclosed, if the condition $x \in D(A)$ holds (but $x' \in Ax$ does not necessarily hold) then A is called almost demiclosed.

From the definition, if A is demiclosed, then A is a closed operator and is almost demiclosed.

LEMMA 2.16. *If A is a maximal dissipative operator on $\overline{D(A)}$, then A is a closed operator.*

PROOF. Let $x_n \in D(A)$, $x_n' \in Ax_n$, $\lim_{n\to\infty} x_n = x$, and $\lim_{n\to\infty} x_n' = x'$. Since A is a dissipative operator, $\langle x_n' - y', x_n - y \rangle_i \leq 0$ $(n = 1, 2, \ldots)$ holds for every $y \in D(A)$ and $y' \in Ay$ ((D$_4$) of Theorem 2.9). Here we let $n \to \infty$; then we obtain $\langle x' - y', x - y \rangle_i \leq 0$ $(y \in D(A), y' \in Ay)$, from the lower semicontinuity of $\langle \cdot, \cdot \rangle_i : X \times X \to (-\infty, \infty)$ (see (v) of Corollary 2.6). Thus, we define $\widetilde{A}$ as follows:

(i) when $x \in D(A)$,

$$\widetilde{A}z = \begin{cases} Az & (z \in D(A)\backslash\{x\}), \\ Ax \cup \{x'\} & (z = x), \end{cases}$$

and

(ii) when $x \notin D(A)$,

$$\widetilde{A}z = \begin{cases} Az & (z \in D(A)), \\ \{x'\} & (z = x). \end{cases}$$

Then $\widetilde{A}$ is a dissipative extension of A. But, $x \in \overline{D(A)}$ and since A is a maximal dissipative operator on $\overline{D(A)}$, it follows that $x \in D(A)$ (so (ii) does not occur). And again, from the maximality of A we obtain $Ax \cup \{x'\} = \widetilde{A}x = Ax$ and thus $x' \in Ax$. $\square$

REMARK. We denote by $\overline{A}$ the operator A that has the closure $\overline{\mathfrak{G}(A)}$ of $\mathfrak{G}(A)$ in $X \times X$ as its graph. $\overline{A}$ is called the closure of the operator A. Obviously $\overline{A}$ is a closed operator and a minimum closed extension of A. As we see from the proof of the above lemma, if A is a dissipative operator, then the closure $\overline{A}$ of A is also a dissipative operator.

LEMMA 2.17. *Let X^* be a uniformly convex Banach space.*

(i) *If A is a maximal dissipative operator on $D(A)$ and is almost demiclosed, then A is demiclosed.*

(ii) *If A is a maximal dissipative operator on $\overline{D(A)}$, then A is demiclosed.*

PROOF. Note that F is a continuous mapping (Theorem 2.3). Let $x_n \in D(A)$, $x_n' \in Ax_n$, $\lim_{n\to\infty} x_n = x$ and w-$\lim_{n\to\infty} x_n' = x'$. Then, from

$$\mathrm{Re}(x_n' - y', F(x_n - y)) \leq 0 \qquad (y \in D(A), y' \in Ay, n = 1, 2, \ldots)$$

and $F(x_n - y) \to F(x - y)$ as $n \to \infty$, we have

$$\mathrm{Re}(x' - y', F(x - y)) \leq 0 \qquad (y \in D(A), y' \in Ay). \qquad (2.32)$$

(i) From the assumption that A is almost demiclosed, we have $x \in D(A)$. If we define $\widetilde{A}$ by $\widetilde{A}z = Az$ $(z \in D(A)\backslash\{x\})$ and $\widetilde{A}z = Ax \cup \{x'\}$ $(z = x)$, then $\widetilde{A}$ is a dissipative extension of A by (2.32). Since A is a maximal dissipative operator on $D(A)$, we have $\widetilde{A}x = Ax$, and obtain $x' \in Ax$.

(ii) If we define $\widetilde{A}$ in the same way as in the proof of Lemma 2.16, then $\widetilde{A}$ is a dissipative extension of A by (2.32). The remaining part of the proof uses the same reasoning as in the proof of Lemma 2.16. We obtain $x \in D(A)$ and $x' \in Ax$. Therefore A is demiclosed. □

LEMMA 2.18. *Let X be a reflexive Banach space and let A be a demiclosed operator. If $x_n \in D(A)$, $\lim_{n\to\infty} x_n = x$, and also $\{ \||Ax_n\|| \}$ is bounded, then $x \in D(A)$ and $\||Ax\|| \leq \liminf_{n\to\infty} \||Ax_n\||$.*

PROOF. Let $\liminf_{n\to\infty} \||Ax_n\|| = M$ and $\varepsilon > 0$ be given arbitrarily. Then by selecting a subsequence $\{n'\}$ from $\{n\}$ and $y_{n'} \in Ax_{n'}$, we can obtain

$$\|y_{n'}\| \leq M + \varepsilon \quad \text{for all } n'.$$

Since X is reflexive, $\{y_{n'}\}$ has a weakly convergent subsequence $\{y_{n_i'}\}$. Set $x' = \text{w-}\lim_{i\to\infty} y_{n_i'}$; then since A is demiclosed, we have $x \in D(A)$ and $x' \in Ax$.

Since $\|x'\| \leq \liminf_{i\to\infty} \|y_{n_i'}\| \leq M + \varepsilon$, we have $\||Ax\|| \leq \|x'\| \leq M + \varepsilon$ and thus obtain $\||Ax\|| \leq M$. □

DEFINITION 2.7. For an operator A, that is not necessarily single-valued, with $D(A) \subset X$ and $R(A) \subset X$ we define A^0x $(x \in D(A))$ by

$$A^0x = \{ x' \in Ax : \|x'\| = \||Ax\|| \}.$$

If Ax $(x \in D(A))$ is an infinite set, it is possible that $A^0x = \varnothing$. Thus, we define the domain $D(A^0)$ of A^0 by

$$D(A^0) = \{ x \in D(A) : A^0x \neq \varnothing \},$$

and call A^0 the canonical restriction of A.

From the definition, we have $A^0 \subset A$, and $A^0 = A$ when A is single valued.

For example, the canonical restriction of the m-dissipative operator $\tilde{g}$ in Example 2.1 is a single-valued function $g^0 : R^1 \to R^1$, given by the following equations:

$$g^0(x) = g_+(x) \quad (\text{if } g_+(x) > 0),$$
$$g^0(x) = 0 \quad (\text{if } g_+(x) \leq 0 \leq g_-(x)),$$
$$g^0(x) = g_-(x) \quad (\text{if } g_-(x) < 0).$$

LEMMA 2.19. *Let X^* be a strictly convex Banach space, and let A be a maximal dissipative operator on $D(A)$.*

(i) *If X is a strictly convex Banach space, then for each $x \in D(A)$, A^0x is a set consisting of a single point or an empty set.*

(ii) *If X is a reflexive Banach space, then $D(A^0) = D(A)$.*

(iii) *If X is a strictly convex and reflexive Banach space, then $D(A^0) = D(A)$ and A^0 is single valued.*

PROOF. Let $x \in D(A)$. Note that Ax is a closed convex set (see Lemma 2.15).

(i) Since the case $\||Ax\|| = 0$ is self-evident, we consider the case $\||Ax\|| \neq 0$. Assume $A^0 x \neq \varnothing$ and let $x', x'' \in A^0 x$ ($\subset Ax$). Then $\|x'\| = \||Ax\|| = \|x''\|$ ($\neq 0$). Since Ax is a convex set, $(x' + x'')/2 \in Ax$. Therefore, we have

$$\||Ax\|| \leq 2^{-1}\|x' + x''\| \leq 2^{-1}(\|x'\| + \|x''\|) = \||Ax\||,$$

and $\|x' + x''\| = \|x'\| + \|x''\|$. Since X is strictly convex, there exists an α (scalar) such that $x' = \alpha x''$. Hence, $|1 + \alpha| \|x''\| = (|\alpha| + 1)\|x''\|$, that is, $|1 + \alpha| = 1 + |\alpha|$, and $|\alpha| = 1$. Therefore, $\alpha = 1$ and $x' = x''$.

(ii) Since Ax is a closed convex set, it follows that Ax is a closed set with respect to the weak topology. From the definition of $\||Ax\||$, we can select a sequence $\{x'_n\}$ in Ax such that $\|x'_n\| \to \||Ax\||$ as $n \to \infty$. Since X is reflexive, we can choose a weakly convergent subsequence $\{x'_{n_i}\}$ of the bounded sequence $\{x'_n\}$. We set $x' = \text{w-}\lim_{i \to \infty} x'_{n_i}$; then $x'_{n_i} \in Ax$ and since Ax is a closed set with respect to the weak topology, we obtain $x' \in Ax$. Furthermore, $\||Ax\|| \leq \|x'\| \leq \lim_{i \to \infty} \|x'_{n_i}\| = \||Ax\||$, that is $\|x'\| = \||Ax\||$. Hence, we have $x' \in A^0 x$ and thus $x \in D(A^0)$. We have shown that $D(A) \subset D(A^0)$ and thus $D(A^0) = D(A)$.

(iii) This is obvious from (i) and (ii). $\square$

THEOREM 2.20. *Let X be a uniformly convex Banach space.*

(i) *If A is a dissipative operator and $R(I - \lambda A) \supset \overline{\text{co}}\, D(A)$ for all $\lambda > 0$, then $\overline{D(A)}$ is a (closed) convex set.*

(ii) *If A is an m-dissipative operator, $\overline{D(A)}$ is a (closed) convex set.*

PROOF. (i) From (ii) of Lemma 2.11, we have

$$\|J_\lambda x - x\| = \lambda \|A_\lambda x\| \leq \lambda \||Ax\|| \qquad (x \in D(A), \ \lambda > 0).$$

From this and the contractiveness of J_λ (see (2.18)), we obtain

$$\lim_{\lambda \to 0+} J_\lambda x = x \qquad (x \in \overline{D(A)}). \tag{2.33}$$

Now, we set $D = \{x \in \overline{\text{co}}\, D(A) : \lim_{\lambda \to 0+} J_\lambda x = x\}$; then we have $\overline{D(A)} = D$ from (2.33). Hence, we need to show that D is a convex set. Let $x, y \in D$. (Since D is a closed set, the result will follow if we can show $(x+y)/2 \in D$.) Since $2^{-1}(x + y) \in \overline{\text{co}}\, D(A)$, note that $J_\lambda(\frac{x+y}{2})$ ($\lambda > 0$) can be defined by the assumption; and

$$\left\|J_\lambda\left(\frac{x+y}{2}\right) - J_\lambda x\right\| \leq \left\|\frac{x-y}{2}\right\|,$$

$$\left\|J_\lambda\left(\frac{x+y}{2}\right) - J_\lambda y\right\| \leq \left\|\frac{x-y}{2}\right\| \qquad (\lambda > 0).$$

From this and (2.33), we have that $\{J_\lambda(\frac{x+y}{2}) : 0 < \lambda < 1\}$ is a bounded set, and thus it is weakly sequentially compact (because a uniformly convex Banach space is reflexive). Therefore, there is a sequence $\{\lambda_n\}$ $(0 < \lambda_n < 1)$ and $z \in X$ such that $\lim_{n \to \infty} \lambda_n = 0$ and

$$\text{w-}\lim_{n \to \infty} J_{\lambda_n} \left(\frac{x+y}{2} \right) = z .$$

From this and (2.33) we have

$$\text{w-}\lim_{n \to \infty} = \left[J_{\lambda_n} \left(\frac{x+y}{2} \right) - J_{\lambda_n} x \right] = z - x ,$$

$$\text{w-}\lim_{n \to \infty} \left[J_{\lambda_n} \left(\frac{x+y}{2} \right) - J_{\lambda_n} y \right] = z - y .$$

Hence, we obtain

$$\|z - x\| \le \liminf_{n \to \infty} \left\| J_{\lambda_n} \left(\frac{x+y}{2} \right) - J_{\lambda_n} x \right\| \le \left\| \frac{x-y}{2} \right\| ,$$

$$\|z - y\| \le \liminf_{n \to \infty} \left\| J_{\lambda_n} \left(\frac{x+y}{2} \right) - J_{\lambda_n} y \right\| \le \left\| \frac{x-y}{2} \right\|$$

and have

$$\|x - y\| = \|(x - z) + (z - y)\| \le \|x - z\| + \|z - y\| \le \|x - y\| .$$

This yields

$$\|x - z\| = \|z - y\| = \|x - y\|/2 \quad \text{and} \quad \|(x - z) + (z - y)\| = \|x - y\| .$$

From the uniform convexity of X, we obtain $z = (x + y)/2$. Thus,

$$\text{w-}\lim_{n \to \infty} J_{\lambda_n} \left(\frac{x+y}{2} \right) = \frac{x+y}{2} .$$

Now, since we can obtain the same result by starting with an arbitrary subsequence $\{J_{\lambda'}(\frac{x+y}{2})\}$ instead of $\{J_\lambda(\frac{x+y}{2}) : 0 < \lambda < 1\}$, where $\lambda' \to 0$ (that is, $\{J_{\lambda'}(\frac{x+y}{2})\}$ has a subsequence that converges weakly to $(x + y)/2$, it follows that

$$\text{w-}\lim_{\lambda \to 0+} J_\lambda \left(\frac{x+y}{2} \right) = \frac{x+y}{2} .$$

Hence,

$$\text{w-}\lim_{\lambda \to 0+} \left[J_\lambda \left(\frac{x+y}{2} \right) - J_\lambda x \right] = \frac{y-x}{2} . \tag{2.34}$$

And also, from

$$\left\| \frac{y-x}{2} \right\| \le \liminf_{\lambda \to 0+} \left\| J_\lambda \left(\frac{x+y}{2} \right) - J_\lambda x \right\|$$

$$\le \limsup_{\lambda \to 0+} \left\| J_\lambda \left(\frac{x+y}{2} \right) - J_\lambda x \right\|$$

$$\le \left\| \frac{y-x}{2} \right\|$$

we have

$$\lim_{\lambda \to 0+} \left\| J_\lambda \left(\frac{x+y}{2} \right) - J_\lambda x \right\| = \left\| \frac{y-x}{2} \right\|.$$

Since X is uniformly convex, from the above equation and (2.34), we obtain

$$\lim_{\lambda \to 0+} \left[J_\lambda \left(\frac{x+y}{2} \right) - J_\lambda x \right] = \frac{y-x}{2}.$$

When we apply (2.33) to this, we obtain

$$\lim_{\lambda \to 0+} J_\lambda \left(\frac{x+y}{2} \right) = \frac{x+y}{2},$$

and hence $(x+y)/2 \in D$. Thus D is a convex set.

(ii) This is obvious from (i). □

We now consider the following problem:

> "Is an m-dissipative operator A determined by its canonical
> restriction A^0? That is, when A and B are m-dissipative
> operators, is $A = B$ if $A^0 = B^0$?"

LEMMA 2.21. *Let C $(\neq \varnothing)$ be a closed convex set of X, and set*

$$\|\|C\|\| = \inf\{\|x\| : x \in C\} \qquad and \quad C^0 = \{x \in C : \|x\| = \|\|C\|\|\}.$$

(i) *A necessary and sufficient condition for x to be an element of C^0 is that $x \in C$ and for every $y \in C$ there is an $f \in F(x)$ such that $\|x\|^2 \leq \mathrm{Re}(y, f)$.*

(ii) *If X^* is a uniformly convex Banach space, then $C^0 \neq \varnothing$ and $F(C^0)$ $(= \{F(x) : x \in C^0\})$ is a set consisting of only one point.* ([5])

PROOF. (i) (Necessity) Let $y \in C$ and $0 < \alpha < 1$. Since $(1-\alpha)x + \alpha y \in C$, $\|x + \alpha(y - x)\| = \|(1-\alpha)x + \alpha y\| \geq \|\|C\|\| = \|x\|$. Hence,

$$\tau_+(y - x, x) = \lim_{\alpha \to 0+} \frac{\|x + \alpha(y - x)\| - \|x\|}{\alpha} \geq 0.$$

From Theorem 2.5, there is an $f \in F(X)$ such that

$$\mathrm{Re}(y - x, f) = \|x\| \tau_+(y - x, x).$$

Thus $0 \leq \mathrm{Re}(y - x, f) = \mathrm{Re}(y, f) - \|x\|^2$; that is, $\|x\|^2 \leq \mathrm{Re}(y, f)$.

(Sufficiency) Since $\|x\|^2 \leq \mathrm{Re}(y, f) \leq \|y\| \|x\|$, $\|x\| \leq \|y\|$ holds for all $y \in C$. Therefore, $x \in C^0$.

(ii) Since X^* is uniformly convex, X^* is strictly convex and a reflexive Banach space; thus X is also a reflexive Banach space. Therefore, by the same method used for the proof of (ii) in Lemma 2.19, we obtain $C^0 \neq \varnothing$. Next, we let $x_i \in C^0$ $(i = 1, 2)$. Then, $\|x_1\| = \|x_2\|$ $(= \|\|C\|\|)$, and from (i) we have

$$\|x_2\|^2 \leq \mathrm{Re}(x_1, F(x_2)) \leq |(x_1, F(x_2))| \leq \|x_1\| \|x_2\| = \|x_2\|^2.$$

(Note that F is a single-valued mapping.) Hence, $(x_1, F(x_2)) = \|x_1\|^2 = \|x_2\|^2 = \|F(x_2)\|^2$, and we have $F(x_2) = F(x_1)$. □

([5]) The unique point in the set $F(C^0)$ is denoted by $F(C^0)$.

LEMMA 2.22. *Let X^* be a uniformly convex Banach space.*

(i) *Let A be a maximal dissipative operator on $\overline{D(A)}$ that satisfies $R(I - \lambda A) \supset D(A)$ $(\lambda > 0)$. Then, for all $x \in D(A)$ we have*

$$\lim_{\lambda \to 0+} F(A_\lambda x) = F(A^0 x), \tag{2.35}$$

where $A_\lambda = \lambda^{-1}(J_\lambda - I)$. Moreover, if X is also a uniformly convex Banach space, then for all $x \in D(A)$ we have

$$\lim_{\lambda \to 0+} A_\lambda x = A^0 x. \tag{2.36}$$

(ii) *If A is an m-dissipative operator, (2.35) holds for all $x \in D(A)$; moreover, if X is also a uniformly convex Banach space, then (2.36) holds for all $x \in D(A)$.*

PROOF. (i) Let $x \in D(A)$. By (ii) of Lemma 2.11,

$$\lambda^{-1}\|J_\lambda x - x\| = \|A_\lambda x\| \leq \|\|Ax\|\| \qquad (\lambda > 0).$$

Take an arbitrary sequence $\{\lambda_n\}$ of positive numbers such that $\lambda_n \to 0$ as $n \to \infty$, and consider the sequence $\{A_{\lambda_n} x\}$. Since $\{A_{\lambda_n} x\}$ is a bounded sequence in a reflexive Banach space X (from the uniform convexity of X^*, X^* is reflexive and so is X), $A_{\lambda_n} x$ contains a weakly convergent subsequence $\{A_{\lambda_n'} x\}$. Now set $y = \text{w-lim}_{n \to \infty} A_{\lambda_n'} x$. Since $A_{\lambda_n'} x \in A J_{\lambda_n'} x$ ((i) of Lemma (2.11)), $\lim_{n \to \infty} J_{\lambda_n'} x = x$, and A is demiclosed ((ii) of Lemma 2.17), and thus we obtain $y \in Ax$. From this we have

$$\|\|Ax\|\| \leq \|y\| \leq \liminf_{n \to \infty} \|A_{\lambda_n'} x\| \leq \limsup_{n \to \infty} \|A_{\lambda_n'} x\| \leq \|\|Ax\|\|.$$

Hence, we have $\|y\| = \|\|Ax\|\| = \lim_{n \to \infty} \|A_{\lambda_n'} x\|$, and obtain $y \in A^0 x$. Next, since $\|F(A_{\lambda_n'} x)\| = \|A_{\lambda_n'} x\| \leq \|\|Ax\|\|$, $\{F(A_{\lambda_n'} x)\}$ is a bounded sequence in X^*, and contains a weakly convergent subsequence $\{F(A_{\lambda_{n_k}'} x)\}$. If we set $\text{w-lim}_{k \to \infty} F(A_{\lambda_{n_k}'} x) = y^*$, then by the dissipativity of A we have

$$\text{Re}(A_{\lambda_{n_k}'} x - y, F(A_{\lambda_{n_k}'} x)) = (\lambda_{n_k'})^{-1} \text{Re}(A_{\lambda_{n_k}'} x - y, F(J_{\lambda_{n_k}'} x - x)) \leq 0,$$

that is, we obtain $\|A_{\lambda_{n_k}'} x\|^2 \leq \text{Re}(y, F(A_{\lambda_{n_k}'} x))$. Now let $k \to \infty$; then $\|y\|^2 \leq \text{Re}(y, y^*)$. From this and

$$\|y^*\| \leq \lim_{k \to \infty} F(A_{\lambda_{n_k}'} x)\| = \lim_{k \to \infty} \|A_{\lambda_{n_k}'} x\| = \|y\|,$$

we have

$$\|y\|^2 \leq \text{Re}(y, y^*) \leq |(y, y^*)| \leq \|y\| \|y^*\| = \|y\|^2.$$

Therefore,

$$(y, y^*) = \|y\|^2 = \|y^*\|^2 \tag{2.37}$$

and we have $y^* = F(y)$. Thus,

$$\text{w-}\lim_{k \to \infty} F(A_{\lambda_{n_k'}} x) = F(y). \tag{2.38}$$

Also, $\lim_{k \to \infty} \|F(A_{\lambda_{n_k'}} x)\| = \|y\| = \|F(y)\|$. Therefore, from the uniform convexity of X^* we obtain

$$\lim_{k \to \infty} F(A_{\lambda_{n_k'}} x) = F(y).$$

Since Ax is a closed convex set (see Lemma 2.15), we have $F(y) = F(A^0 x)$ from part (ii) of Lemma 2.21. Therefore

$$\lim_{k \to \infty} F(A_{\lambda_{n_k'}} x) = F(A^0 x). \tag{2.39}$$

Thus, we have shown that for every sequence $\{\lambda_n\}$ of positive numbers that converges to 0 the sequence $\{F(A_{\lambda_n} x)\}$ contains a subsequence that converges to the same limit $F(A^0 x)$ Therefore, (2.35) holds.

Next, let X be a uniformly convex Banach space with $x \in D(A)$, and let $\{\lambda_n\}$ be an arbitrary sequence of positive numbers which converges to 0. Then, as shown above, $\{A_{\lambda_n} x\}$ contains a weakly convergent subsequence $\{A_{\lambda_n'} x\}$. If we set $y = \text{w-}\lim_{n \to \infty} A_{\lambda_n'} x$, then $\|y\| = \lim_{n \to \infty} \|A_{\lambda_n'} x\|$ and $y \in A^0 x$. But from the uniform convexity of X, A^0 is a single-valued operator (see (iii) of Lemma 2.19). Therefore $y = A^0 x$. Thus, $A^0 x = \text{w-}\lim_{n \to \infty} A_{\lambda_n'} x$ and $\|A^0 x\| = \lim_{n \to \infty} \|A_{\lambda_n'} x\|$. From the uniform convexity of X we obtain

$$A^0 x = \lim_{n \to \infty} A_{\lambda_n'} x.$$

Therefore, we have shown (2.36).

(ii) This is obvious from (i) of Lemma 2.12. $\quad\square$

THEOREM 2.23. *Let X^* be a uniformly convex Banach space, let A be an m-dissipative operator, and let A' be a single-valued operator that satisfies $A'u \in Au$ and $\|A'u\| \leq \varphi(\||Au\||)$ $(u \in D(A))$ and maps $D(A)$ to X. Here, $\varphi: [0, \infty) \to [0, \infty)$ is a bounded function on every bounded interval of $[0, \infty)$.*

If $x' \in X$ and $x \in D(A)$ satisfy

$$\text{Re}(x' - A'u, F(x - u)) \leq 0 \qquad (u \in D(A)), \tag{2.40}$$

then $x' \in Ax$.

PROOF. (I) The case $x' = 0$. In this case (2.40) becomes

$$\text{Re}(A'u, F(x - u)) \geq 0 \qquad (u \in D(A)). \tag{2.40}'$$

If we set $u = J_\lambda x$ in this inequality, then for all $\lambda > 0$ we have

$$\text{Re}(A' J_\lambda x, F(A_\lambda x)) \; (= -\lambda^{-1} \text{Re}(A' J_\lambda x, F(x - J_\lambda x))) \leq 0. \tag{2.41}$$

Since $\|A'J_\lambda x\| \leq \varphi(\||AJ_\lambda x\||)$ and $\||AJ_\lambda x\|| \leq \|A_\lambda x\| \leq \||Ax\||$ (see Lemma 2.11), $\{A'J_\lambda x : 0 < \lambda \leq 1\}$ is a bounded subset of X (reflexive Banach space). Therefore, there is a sequence $\{\lambda_n\}$ of positive numbers with $\lim_{n\to\infty} \lambda_n = 0$ and there is a point $y \in X$ such that $\text{w-}\lim_{n\to\infty} A'J_{\lambda_n}x = y$. Also,

$$\|J_\lambda x - x\| = \lambda\|A_\lambda x\| \leq \lambda\||Ax\|| \to 0 \quad \text{as } \lambda \to 0+.$$

Then, since A is demiclosed (see Lemmas 2.12 and 2.17), we have $y \in Ax$. From (ii) of Lemma 2.22, we have $\lim_{\lambda\to 0+} F(A_\lambda x) = F(A^0 x)$. From this and $\text{w-}\lim_{n\to\infty} A'J_{\lambda_n}x = y$ and (2.41), we obtain $\text{Re}(y, F(A^0 x)) \leq 0$, that is, $\text{Re}(y, F(z)) \leq 0$ $(z \in A^0 x)$. (Note that since Ax is a closed convex set, $F(z) = F(A^0 x)$ for all $z \in A^0 x$. See part (ii) of Lemma 2.21 and its footnote.) Also, we have from (i) of Lemma 2.21 that

$$\|z\|^2 \leq \text{Re}(y, F(z)) \quad (z \in A^0 x).$$

(Here we use that $y \in Ax$.) As a result, we have $\|z\|^2 \leq 0$ $(z \in A^0 x)$ and obtain $A^0 x = \{0\}$. Thus $x' = 0 \in Ax$.

(II) The case $x' \neq 0$. Define operators B and B' by

$$Bu = Au - x' \ (= \{u' - x' : u' \in Au\}), \quad B'u = A'u - x' \quad (u \in D(A)).$$

Then, B is an m-dissipative operator with $D(B) = D(B') = D(A)$ and B' is a single-valued operator that satisfies $B'u \in Bu$ $(u \in D(B))$. Now we set

$$\psi_0(t) = \sup\{\varphi(s) : 0 \leq s \leq t\} \quad (t \geq 0);$$

then $\psi^0 \colon [0, \infty) \to [0, \infty)$ is an increasing function and $\varphi(t) \leq \psi_0(t)$ $(t \geq 0)$. Since $\||Au\|| \leq \||Bu\|| + \|x'\|$ $(u \in D(B))$, for $u \in D(B)$ we have

$$\|B'u\| \leq \|A'u\| + \|x'\| \leq \varphi(\||Au\||) + \|x'\|$$
$$\leq \psi_0(\||Au\||) + \|x'\| \leq \psi_0(\||Bu\|| + \|x'\|) + \|x'\|.$$

Hence, if we set $\psi(t) = \psi_0(t+\|x'\|)+\|x'\|$ $(t \geq 0)$, then $\psi \colon [0, \infty) \to [0, \infty)$ is an increasing function and

$$\|B'u\| \leq \psi(\||Bu\||) \quad (u \in D(B)).$$

And, the assumption (2.40) becomes

$$\text{Re}(B'u, F(x - u)) \geq 0 \quad (u \in D(B)). \tag{2.40}''$$

Thus, if we consider B, B' and ψ, instead of A, A', and φ, we have case (I) and obtain $0 \in Bx$ $(= Ax - x')$. That is, $x' \in Ax$. $\square$

COROLLARY 2.4. *Let X^* be a uniformly convex Banach space, and let both A and B be m-dissipative operators.*
(i) *If $D(A) = D(B)$ and $A^0 x \cap B^0 x \neq \varnothing$ for each $x \in D(A)$, then $A = B$.*
(ii) *If $A^0 = B^0$, then $A = B$.*

PROOF. (i) Consider a single-valued operator A' with $D(A') = D(A)$ such that $A'u \in A^0 u \cap B^0 u$ for each $u \in D(A)$ and set $B' = A'$. Then

$$\|A'u\| = \|B'u\| = \||Au\|| = \||Bu\|| \qquad (u \in D(A) = D(B)).$$

Hence, by considering the function $\varphi(t) = t$ $(t \geq 0)$, we see that A, A', B, and B' satisfy the conditions of Theorem 2.23.

Now, let $x \in D(A) = D(B)$. If $x' \in Bx$, then from the dissipativity of the operator B, we have

$$\mathrm{Re}(x' - A'u, F(x - u)) = \mathrm{Re}(x' - B'u, F(x - u)) \leq 0 \qquad (u \in D(A)).$$

That is, (2.40) holds, and hence, we obtain from the previous theorem that $x' \in Ax$. Therefore, we have $Bx \subset Ax$. Similarly, we obtain $Ax \subset Bx$ and hence $Ax = Bx$.

(ii) When we note that $D(A^0) = D(A)$ and $D(B^0) = D(B)$ (see (ii) of Lemma 2.19), (ii) is self-evident from (i). □

Finally, we consider the following problem:

> "Is a maximal dissipative operator an m-dissipative operator?"

(As we saw in Lemma 2.12, an m-dissipative operator is always a maximal dissipative operator.)

THEOREM 2.25. *Let X be a Hilbert space and let X_0 $(\subset X)$ be a closed convex set. If A is a dissipative operator with $D(A) \subset X_0$, then for every $y \in X$ there exists an $x \in X_0$ such that*

$$\mathrm{Re}(-u' + x, u - x) \geq \mathrm{Re}(y, u - x) \qquad (u \in D(A),\ u' \in Au). \qquad (2.42)$$

PROOF. (I) The case $y = 0$. We will show that there exists an $x \in X_0$ that satisfies

$$\mathrm{Re}(-u' + x, u - x) \geq 0 \qquad (u \in D(A),\ u' \in Au). \qquad (2.42)'$$

For each $u \in D(A)$ and $u' \in Au$ we set

$$X_0(u, u') = \{x \in X_0 : \mathrm{Re}(-u' + x, u - x) \geq 0\}.$$

Since $u \in X_0(u, u')$, we have $X_0(u, u') \neq \varnothing$. We see that $X_0(u, u')$ is a convex set from the fact that $\mathrm{Re}(-u' + x, u - x) = \mathrm{Re}(-u', u) + \mathrm{Re}(-u', -x) + \mathrm{Re}(x, u) - \|x\|^2$ and $\|\cdot\|^2 \colon X \to [0, \infty)$ is a convex function (that is, $\|tx_1 + (1 - t)x_2\|^2 \leq t\|x_1\|^2 + (1 - t)\|x_2\|^2$ $(0 < t < 1)$). Also, $X_0(u, u')$ is obviously a closed set. Further, from $\|x\|^2 \leq \|u\|\|u'\| + (\|u\| + \|u'\|)\|x\|$ $(x \in X_0(u, u'))$, we have that $X_0(u, u')$ is a bounded set. Since a closed convex subset of X is a weakly closed set (a closed set with respect to the weak topology, see Theorem 1.7), we have that $X_0(u, u')$ is a bounded weakly closed set. Therefore, it is compact in the weak topology. [6]

Now, to show that there is an $x \in X_0$ that satisfies $(2.42)'$, we prove an equivalent proposition

$$\bigcap \{X_0(u, u') : u \in D(A), \ u' \in Au\} \neq \varnothing. \tag{2.43}$$

Since $X_0(u, u')$ is weakly compact, in order to prove (2.43) it suffices to show that for every finite number of $u_i \in D(A)$ and $u_i' \in Au$ $(i = 1, 2, \ldots, n)$, we have

$$\bigcap_{i=1}^{n} X_0(u_i, u_i') \neq \varnothing, \tag{2.44}$$

(that is, the family of sets $\{X_0(u, u') : u \in D(A) \text{ and } u' \in Au\}$ has a finite intersection property.)

Thus we need to show that (2.44) holds. Let R^n be n-dimensional Euclidean space and set

$$K = \left\{ \boldsymbol{\lambda} = (\lambda_1, \lambda_2, \ldots, \lambda_n) \in R^n : \lambda_i \geq 0 \ (i = 1, 2, \ldots, n), \ \sum_{i=1}^{n} \lambda_i = 1 \right\}.$$

Then K is a bounded closed convex subset of R^n and hence K is compact. Now we define a function (functional) $\psi : K \times K \to R^1 \ (= (-\infty, \infty))$ by

$$\psi(\boldsymbol{\lambda}, \boldsymbol{\mu}) = \sum_{i=1}^{n} \mu_i \operatorname{Re}(x(\boldsymbol{\lambda}) - u_i', x(\boldsymbol{\lambda}) - u_i)$$

$$(\boldsymbol{\lambda} = (\lambda_1, \lambda_2, \ldots, \lambda_n), \ \boldsymbol{\mu} = (\mu_1, \mu_2, \ldots, \mu_n) \in K),$$

where $x(\boldsymbol{\lambda}) = \sum_{j=1}^{n} \lambda_j u_j \ (\in X_0$ should be noted). ψ is continuous, convex with respect to $\boldsymbol{\lambda}$, and a linear function with respect to $\boldsymbol{\mu}$. From the Minimax theorem (see Appendix) it follows that there exists a $\boldsymbol{\lambda}^0 \in K$ such that

$$\psi(\boldsymbol{\lambda}^0, \boldsymbol{\mu}) \leq \max\{\psi(\boldsymbol{\lambda}, \boldsymbol{\lambda}) : \boldsymbol{\lambda} \in K\} \qquad (\boldsymbol{\mu} \in K). \tag{2.45}$$

Since $u_i \in D(A)$ and $u_i' \in Au_i$ $(i = 1, 2, \ldots, n)$, we have from the dissipativity of A that

$$\operatorname{Re}(u_j' - u_i', u_j - u_i) \leq 0 \qquad (i, j = 1, 2, \ldots, n).$$

---

[6] We can obtain this from Theorem 1.4 (see, for example, p. 425 of [45]).

Therefore, for all $\lambda = (\lambda_1, \lambda_2, \ldots, \lambda_n) \in K$ we have

$$\psi(\lambda, \lambda) = \sum_{i=1}^{n} \lambda_i \, \mathrm{Re} \left(\sum_{j=1}^{n} \lambda_j(u_j - u_i'), \sum_{k=1}^{n} \lambda_k(u_k - u_i) \right)$$

$$= \sum_{i,j=1}^{n} \lambda_i \lambda_j \, \mathrm{Re}(-u_i', u_j - u_i)$$

$$= \frac{1}{2} \sum_{i,j=1}^{n} \lambda_i \lambda_j \, \mathrm{Re}(u_j' - u_i', u_j - u_i) \leq 0 .$$

From this and (2.45), we have $\psi(\lambda_0, \mu) \leq 0$ $(\mu \in K)$; that is, we obtain

$$\sum_{i=1}^{n} \mu_i \, \mathrm{Re}(-u_i' + x(\lambda^0), u_i - x(\lambda^0)) \geq 0 \qquad (\mu = (\mu_1, \mu_2, \ldots, \mu_n) \in K) .$$

When we set

$$\mu = \mu^{(i)} \equiv (0, \ldots, 0, \overset{(ith)}{1}, 0, \ldots, 0) \qquad (i = 1, 2, \ldots, n)$$

in the above inequality, we have

$$\mathrm{Re}(-u_i' + x(\lambda^0), u_i - x(\lambda^0)) \geq 0 \qquad (i = 1, 2, \ldots, n) .$$

Therefore, we have $x(\lambda^0) \in X_0(u_i, u_i')$ $(i = 1, 2, \ldots, n)$, and (2.44) holds.

(II) The case $y \neq 0$. We set $X_1 = X_0 - \{y\}$ $(\{x - y : x \in X_0\})$; then X_1 is a closed convex set. Next, set $D(B) = D(A) - \{y\}$ $(= \{x - y : x \in D(A)\})$ and define an operator B (not necessarily singled valued) with domain $D(B)$ by

$$B(x - y) = Ax \qquad (x - y \in D(B)) .$$

Then B is a dissipative operator and $D(B) \subset X_1$. Therefore, we can apply the result of case (I) to X_1 and B, and obtain that there exists $(x - y) \in X_1$ such that

$$\mathrm{Re}(-u' + (x-y), (u-y) - (x-y)) \geq 0 \qquad (u-y \in D(B), \ u' \in B(u-y) = Au) . \tag{2.42''}$$

In other words, there exists an $x \in X_0$ such that

$$\mathrm{Re}(-u' + x, u - x) \geq \mathrm{Re}(y, u - x) \qquad (u \in D(A), \ u' \in Au) . \qquad \square$$

THEOREM 2.26. *Let X_0 be a closed convex set of a Hilbert space X and let B be a dissipative operator with $D(B) \subset X_0$. Then, there is an m-dissipative operator A such that $D(A) \subset X_0$ and $B \subset A$.*

PROOF. Set

$$\mathfrak{A} = \{\text{dissipative operator } C : B \subset C, \ D(C) \subset X_0\} .$$

Then, $\mathfrak{A}$ $(\neq \varnothing)$ forms an ordered set with relation "$\subset$", and every totally ordered subset of $\mathfrak{A}$ always has an upper bound. (In fact, for a totally ordered subset $\{C_\alpha\}$ of $\mathfrak{A}$ we define $D(C) = \bigcup_\alpha D(C_\alpha)$ and $Cx = \bigcup_{\alpha \in \Lambda(x)} C_\alpha x$ $(x \in D(C))$, where $\Lambda(x) = \{\alpha : x \in D(C_\alpha)\}$. Then C is an upper bound of $\{C_\alpha\}$.) Therefore, by Zorn's lemma $\mathfrak{A}$ contains a maximal element A. Obviously, A is a dissipative operator with $B \subset A$ and $D(A) \subset X_0$. Moreover,

$$R(I - A) = X. \tag{2.46}$$

In fact, by Theorem 2.25 for every $y \in X$ there is an $x \in X_0$ such that

$$\mathrm{Re}(u' - (x - y), u - x) \le 0 \qquad (u \in D(A), \ u' \in Au).$$

Since A is a maximal element of $\mathfrak{A}$, A is a maximal dissipative operator on X_0. Hence, we obtain

$$x \in D(A) \qquad \text{and} \qquad (x - y) \in Ax \quad (\text{that is, } y \in (I - A)x),$$

and (2.46) is proved. Therefore, by Lemma 2.13, A is an m-dissipative operator. $\square$

As a corollary of the above theorem, we obtain the following result:

COROLLARY 2.27. *The maximal dissipative operator in a Hilbert space is always an m-dissipative operator.*

PROOF. Let A be the maximal dissipative operator in a Hilbert space X. If we let $X_0 = X$ and apply Theorem 2.26, it follows that there is an m-dissipative operator $\tilde{A}$ such that $A \subset \tilde{A}$. Since A is maximal, we have $A = \tilde{A}$ and thus A is an m-dissipative operator. $\square$

The above corollary does not hold in a Banach space. Next we give such an example.

EXAMPLE 2.6. Let $1 < p < \infty$ and $p \ne 2$. A set consisting of all ordered pairs (a, b) of real numbers forms a uniformly convex (real) Banach space through ordinary addition and scalar multiplication and the norm $\|(a, b)\| = (|a|^p + |b|^p)^{1/p}$. We denote this space by $l^p(2)$.([7])

Now we define an (single-valued) operator A with $D(A) \subset l^p(2)$ and $R(A) \subset l^p(2)$ as follows:

$$D(A) = \{(0, 0), (0, 1), (1, 0)\},$$
$$A(0, 0) = (0, 0), \qquad A(0, 1) = (-1, 0), \qquad A(1, 0) = (0, 1).$$

Then A is a dissipative operator. We can show this by investigating condition (D_2) of Theorem 2.9. Set

$$\Delta_1 = \{(a, b) : 0 < a < b < 1/2\}, \qquad \Delta_2 = \{(a, b) : 0 < b < a < 1/2\}.$$

(See Figure 2.1.)

([7]) $l^p(2)$ is the so-called two-dimensional Minkowski space.

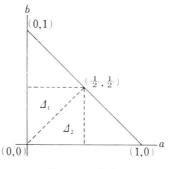

FIGURE 2.1

(i) Let B be a dissipative extension of A. If $p > 2$, then $D(B) \cap \Delta_1 = \varnothing$; if $1 < p < 2$, then $D(B) \cap \Delta_2 = \varnothing$.

In fact, if $(a, b) \in D(B) \cap (\Delta_1 \cup \Delta_2)$ and $(a', b') \in B(a, b)$, then since B is a dissipative extension of A, we obtain

$$\tau_-((0, 0) - (a', b'), (0, 0) - (a, b))$$
$$= (a^p + b^p)^{(1/p)-1}(a' a^{p-1} + b' b^{p-1}) \leq 0,$$
$$\tau_-((-1, 0) - (a', b'), (0, 1) - (a, b))$$
$$= -[a^p + (1 - b)^p]^{(1/p)-1}[-(1 + a')a^{p-1} + b'(1 - b)^{p-1}] \leq 0,$$
$$\tau_-((0, 1) - (a', b'), (1, 0) - (a, b))$$
$$= -[(1 - a)^p + b^p]^{(1/p)-1}[a'(1 - a)^{p-1} + (1 - b')b^{p-1}] \leq 0$$

(condition (D_3) of Theorem 2.9). Hence, we have $a' a^{p-1} + b' b^{p-1} \leq 0$, $b'(1 - b)^{p-1} \geq (1 + a')a^{p-1}$, and $a'(1 - a)^{p-1} + (1 - b')b^{p-1} \geq 0$. From this we obtain

$$b^{p-1} + (1 - b)^{p-1} \geq a^{p-1} + (1 - a)^{p-1}. \tag{2.47}$$

We consider the function

$$\varphi(t) = t^{p-1} + (1 - t)^{p-1};$$

if $p > 2$, then φ is strictly decreasing on the interval $(0, 1/2)$, and if $1 < p < 2$, then φ is a strictly increasing function on the interval $(0, 1/2)$. Therefore, there is no point $(a, b) \in D(B) \cap \Delta_1$ that satisfies (2.47) when $p > 2$. That is, $D(B) \cap \Delta_1 = \varnothing$. Also, there is no point $(a, b) \in D(B) \cap \Delta_2$ that satisfies (2.47) when $1 < p < 2$. That is, $D(B) \cap \Delta_2 = \varnothing$.

(ii) Let $\widetilde{A}$ be a maximal dissipative extension of A; that is, $\widetilde{A}$ is a maximal dissipative operator such that $\widetilde{A} \supset A$ (see (ii) of Lemma 2.12). Then $\widetilde{A}$ is not an m-dissipative operator.

In fact, if $\widetilde{A}$ is an m-dissipative operator, then by (ii) of Theorem 2.20, $\overline{D(\widetilde{A})}$ is a closed convex set. From this and $D(\widetilde{A}) \supset D(A)$, it must be that

$$\overline{D(\widetilde{A})} \supset \Delta_1 \cup \Delta_2.$$

On the other hand, from (i), if $p > 2$, then $\overline{D(\widetilde{A})} \cap \Delta_1 = \varnothing$, and also, if $1 < p < 2$, then $\overline{D(\widetilde{A})} \cap \Delta_2 = \varnothing$. This is a contradiction. Thus, $\widetilde{A}$ is a maximal dissipative operator but is not an m-dissipative operator.

REMARK. Let n be a fixed arbitrary natural number. The set consisting of all ordered n pairs of real numbers $(a_1, a_2, \ldots, a_n)$ forms a (real) Banach space with the ordinary addition and scalar multiplication and norm given by $\|(a_1, \ldots, a_n)\| = \max\{|a_i| : 1 \le i \le n\}$. We denote this space by $l^{\infty}(n)$. Then, it is known that a maximal dissipative operator in the space $l^{\infty}(n)$ is always an m-dissipative operator. (See [37] of the bibliography.)

CHAPTER 3

Semigroups of Nonlinear Contractions

As in Chapter 2, X is a real or complex Banach space, unless specified otherwise. When X_0 ($\neq \varnothing$) is a subset of X, we use the notation $T \in \text{Cont}(X_0)$ for a contraction operator T that maps X_0 into itself. That is, for $T \in \text{Cont}(X_0)$ we have $T: X_0 \to X_0$ and

$$\|Tx - Ty\| \leq \|x - y\| \qquad (x, y \in X_0). \tag{3.1}$$

§1. Semigroups of contractions

DEFINITION 3.1. Let $X_0 (\neq \varnothing)$ be a subset of X. $\{T(t): t \geq 0\}$ is called a semigroup of contractions on X_0 if (i) through (iii) hold:

(i) $T(t) \in \text{Cont}(X_0)$ for each $t > 0$ and $T(0)x = x$ $(x \in X_0)$;

(ii) $T(t + s) = T(t)T(s)$ $(t, s \geq 0)$;

(iii) $\lim_{t \to 0+} T(t)x = x$ $(x \in X_0)$.

It is easy to see that $T(t)x: [0, \infty) \to X_0$ is continuous for each $x \in X_0$.

In particular, when X_0 is a linear set and each $T(t)$ is a linear operator, a semigroup of contractions $\{T(t): t \geq 0\}$ on X_0 is called a semigroup of linear contractions on X_0.

REMARK. It is easy to see that a semigroup of contractions on X_0 can be extended uniquely to a semigroup of contractions on $\overline{X_0}$. Therefore, when we consider a semigroup of contractions on X_0, we may assume, without loss of generality, that X_0 is closed.

LEMMA 3.1. Let $\{T(t): t \geq 0\}$ be a semigroup of contractions on X_0 and let $x \in X_0$. If $\liminf_{t \to 0+} \|T(t)x - x\|/t$ (denoted below by L) $< \infty$, then

(i) $\|T(t + h)x - T(t)x\| \leq hL$ $(t, h \geq 0)$;

(ii) for each $t \geq 0$, the limit

$$\lim_{h \to 0+} \|T(t + h)x - T(t)x\|/h$$

exists and is finite-valued. (We denote this limit by $\varphi(t)$.) In addition, $\varphi(t): [0, \infty) \to [0, \infty)$ is a monotone decreasing function.

PROOF. (i) For an arbitrary $\varepsilon > 0$, we can select a sequence $\{h_k\}$ of positive numbers that converges to 0 such that

$$\|T(h_k)x - x\| < (L + \varepsilon)h_k \qquad (k = 1, 2, \dots).$$

45

For every $r_i \geq 0$ $(i = 1, 2, \ldots, n)$ and $\tau \geq 0$, we have

$$\left\| T\left(\tau + \sum_{i=1}^{n} r_i\right)x - x \right\| \leq \|T(\tau)x - x\| + \sum_{i=1}^{n} \|T(r_i)x - x\|. \tag{3.2}$$

Also, we note that for $t, h \geq 0$

$$\|T(t+h)x - T(t)x\| \leq \|T(h)x - x\|. \tag{3.3}$$

Let $h > 0$, and $n_k = [h/h_k]\,(^1)$, $\tau = h - n_k h_k$, $r_i = h_k$ $(i = 1, 2, \ldots, n_k)$, and apply (3.2). Then,

$$\|T(h)x - x\| \leq \|T(h - n_k h_k)x - x\| + n_k\|T(h_k)x - x\|$$
$$\leq \|T(h - n_k h_k)x - x\| + (L + \varepsilon)h.$$

Here we let $k \to \infty$; then since $0 \leq h - n_k h_k < h_k \to 0$, we have

$$\|T(h)x - x\| \leq (L + \varepsilon)h.$$

Thus $\|T(h)x - x\| \leq Lh$. From this and (3.3), we obtain (i).

(ii) From (i) we have

$$\limsup_{h \to 0+} \|T(h)x - x\|/h \leq L$$

$(= \liminf_{h \to 0+} \|T(h)x - x\|/h)$. Hence, we have obtained that

"$\lim_{h \to 0+} \|T(h)x - x\|/h$ $(= L)$ exists and is finite-valued."

Next, for every $t > 0$

$$\|T(h)T(t)x - T(t)x\| \leq \|T(h)x - x\| \qquad (h > 0);$$

therefore, $\liminf_{h \to 0+} \|T(h)T(t)x - T(t)x\|/h \leq L < \infty$. Hence the statement above in quotes holds even if we replace x by $T(t)x$. That is, for each $t > 0$

$$\lim_{h \to 0+} \|T(t+h)x - T(t)x\|/h \qquad \left(= \lim_{h \to 0+} \|T(h)T(t)x - T(t)x\|/h\right)$$

exists and has finite value. Finally, from

$$\|T(t+h)x - T(t)x\| \leq \|T(s+h)x - T(s)x\| \qquad (s \leq t),$$

$\varphi: [0, \infty) \to [0, \infty)$ is monotone decreasing. $\square$

DEFINITION 3.2. Let $\{T(t): t \geq 0\}$ be a semigroup of contractions on X_0. Set

$$A_h = [T(h) - I]/h \qquad (h > 0).$$

Define operators A_0 and A' with domains

$$D(A_0) = \left\{x \in X_0: \lim_{h \to 0+} A_h x \text{ exists}\right\},$$

$$D(A') = \left\{x \in X_0: \text{w-}\lim_{h \to 0+} A_h x \text{ exists}\right\}$$

$(^1)$ $[s]$ denotes the largest integer for an arbitrary real number s that is not larger than s.

by

$$A_0 x = \lim_{h \to 0+} A_h x \qquad (x \in D(A_0)),$$

$$A' x = \text{w-}\lim_{h \to 0+} A_h x \qquad (x \in D(A')).$$

A_0 and A' are called the infinitesimal generator and the weak infinitesimal generator of $\{T(t): t \geq 0\}$, respectively.

From the definition, $A_0 \subset A'$, that is, A' is an extension of A_0.

LEMMA 3.2. *Let* $\{T(t): t \geq 0\}$ *be a semigroup of contractions on* X_0.
(i) *For each* $h > 0$, A_h *is a dissipative operator.*
(ii) *The weak infinitesimal generator* A' *is a dissipative operator.*
Therefore, the infinitesimal generator A_0 *is also a dissipative operator.*

PROOF. (i) Let $h > 0$ and $x, y \in X_0$. For every $f \in F(x - y)$ we have

$$\text{Re}(A_h x - A_h y, f) = h^{-1}[\text{Re}(T(h)x - T(h)y, f) - \|x - y\|^2]$$
$$\leq h^{-1}(\|x - y\|^2 - \|x - y\|^2) = 0.$$

(ii) Let $x, y \in D(A')$. From (i), for every $f \in F(x - y)$ we have

$$\text{Re}(A_h x - A_h y, f) \leq 0.$$

Now let $h \to 0+$; then we have

$$\text{Re}(A' x - A' y, f) \leq 0. \quad \square$$

REMARK. As we see from the above proof, A_h $(h > 0)$, A', and A_0 are all strictly dissipative operators.

DEFINITION 3.3. Let $\{T(t): t \geq 0\}$ be a semigroup of contractions on X_0 and set

$$\widehat{D} = \left\{ x \in X_0 : \liminf_{h \to 0+} \|A_h x\| < \infty \right\}.$$

An operator A (not necessarily single-valued) is called the (g)-operator of $\{T(t): t \geq 0\}$ if $A_0 \subset A$, $D(A) \subset \widehat{D}$, and A is a maximal dissipative operator on $\widehat{D}$.

From Lemma 3.1 we have

$$\widehat{D} = \{x \in X_0; T(t)x: [0, \infty) \to X_0 \text{ is Lipschitz continuous}\}. \tag{3.4}$$

Also, if w-$\lim_{h \to 0+} A_h x$ exists, then from the uniform boundedness theorem $\limsup_{h \to 0+} \|A_h x\| < \infty$. Therefore, we have

$$D(A_0) \subset D(A') \subset \widehat{D}. \tag{3.5}$$

REMARK. If $D(A_0) \neq \varnothing$, then the (g)-operator of $\{T(t): t \geq 0\}$ always exists (see the proof of part (ii) of Lemma 2.12).

We will now investigate relations among the infinitesimal generator of a semigroup of contractions, the weak infinitesimal generator, and the (g)-operator.

LEMMA 3.3. *Let $\{T(t): t \geq 0\}$ be a semigroup of contractions on X_0.*
(i) *For each $t \geq 0$, $T(t)\widehat{D}$ $(= \{T(t)x: x \in \widehat{D}\}) \subset \widehat{D}$.*
(ii) *If X is a reflexive Banach space, then for each $x \in \widehat{D}$ we have*

$$T(t)x \in D(A_0), \qquad (d/dt)T(t)x = A_0 T(t)x \quad (a.e.\ t),$$

and hence $\widehat{D} \subset \overline{D(A_0)}$.

PROOF. (i) Let $t \geq 0$ and $x \in \widehat{D}$. Since

$$\|A_h T(t)x\| = h^{-1}\|T(t)T(h)x - T(t)x\| \leq \|A_h x\| \qquad (h > 0),$$

we have

$$\liminf_{h \to 0+} \|A_h T(t)x\| \leq \liminf_{h \to 0+} \|A_h x\| < \infty.$$

Therefore, $T(t)x \in \widehat{D}$.

(ii) Let $x \in \widehat{D}$. Since $T(t)x: [0, \infty) \to X_0$ is Lipschitz continuous, $T(x)$ is (strongly) absolutely continuous. Thus, since X is reflexive, a theorem of Radon-Nikodym type holds (see Theorem 1.17); that is, $T(t)x$ is strongly differentiable for almost every t (a.e. t). And, at the point t_0 where $T(t)x$ is strongly differentiable, we have

$$\begin{aligned}
[(d/dt)T(t)x]_{t=t_0} &= \lim_{h \to 0+} h^{-1}[T(t_0 + h)x - T(t_0)x] \\
&= \lim_{h \to 0+} A_h T(t_0)x = A_0 T(t_0)x.
\end{aligned}$$

Thus we obtain (3.6). Next, from $T(t)x \in D(A_0)$ (a.e. t) and from $\lim_{t \to 0+} T(t)x = x$, we obtain $x \in \overline{D(A_0)}$. Therefore, $\widehat{D} \subset \overline{D(A_0)}$. □

LEMMA 3.4. *Let X be a reflexive Banach space and let $\{T(t): t \geq 0\}$ be a semigroup of contractions on X_0. If B is a dissipative extension of A_0 and $D(B) \subset \overline{D(A_0)} \cap X_0$, then*
(i) *for every $x \in \overline{D(A_0)} \cap X_0$, $x_0 \in D(B)$, and $y_0 \in Bx_0$, we have*

$$\|T(t)x - x_0\|^2 - \|x - x_0\|^2 \leq 2\int_0^t \langle y_0, T(\tau)x - x_0\rangle_s \, d\tau \qquad (t \geq 0);$$

(ii) *$D(B) \subset \widehat{D}$, and for every $x_0 \in D(B)$ and $t \geq 0$, we have*

$$\|T(t)x_0 - x_0\| \leq t\|\,|Bx_0|\,\|.$$

PROOF. (i) Let $x_0 \in D(B)$ and $y_0 \in Bx_0$. First, we show that the required inequality holds when $x \in D(A_0)$. By (ii) of Lemma 3.3

$$(d/d\tau)T(\tau)x = A_0 T(\tau)x \in BT(\tau)x \quad (a.e.\ \tau). \tag{3.7}$$

Also, since $\|T(\tau)x - x_0\|: [0, \infty) \to [0, \infty)$ is Lipschitz continuous, it is differentiable a.e. τ. Hence by Lemma 2.8, we have

$$(d/d\tau)\|T(\tau)x - x_0\|^2 = 2\langle (d/d\tau)T(\tau)x, T(\tau)x - x_0\rangle_i \quad (a.e.\ \tau).$$

From (3.7) and the dissipativity of B,

$$\langle (d/d\tau)T(\tau)x - y_0, T(\tau)x - x_0 \rangle_i \leq 0 \quad \text{(a.e. } \tau\text{)}.$$

Therefore, we have

$$\langle (d/d\tau)T(\tau)x, T(\tau)x - x_0 \rangle_i$$
$$\leq \langle (d/d\tau)T(\tau)x - y_0, T(\tau)x - x_0 \rangle_i + \langle y_0, T(\tau)x - x_0 \rangle_s$$
$$\leq \langle (y_0, T(\tau)x - x_0 \rangle_s$$

and obtain

$$(d/d\tau)\|T(\tau)x - x_0\|^2 \leq 2\langle y_0, T(\tau)x - x_0 \rangle_s \quad \text{(a.e. } \tau\text{)}.$$

We integrate this on $[0, t]$ ($t \geq 0$ is arbitrary) and have

$$\|T(t)x - x_0\|^2 - \|x - x_0\|^2 \leq 2\int_0^t \langle y_0, T(\tau)x - x_0 \rangle_s \, d\tau. \tag{3.8}$$

Next, let $x \in \overline{D(A_0)} \cap X_0$ and take a sequence $\{x_n\}$ in $D(A_0)$ such that $\lim_{n\to\infty} x_n = x$. Since (3.8) holds for every x_n ($\in D(A_0)$), we have

$$\|T(t)x_n - x_0\|^2 - \|x_n - x_0\|^2 \leq 2\int_0^t \langle y_0, T(\tau)x_n - x_0 \rangle_s \, d\tau \quad (t \geq 0).$$

Let $n \to \infty$. Note

$$|\langle y_0, T(\tau)x_n - x_0 \rangle_s| \leq \|y_0\| \|T(\tau)x_n - x_0\|$$
$$\leq \|y_0\|(\|T(\tau)x_0\| + 2\|x_0\| + \sup_{n\geq 1}\|x_n\|) \in L^1(0, t).$$

Then by the Lebesgue convergence theorem and from the upper semicontinuity of $\langle \cdot, \cdot \rangle_s : X \times X \to (-\infty, \infty)$ (see (iv) of Corollary 2.6) we have

$$\|T(t)x - x_0\|^2 - \|x - x_0\|^2 \leq 2\limsup_{n\to\infty}\int_0^t \langle y_0, T(\tau)x_n - x_0 \rangle_s \, d\tau$$
$$\leq 2\int_0^t \limsup_{n\to\infty}\langle y_0, T(\tau)x_n - x_0 \rangle_s \, d\tau$$
$$\leq 2\int_0^t \langle y_0, T(\tau)x - x_0 \rangle_s \, d\tau.$$

(ii) Since $D(B) \subset \overline{D(A_0)} \cap X_0$, if we set $x = x_0$ in (i), it follows that $\|T(t)x_0 - x_0\|^2 \leq 2\|y_0\|\int_0^t \|T(\tau)x_0 - x_0\| \, d\tau$ ($t \geq 0$) for every element $y_0 \in Bx_0$. Therefore

$$\|T(t)x_0 - x_0\|^2 \leq 2\||Bx_0\||\int_0^t \|T(\tau)x_0 - x_0\| \, d\tau \quad (t \geq 0).$$

From this we obtain (see the next remark)

$$\|T(t)x_0 - x_0\| \leq t\||Bx_0\|| \quad (t \geq 0). \quad \square$$

REMARK. Let $\sigma(t)$ and $\varphi(t)$ be nonnegative measurable functions defined on $[0, T]$, and let $\varphi(t)\sigma(t)$ be integrable on $[0, T]$. Let a be a nonnegative constant. If $\sigma(t)^2 \leq a^2 + 2\int_0^t \varphi(s)\sigma(s)\,ds$ $(0 \leq t \leq T)$, then $\sigma(t) \leq a + \int_0^t \varphi(s)\,ds$ $(0 \leq t \leq T)$. Next, we prove this. If we set $g(t) = a^2 + 2\int_0^t \varphi(s)\sigma(s)\,ds$ $(0 \leq t \leq T)$, then g is nonnegative, increasing, and absolutely continuous on $[0, T]$. Let $T_0 = \sup\{t: 0 \leq t \leq T, g(t) = a^2\}$. Obviously, $g(T_0) = a^2$. If $T_0 = T$, then $g(t) = a^2$ for all $t \in [0, T]$. Since $\sigma(t)^2 \leq g(t)$, $\sigma(t) \leq a \leq a + \int_0^t \varphi(s)\,ds$ $(0 \leq t \leq T)$. Next, we consider the case $T_0 < T$. If $0 \leq t \leq T_0$, then $\sigma(t) \leq g(t)^{1/2} = a \leq a + \int_0^t \varphi(s)\,ds$. In order to show that the required estimation equation may also be obtained for the case $T_0 < t$, we take an arbitrary τ with $T_0 < \tau < T$. Since $g(\tau) > a^2$ and $\sqrt{g}$ is absolutely continuous on $[\tau, T]$, and $(d/ds)\sqrt{g(s)} = g'(s)/(2\sqrt{g(s)}) = \varphi(s)\sigma(s)/\sqrt{g(s)} \leq \varphi(s)$ (a.e. $s \in [\tau, T]$), by integrating $(d/ds)\sqrt{g(s)} \leq \varphi(s)$ on $[\tau, t]$, we have $\sqrt{g(t)} - \sqrt{g(\tau)} \leq \int_\tau^t \varphi(s)\,ds$ $(\tau \leq t \leq T)$. Here we let $\tau \to T_0$; then $\sqrt{g(t)} - \sqrt{g(T_0)} \leq \int_{T_0}^t \varphi(s)\,ds \leq \int_0^t \varphi(s)\,ds$ $(T_0 < t \leq T)$. Since $\sigma(t) \leq \sqrt{g(t)}$ and $g(T_0) = a^2$, we have $\sigma(t) \leq a + \int_0^t \varphi(s)\,ds$ $(T_0 < t \leq T)$.

THEOREM 3.5. *Let X be a reflexive Banach space, and let A be the (g)-operator of a semigroup of contractions $\{T(t): t \geq 0\}$ on X_0. Then, the following* (i) *through* (v) *hold:*

(i) *Let $x \in X_0$. If* w-$\lim_{n\to\infty} t_n^{-1}(T(t_n)x - x)$ *exists (we denote this limit by x') for some sequence of positive numbers $\{t_n\}$ with $\lim_{n\to\infty} t_n = 0$, then $x \in D(A^0)$ and $x' \in A^0 x$, where A^0 is the canonical restriction of A (see Theorem* 2.7).

(ii) $A_0 \subset A' \subset A^0$.

(iii) $D(A^0) = D(A) = \widehat{D}$ *and for arbitrary $x \in \widehat{D}$*

$$\lim_{h\to 0+} \|A_h x\| = \|\,|Ax|\,\|. \tag{3.9}$$

(iv) *A is a maximal dissipative operator on $\overline{D(A)} \cap X_0$.*

(v) *If X_0 is a closed set, then A is a maximal dissipative operator on $\overline{D(A)}$ and also is almost demiclosed. In particular, if X^* is a uniformly convex Banach space, then A is demiclosed.*

PROOF. Note that $A_0 \subset A$, $D(A) \subset \widehat{D} \subset \overline{D(A_0)} = \overline{D(A)}$ (see (ii) of Lemma 3.3).

(i) From $\liminf_{h\to 0+} \|h^{-1}(T(h)x - x)\| \leq \liminf_{n\to\infty} \|t_n^{-1}(T(t_n)x - x\| < \infty$, we have $x \in \widehat{D}$. Take arbitrary elements $x_0 \in D(A)$ and $y_0 \in Ax_0$. Then by (i) of Lemma 3.4

$$\|T(t)x - x_0\|^2 - \|x - x_0\|^2 \leq 2\int_0^t \langle y_0, T(\tau)x - x_0 \rangle_s \, d\tau \qquad (t \geq 0).$$

Since $\|T(t)x-x_0\|^2-\|x-x_0\|^2 \geq 2\operatorname{Re}(T(t)x-x, f)$ for every $f \in F(x-x_0)$, we obtain

$$\operatorname{Re}(T(t)x - x, f) \leq \int_0^t \langle y_0, T(\tau)x - x_0 \rangle_s d\tau \qquad (t \geq 0). \qquad (3.10)$$

From the upper semicontinuity of $\langle \cdot, \cdot \rangle_s : X \times X \to (-\infty, \infty)$, for every $\varepsilon > 0$, we can choose a $\delta > 0$ such that if $0 \leq \tau < \delta$, then

$$\langle y_0, T(\tau)x - x_0 \rangle < \langle y_0, x - x_0 \rangle_s + \varepsilon.$$

Thus from (3.10), if $0 < t < \varepsilon$, then

$$\operatorname{Re}(t^{-1}(T(t)x - x), f) \leq \langle y_0, x - x_0 \rangle_s + \varepsilon.$$

Therefore, it follows that for every $f \in F(x - x_0)$ we have

$$\operatorname{Re}(x', f) \leq \langle y_0, x - x_0 \rangle_s.$$

Since there exists an element $g \in F(x - x_0)$ that satisfies $\langle y_0, x - x_0 \rangle_s = \operatorname{Re}(y_0, g)$ ((i) of Theorem 2.5), we have

$$\operatorname{Re}(x' - y_0, g) \leq 0.$$

This shows that the operator $\widetilde{A}$ defined by the following equation is a dissipative operator:

$$\widetilde{A}z = \begin{cases} Az & (z \in D(A)\backslash\{x\}), \\ Ax \cup \{x'\} & (z = x)(^2). \end{cases}$$

Obviously $\widetilde{A} \supset A$ and thus $\widetilde{A}$ is the dissipative extension of A. Since $x \in \widehat{D}$ and A is a maximal dissipative operator on $\widehat{D}$, we obtain

$$x \in D(A), \qquad x' \in Ax. \qquad (3.11)$$

Next, from (ii) of Lemma 3.4 (we apply this with $B = A$) and (ii) of Lemma 3.1, we have

$$\lim_{h \to 0+} h^{-1}\|T(h)x - x\| \leq \||Ax\||.$$

On the other hand, since $x' \in Ax$ and $x' = \text{w-}\lim_{n \to \infty} t_n^{-1}(T(t_n)x - x)$, we have

$$\||Ax\|| \leq \|x'\| \leq \lim_{n \to \infty} t_n^{-1}\|T(t_n)x - x\| = \lim_{h \to 0+} h^{-1}\|T(h)x - x\|.$$

Hence

$$\||Ax\|| = \|x'\| = \lim_{h \to 0+} h^{-1}\|T(h)x - x\| \quad \left(= \lim_{h \to 0+} \|A_h x\| \right). \qquad (3.12)$$

From this and (3.11), we obtain

$$x \in D(A^0), \qquad x' \in A^0 x.$$

(ii) is obvious from (i).

$(^2)$ When $x \notin D(A)$, set $Ax = \varnothing$.

(iii) Let $x \in \hat{D}$. Then since $\liminf_{t \to 0+} t^{-1}\|T(t)x - x\| < \infty$, by the reflexivity of X we can choose a sequence $\{t_n\}$ of positive numbers that converges to 0, such that $\text{w-}\lim_{n \to \infty} t_n^{-1}(T(t_n)x - x)$ exists. Therefore, from (i) we have

$$x \in D(A^0) \quad \text{and} \quad |||Ax||| = \lim_{h \to 0+} \|A_h x\| \quad (\text{see } (3.12)).$$

Hence $\hat{D} \subset D(A^0)$. On the other hand, since $D(A^0) \subset D(A) \subset \hat{D}$, we have shown (iii).

(iv) Let $\tilde{A}$ be a dissipative extension of A, and define an operator B (not necessarily single-valued) by $D(B) = D(\tilde{A}) \cap (\overline{D(A)} \cap X_0)$ and $Bx = \tilde{A}x$ $(x \in D(B))$. Then B is a dissipative extension of A (thus of A_0) and $D(B) \subset \overline{D(A)} \cap X_0 = \overline{D(A_0)} \cap X_0$. Therefore, by part (ii) of Lemma 3.4, we have that $D(B) \subset \hat{D}$. Also, because A is a maximal dissipative operator on $\hat{D}$, we have

$$D(B) \ (= D(B) \cap \hat{D} = D(A) \cap \hat{D}) = D(A), \qquad Bx = Ax \quad (x \in D(A)).$$

Thus, for the dissipative extension $\tilde{A}$ of A

$$D(\tilde{A}) \cap (\overline{D(A)} \cap X_0) = D(A) \ (= D(A) \cap (\overline{D(A)} \cap X_0)),$$
$$\tilde{A}x = Ax \quad (x \in D(A)).$$

That is, A is a maximal dissipative operator on $\overline{D(A)} \cap X_0$.

(v) If X_0 is a closed set, then since $\overline{D(A)} \subset X_0$, from (iv), A is a maximal dissipative operator on $\overline{D(A)}$. To show A is almost demiclosed, we let $x_n \in D(A)$, $x_n' \in Ax_n$, $\lim_{n \to \infty} x_n = x$ and $\text{w-}\lim_{n \to \infty} x_n' = x'$. By part (ii) of Lemma 3.4, we have

$$\|T(t)x_n - x_n\| \leq t|||Ax_n||| \leq t\|x_n'\| \leq Mt \qquad (t \geq 0),$$

where M is a constant such that $\|x_n'\| \leq M$ $(n \geq 1)$. As $n \to \infty$, $\|T(t)x - x\| \leq Mt$ $(t \geq 0)$. Thus, $x \in \hat{D} = D(A)$, and A is almost demiclosed. Finally, if X^* is uniformly convex, then by Lemma 2.17 A is demiclosed. $\square$

COROLLARY 3.6. *Let X^* be a strictly convex Banach space, and let A be the (g)-operator of a semigroup of contractions $\{T(t): t \geq 0\}$ on X_0.*

(i) *If X is a strictly convex and reflexive Banach space, then $A^0 = A'$ (the weak infinitesimal generator) and $D(A') = \hat{D}$.*

(ii) *If X is a uniformly convex Banach space, then $A^0 = A' = A_0$ (the infinitesimal generator) and $D(A_0) = \hat{D}$.*

PROOF. (i) From (ii) and (iii) of Theorem 3.5,

$$A' \subset A^0, \quad \text{and} \quad D(A^0) = D(A) = \hat{D}.$$

We let $x \in \hat{D}$ $(= D(A^0))$; that is, $\|A_h x\| = h^{-1}\|T(h)x - x\| = O(1)$ $(h \to 0+)$. For every sequence $\{h_n\}$ of positive numbers with $\lim_{n\to\infty} h_n = 0$, we can select a subsequence $\{h_n'\}$ such that w-$\lim_{n\to\infty} A_{h_n'} x = $ w-$\lim_{n\to\infty}(T(h_n')x - x)/h_n'$ exists (we use x' to denote this limit.) Then, by part (i) of Theorem 3.5, we have that $x' \in A^0 x$. Since A^0 is a single-valued operator by Lemma 2.19, we have $x' = A^0 x$ and w-$\lim_{n\to\infty} A_{h_n'} x = A^0 x$. In conclusion, we have shown that every sequence $\{A_{h_n} x\}$ (with $h_n > 0$ and $\lim_{n\to\infty} h_n = 0$) has a subsequence that converges weakly to the same limit $A^0 x$. Hence,

$$\text{w-}\lim_{h\to 0+} A_h x = A^0 x.$$

From this we obtain $x \in D(A')$ and $A^0 x = A'x$; thus $D(A^0) = \hat{D} \subset D(A')$. From this and $D(A') \subset \hat{D}$, we obtain $D(A') = \hat{D} = D(A^0)$ and $A^0 = A'$.

(ii) From (i), $A_0 \subset A' = A^0$. If $x \in D(A')$ $(= D(A^0) = \hat{D})$, then w-$\lim_{h\to 0+} A_h x = A'x$, and furthermore, from (3.9)

$$\lim_{h\to 0+} \|A_h x\| = \|A^0 x\| = \|A'x\|.$$

Then, from the uniform convexity of X we obtain

$$\lim_{h\to 0+} A_h x = A'x \qquad (x \in D(A'))$$

and $A' = A_0$. □

REMARK. Furthermore, we obtain the following corollary from Theorem 3.5:

COROLLARY. *Let X be a uniformly convex Banach space and $\{T(t): t \ge 0\}$ a semigroup of contractions on X_0. Then $A' = A_0$. That is, for a semigroup of contractions defined on a subset of a uniformly convex Banach space, the weak infinitesimal generator and the infinitesimal generator coincide.*

In fact, if $x \in D(A')$, then since $A'x = $ w-$\lim_{h\to 0+} A_h x$ and $\|A'x\| = \lim_{h\to 0+}\|A_h x\|$ (we set $x' = A'x$ and apply (3.12)), we obtain from the uniform convexity of X that $\lim_{h\to 0+} A_h x = A'x$.

COROLLARY 3.7. *Assume both X and X^* are uniformly convex Banach spaces. Let $\{T(t): t \ge 0\}$ be a semigroup of contractions on X_0 and $x \in \hat{D}$. Then the following hold*:

(i) *$T(t)x \in D(A)$ for every $t \ge 0$, and $T(t)x$ is strongly right differentiable and the strong right derivative $D^+ T(t)x$ $(= A_0 T(t)x)$ is (strongly) right continuous at every $t \ge 0$.*

(ii) *$\|D^+ T(t)x\|$ $(= \|A_0 T(t)x\|): [0, \infty) \to [0, \infty)$ is monotone decreasing; further, at the point of continuity of $\|D^+ T(t)x\|$, $D^+ T(t)x$ is continuous and $T(t)x$ is strongly differentiable. (Therefore, except at most a countable number of points (of t), $D^+ T(t)x$ is continuous and $T(t)x$ is strongly differentiable.)*

PROOF. Let A be the (g)-operator of $\{T(t); t \geq 0\}$. Note that $D(A^0) = D(A) = \hat{D} = D(A_0)$ and $A^0 = A_0$ (see (iii) of Theorem 3.5 and (ii) of Corollary 3.6).

(i) From $T(t)\hat{D} \subset \hat{D}$ ((i) of Lemma 3.3), we have $T(t)x \in \hat{D} = D(A_0)$ for every $t \geq 0$. And from this,

$$D^+ T(t)x = \lim_{h \to 0+} h^{-1}[T(t+h)x - T(t)x] = \lim_{h \to 0+} A_h T(t)x = A_0 T(t)x.$$

Next, we show that $D^+ T(t)x : [0, \infty) \to X$ is right continuous. Let $t \geq 0$ and $\{t_n\}$ be an arbitrary sequence such that $\lim_{n \to \infty} t_n = t$ and $t_n > t$. From

$$\|D^+ T(t_n)x\| = \lim_{h \to 0+} h^{-1}\|T(t_n + h)x - T(t_n)x\|$$

$$\leq \lim_{h \to 0+} h^{-1}\|T(t+h)x - T(t)x\| = \|D^+ T(t)x\|$$

$\{D^+ T(t_n)x\}$ is a bounded sequence. Since X is a reflexive Banach space, $\{D^+ T(t_n)x\}$ contains a weakly convergent subsequence $\{D^+ T(t_{n'})x\}$. Now we set

$$\text{w-}\lim_{n' \to \infty} D^+ T(t_{n'})x = y.$$

Then we have

$$y \in AT(t)x. \tag{3.13}$$

In fact, from $D^+ T(t_{n'})x = A_0 T(t_{n'})x \; (= A^0 T(t_{n'})x) \in AT(t_{n'})x$ and the dissipativity of A, we have

$$\text{Re}(D^+ T(t_{n'})x - z', F(T(t_{n'})x - z)) \leq 0$$

for every $z \in D(A)$ and $z' \in Az$. Here, if we let $n' \to \infty$ (note the duality mapping F is continuous (see Theorem 2.3)), then we obtain

$$\text{Re}(y - z', F(T(t)x - z)) \leq 0 \qquad (z \in D(A), z' \in Az).$$

Since $T(t)x \in D(A) = \hat{D}$ and A is a maximal dissipative operator on $\hat{D}$ (by the definition of (g)-operator), we obtain (3.13) from the above inequality. Then, from (3.13) we have

$$\|y\| \leq \liminf_{n' \to \infty} \|D^+ T(t_{n'})x\| \leq \limsup_{n' \to \infty} \|D^+ T(t_{n'})x\| \leq \|D^+ T(t)x\|$$

$$= \|A_0 T(t)x\| = \|A^0 T(t)x\| = \||AT(t)x\|| \leq \|y\|.$$

Hence $\|y\| = \||AT(t)x\||$. From this and (3.13), we have $y = A^0 T(t)x$. Thus

$$\text{w-}\lim_{n' \to \infty} D^+ T(t_{n'})x = A^0 T(t)x, \qquad \lim_{n' \to \infty} \|D^+ T(t_{n'})x\| = \|A^0 T(t)x\|.$$

Therefore, from the uniform convexity of X, we obtain

$$\lim_{n' \to \infty} D^+ T(t_{n'})x = A^0 T(t)x \; (= A_0 T(t)x) = D^+ T(t)x,$$

and $D^+T(s)x$ is right continuous at the point $s = t$.

(ii) That $\|D^+T(s)x\|: [0, \infty) \to [0, \infty)$ is monotone decreasing is shown in the proof of (i). Let $t > 0$ be a point of continuity of $\|D^+T(s)x\|$. First, we show that $D^+T(s)x$ is continuous at $s = t$. Since we have shown in (i) that it is right continuous, we now prove that it is also left continuous at t. Take an arbitrary sequence $\{t_n\}$ of positive numbers such that $\lim_{n\to\infty} t_n = t$ and $t_n < t$. Since $\lim_{n\to\infty} \|D^+T(t_n)x\| = \|D^+T(t)x\|$, $\{D^+T(t_n)x\}$ is a bounded sequence, and thus it contains a weakly convergent subsequence $\{D^+T(t_{n'})x\}$. Now if we set

$$\text{w-}\lim_{n'\to\infty} D^+T(t_{n'})x = u,$$

then we obtain $u \in AT(t)x$ by the same method as is used in proving (3.13). Thus

$$\|u\| \leq \lim_{n'\to\infty} \|D^+T(t_{n'})x\| = \|D^+T(t)x\| = \|A_0T(t)x\|$$

$$= \|A^0T(t)x\| = \|\|AT(t)x\|\| \leq \|u\|.$$

Applying the same method as in proving (i), we obtain $u = A^0T(t)x$ and

$$\lim_{n'\to\infty} D^+T(t_{n'})x = A^0T(t)x = D^+T(t)x.$$

Thus we have shown $\lim_{s\to t-0} D^+T(s)x = D^+T(t)x$. Next, we prove that $T(s)x$ is strongly differentiable at $s = t$. Since $T(s)x: [0, \infty) \to X$ is Lipschitz continuous, it is (strongly) absolutely continuous. Then, from the reflexivity of X, the theorem of Radon-Nikodym type holds. That is, $T(s)x$ is strongly differentiable a.e. s and its strong derivative $(d/ds)T(s)x$ is (Bochner) integrable on every bounded interval of $[0, \infty)$, and for arbitrary $s_1, s_2 \geq 0$ we have (Theorem 1.17)

$$T(s_2)x - T(s_1)x = \int_{s_1}^{s_2} (d/ds)T(s)x \, ds = \int_{s_1}^{s_2} D^+T(s)x \, ds.$$

In particular, for every $\tau \geq 0$ we have

$$T(\tau)x - T(t)x = \int_t^{\tau} D^+T(s)x \, ds.$$

Since $D^+T(s)x$ is continuous at $s = t$, from the above equation we obtain

$$\left\| \frac{T(\tau)x - T(t)x}{\tau - t} - D^+T(t)x \right\| \leq \frac{1}{|\tau - t|} \left| \int_t^{\tau} \|D^+T(s)x - D^+T(t)x\| \, ds \right|$$

$$\to 0 \quad (\tau \to t).$$

Hence, $T(s)x$ is strongly differentiable at $s = t$. $\square$

By the Hille-Yosida theory, the infinitesimal generator of a semigroup of linear contractions on a Banach space X is an m-dissipative operator with a dense domain in X. But, as we see from the next example, this does not hold for semigroups of nonlinear contractions. Also, for the semigroup of linear contractions on X, it is known that the weak infinitesimal generator and the infinitesimal generator always coincide (see Theorem 10.5.4, p. 318, [49]). However, with regard to semigroups of nonlinear contractions, as a general theory, nothing appears to be known other than the results mentioned in the remark after Corollary 3.6.

EXAMPLE 3.1. Let $X = R^1 = (-\infty, \infty)$ and define an operator $T(t)$: $X \to X$ for each $t \geq 0$ by

$$T(t)x = \begin{cases} \max\{0, x - t\}, & x > 0, \\ x, & x \leq 0. \end{cases}$$

Then, $\{T(t): t \geq 0\}$ is a semigroup of (nonlinear) contractions on X and its infinitesimal generator A_0 is

$$A_0 x = \begin{cases} -1, & x > 0, \\ 0, & x \leq 0 \end{cases}$$

on $D(A_0) = X$. A_0 is a dissipative operator, but because

$$R(I - \lambda A_0) = (-\infty, 0] \cup (\lambda, \infty) \neq X \qquad (\lambda > 0)$$

A_0 is not an m-dissipative operator. Also, for $x > 0$

$$D^+ T(t)x = \begin{cases} -1, & 0 \leq t < x, \\ 0, & x \leq t; \end{cases}$$

$$D^- T(t)x = \begin{cases} -1, & 0 < t \leq x, \\ 0, & x < t \end{cases}$$

(where $D^- T(t)x$ is the left derivative of the function of s, $T(s)x$, at $s = t$). Therefore, at the point of discontinuity $t = x > 0$ ([3]) of $\|D^+ T(t)x\|$ ($= |D^+ T(t)x|$) (the function of t), $T(t)x$ is always not differentiable.

EXAMPLE 3.2. Let $X = C[0, 1]$ and define a function $\varphi: (-\infty, \infty) \to (-\infty, \infty)$ by

$$\varphi(\eta) = \begin{cases} \eta, & \eta \geq 0, \\ 2\eta, & \eta < 0. \end{cases}$$

And define an operator $T(t): X \to X$ for each $t \geq 0$ by

$$[T(t)x](\xi) = \varphi(t + \varphi^{-1}(x(\xi))), \qquad x \in X, \quad 0 \leq \xi \leq 1.$$

Then, we see that $\{T(t): t \geq 0\}$ is a semigroup of (nonlinear) contractions on X. Let us examine the weak infinitesimal generator A' of this semigroup of contractions.

Let $x \in D(A')$. Then since w-$\lim_{h \to 0+} A_h x = A'x$, we have

$$\lim_{h \to 0+} h^{-1} \int_0^1 ([T(h)x](\xi) - x(\xi)) \, dv(\xi) = \int_0^1 (A'x)(\xi) \, dv(\xi)$$

([3]) When $x \leq 0$, since $D^+ T(t)x = 0$ $(0 \leq t < \infty)$ $\|D^+ T(t)x\|$ does not have points of discontinuity.

for every $v \in BV[0, 1]$ (the set of bounded variational functions defined on $[0, 1]$). In particular, if v is the characteristic function of the interval $(\zeta, 1]$, where $0 \le \zeta \le 1$; then we obtain

$$\lim_{h \to 0+} h^{-1}([T(h)x](\zeta) - x(\zeta)) = (A'x)(\zeta) \qquad (0 \le \zeta \le 1).$$

From this we obtain

$$(A'x)(\zeta) = \begin{cases} 1, & \zeta \in \{\xi : x(\xi) \ge 0\}, \\ 2, & \zeta \in \{\xi : x(\xi) < 0\}. \end{cases}$$

However, since $A'x \in C[0, 1]$, it must be that either $x(\xi) \ge 0$ for all $\xi \in [0, 1]$ (and then $(A'x)(\xi) \equiv 1$), or $x(\xi) < 0$ for all $\xi \in [0, 1]$ (and then $(A'x)(\xi) \equiv 2$). Conversely, it is easy to see that for $x \in C[0, 1]$, if either $x(\xi) \ge 0$ for all $\xi \in [0, 1]$, or $x(\xi) < 0$ for all $\xi \in [0, 1]$, then $x \in D(A')$. (In fact, this follows because if $x(\xi) \ge 0$ $(\xi \in [0, 1])$, then $[A_h x](\xi) \equiv 1$ $(h > 0)$. Also, assume $x(\xi) < 0$ $(\xi \in [0, 1])$, and let M (< 0) be a maximum value of the function x; then $[A_h x](\xi) \equiv 2$ when $0 < h < -\frac{1}{2}M$.)

Therefore, $D(A') = \{x \in C[0, 1]: x(\xi) \ge 0$ $(\forall \xi \in [0, 1])$, or $x(\xi) < 0$ $(\forall \xi \in [0, 1])\}$, and $A'x$ $(x \in D(A'))$ is

$$[A'x](\xi) \equiv \begin{cases} 1, & x(\xi) \ge 0 \ (\forall \xi \in [0, 1]), \\ 2, & x(\xi) < 0 \ (\forall \xi \in [0, 1]). \end{cases}$$

Also, as we see from the above consideration, the infinitesimal generator of this semigroup of contractions is $A_0 = A'$, and its domain $D(A_0)$ $(= D(A'))$ is not dense in X. Moreover, A_0 is not an m-dissipative operator.

Furthermore, for this semigroup of contractions, since $\|A_h x\| \le 2$ for all $x \in X$ and $h > 0$, we have $\hat{D} = X$.

As we have seen in the example above, for the semigroup of nonlinear contractions in a Banach space, the domain of the infinitesimal generator is not necessarily dense (in the domain of semigroup); but we have not yet touched on anything concerning the existence of the infinitesimal generator. After all, does a semigroup of nonlinear contractions always have an infinitesimal generator A_0? That is, is it true that $D(A_0) \ne \varnothing$? In fact, as we will show later, we can construct a semigroup of nonlinear contractions that can be defined on all of the space $C[-1, 1]$ and does not have the weak infinitesimal generator (therefore, the infinitesimal generator) (see Example 4.7 of §4.4). As we see from the definition of infinitesimal generator, the problem of existence of the infinitesimal generator is closely related to the problem of differentiability of a semigroup of contractions. In fact, when we let $\{T(t): t \ge 0\}$ be a semigroup of contractions and let A_0 be its infinitesimal generator, a necessary and sufficient condition for $T(t)x \in D(A_0)$ is that $T(s)x$ is strongly right differentiable at $s = t$ (≥ 0). The differentiability of a function which is defined on an interval of the real axis and which has values in a Banach space X depends largely upon whether X is reflexive or not. In fact, this follows because when X is a reflexive Banach space, the possibility for a function to be strongly differentiable almost everywhere can be derived from the (strongly) absolute continuity of the function, but when X is not reflexive, this does not hold (there exists a function which is not strongly right differentiable everywhere and yet is Lipschitz continuous). (See, for example, §15.2 of [97].) With this kind of reasoning (of course, it is a general argument) the existence of the infinitesimal generator of a semigroup of nonlinear contractions is thought to be greatly different, depending upon whether X is reflexive or not. The following related problem also arises:

"If $\{T(t): t \geq 0\}$ is a semigroup of nonlinear contractions on X_0 $(\subset X)$, is

$$\hat{D} = \{x \in X_0 \, ; \, T(t)x : [0, \infty) \to X_0 \text{ is Lipschitz continuous}\}$$

dense in X_0?"

If $\hat{D}$ is dense in X_0, when X is a reflexive Banach space, as we have shown in Lemma 3.3, $\{T(t): t \geq 0\}$ has the infinitesimal generator A_0 whose domain is dense in X_0, because $D(A_0)$ is dense in $\hat{D}$.

As we describe in the next section, if X is a Hilbert space or Banach space of finite dimension, then $\hat{D}$ is dense in X_0 (a closed convex set), but questions other than this remain unsolved. (Baillon, J. Funct. Anal. (1978), proved that if X^* is uniformly convex then Theorem 3.13 of §2 remains true and hence $\hat{D}$ is dense in X_0. See also Reich, J. Funct. Anal. (1980).)

§2. Denseness of $\hat{D}$

In this section, we assume X_0 $(\neq \varnothing)$ is a closed convex subset of a Banach space X and we let $\{T(t): t \geq 0\}$ be a semigroup of contractions on X_0. We define A_h $(h > 0)$, A_0, $\hat{D}$, etc., for this semigroup of contractions in the same way as in the previous section. The objective of this section is to investigate the denseness of $\hat{D}$ in X_0.

A_h $(h > 0)$ is a dissipative operator that satisfies.

$$R(I - \lambda A_h) \supset X_0 = D(A_h) \qquad (\lambda > 0). \qquad (3.14)$$

(See Example 2.4.) Thus, for every $h > 0$ and $\lambda > 0$, we set

$$J_{\lambda, h} x = (I - \lambda A_h)^{-1} x \qquad (x \in R(I - \lambda A_h));$$

then, $J_{\lambda, h} : R(I - \lambda A_h) \to X_0$ is a contraction operator (see Corollary 2.10), and from Lemma 2.11 we have

$$\|J_{\lambda, h} x - x\| \leq \lambda \|A_h x\| \qquad (x \in X_0), \qquad (3.15)$$

$$J_{\lambda, h} x = J_{\mu, h}\left(\frac{\mu}{\lambda} x + \frac{\lambda - \mu}{\lambda} J_{\lambda, h} x\right) \qquad (x \in X_0, \mu > 0). \qquad (3.16)$$

Also we easily obtain from the definition of $J_{\lambda,h}$ that

$$J_{\lambda,h}x = \frac{h}{\lambda+h}x + \frac{\lambda}{\lambda+h}T(h)J_{\lambda,h}x \qquad (x \in X_0, \lambda > 0, h > 0). \qquad (3.17)$$

LEMMA 3.8. *Let* $x \in X_0$, $\lambda > 0$, *and* $h > 0$. *Then for every natural number* n, *we have*

$$\sum_{i=1}^{n}(\langle y_{nh} - x, y_{nh} - y_h\rangle_s + \langle y_h - x, y_h - T(ih)y_{nh}\rangle_s) \leq 0,$$

where $y_\tau = J_{\lambda,\tau}x$ $(\tau > 0)$.

PROOF. From (3.17) we have

$$T(h)y_h = y_h + \frac{h}{\lambda}(y_h - x). \qquad (3.18)$$

From this, for every $z \in X_0$ we have

$$\|y_h - z\|^2 \geq \|T(h)y_h - T(h)z\|^2 = \left\|y_h - T(h)z + \frac{h}{\lambda}(y_h - x)\right\|^2$$

$$\geq \|y_h - T(h)z\|^2 + \frac{2h}{\lambda}\langle y_h - x, y_h - T(h)z\rangle_s.$$

(Here we apply the inequality $\|u + v\|^2 \geq \|u\|^2 + 2\langle v, u\rangle_s$.) Furthermore, we replace z in the above inequality by $T(h)z$, $T(2h)z$, $\ldots$, $T((n-1)h)z$, and add the respective sides of the n inequality to obtain

$$\|y_h - z\|^2 \geq \|y_h - T(nh)z\|^2 + \frac{2h}{\lambda}\sum_{i=1}^{n}\langle y_h - x, y_h - T(ih)z\rangle_s. \qquad (3.19)$$

Set $z = y_{nh}$ in (3.19), and note that $T(nh)y_{nh} = y_{nh} + \frac{nh}{\lambda}(y_{nh} - x)$ (this equation is obtained by replacing h by nh in (3.18)); then

$$\|y_h - y_{nh}\|^2 \geq \left\|y_h - y_{nh} - \frac{nh}{\lambda}(y_{nh} - x)\right\|^2 + \frac{2h}{\lambda}\sum_{i=1}^{n}\langle y_h - x, y_h - T(ih)y_{nh}\rangle_s$$

$$\geq \|y_h - y_{nh}\|^2 + \frac{2nh}{\lambda}\langle y_{nh} - x, y_{nh} - y_h\rangle_s$$

$$+ \frac{2h}{\lambda}\sum_{i=1}^{n}\langle y_h - x, y_h - T(ih)y_{nh}\rangle_s.$$

Therefore,

$$\sum_{i=1}^{n}(\langle y_{nk} - x, y_{nk} - y_h\rangle_s + \langle y_h - x, y_h - T(ih)y_{nh}\rangle_s) \leq 0. \quad \square$$

LEMMA 3.9. *For an arbitrary* $x \in X_0$

$$\lim_{(\lambda,h)\to(0,0)} J_{\lambda,h}x = x.$$

PROOF. Let $x \in X_0$. From Lemma 3.8, for every $\lambda > 0$, $h > 0$, and every natural number n, there exists a natural number i $(1 \leq i \leq n)$ such that

$$\langle y_{nh} - x, y_{nh} - y_h \rangle_s + \langle y_h - x, y_h - T(ih)y_{nh} \rangle_s \leq 0.$$

Because

$$\langle y_{nh} - x, y_{nh} - y_h \rangle_s \geq -\|y_{nh} - x\|(\|y_h - x\| + \|y_{nh} - x\|),$$

$$\langle y_h - x, y_h - T(ih)y_{nh} \rangle_s$$
$$\geq \|y_h - T(ih)y_{nh}\|^2 - \|T(ih)y_{nh} - x\| \, \|y_h - T(ih)y_{nh}\|$$
$$\geq (\|y_h - x\| - \|x - T(ih)y_{nh}\|)^2$$
$$\quad - \|T(ih)y_{nh} - x\|(\|y_h - x\| + \|x - T(ih)y_{nh}\|)$$

we have

$$0 \geq \|y_h - x\|^2 - \|y_h - x\|(3\|x - T(ih)y_{nh}\| + \|y_{nh} - x\|) - \|y_{nh} - x\|^2$$
$$\geq \|y_h - x\|^2 - \|y_h - x\|(4\|y_{nh} - x\| + 3\|x - T(ih)x\|) - \|y_{nh} - x\|^2.$$

From this,

$$\|y_h - x\| \leq \frac{7 + \sqrt{53}}{2} \max\{\|y_{nh} - x\|, \|x - T(ih)x\|\}.$$

And also, from (3.15)

$$\|y_{nh} - x\| = \|J_{\lambda, nh}x - x\| \leq \frac{\lambda}{nh}\|T(nh)x - x\|.$$

Therefore we obtain

$$\|J_{\lambda, h}x - x\| \leq K \max\left\{\frac{\lambda}{nh}\|T(nh)x - x\|, \|x - T(ih)x\|\right\}, \qquad (3.20)$$

where $K = (7 + \sqrt{53})/2$.

Now, since $\lim_{t \to 0+} T(t)x = x$, for every $\varepsilon > 0$, we can choose a $\delta > 0$ such that if $0 \leq t \leq \delta$, then $\|T(t)x - x\| < \varepsilon$. Let $0 < h < \delta/2$ and set $n = [\delta/h]$. Then since $\delta/2 < nh \leq \delta$ and $0 < ih \leq nh \leq \delta$, we obtain from (3.20)

$$\|J_{\lambda, h}x - x\| \leq K \max\left\{\frac{2\lambda\varepsilon}{\delta}, \varepsilon\right\}. \qquad (3.20)'$$

Hence, if $0 < h < \delta/2$ and $0 < \lambda < \delta/2$, then we have $\|J_{\lambda, h}x - x\| \leq K\varepsilon$, and thus $\lim_{(\lambda, h) \to (0, 0)} J_{\lambda, h}x = x$. $\square$

REMARK. From (3.20)$'$ and $\|J_{\lambda, h}x - x\| \leq \lambda\|A_h x\| \leq (2\lambda/\delta)\|T(h)x - x\|$ $(h \geq \delta/2)$, $\{\|J_{\lambda, h}x - x\| : 0 < h \leq \omega\}$ is a bounded set for every $\lambda > 0$, $\omega > 0$, and $x \in X_0$.

LEMMA 3.10. *Let $\{h_n\}$ be a sequence of positive numbers with $\lim_{n \to \infty} h_n = 0$. If for a certain $\lambda > 0$, $\lim_{n \to \infty} J_{\lambda, h_n} x$ exists for all $x \in X_0$, then $\lim_{n \to \infty} J_{\mu, h_n} x$ exists for all $\mu > \lambda$ and all $x \in X_0$.*

PROOF. Let $J_\lambda x = \lim_{n \to \infty} J_{\lambda, h_n} x$ $(x \in X_0)$. Then J_λ is a contraction operator with $X_0 \to X_0$. If we let $\mu > \lambda$, then we have

$$R\left(\frac{\mu}{\lambda} I + \frac{\lambda - \mu}{\lambda} J_\lambda\right) \supset X_0. \tag{3.21}$$

In fact, we take and fix an arbitrary $z \in X_0$, and define an operator U by

$$Uy = \frac{\lambda}{\mu} z + \frac{\mu - \lambda}{\mu} J_\lambda y \qquad (y \in X_0).$$

Since X_0 is a closed convex set and U is an operator of $X_0 \to X_0$, and also $\|Uy_1 - Uy_2\| \leq (1 - \lambda/\mu)\|y_1 - y_2\|$ $(y_1, y_2 \in X_0)$, $U: X_0 \to X_0$ is a strictly contraction operator. Therefore, U has a fixed point y, that is, $y = Uy$. From this we obtain

$$\frac{\mu}{\lambda} y + \frac{\lambda - \mu}{\lambda} J_\lambda y = z,$$

and (3.21) is proved.

Let $x \in X_0$. Then from (3.12) there exists $y \in X_0$ such that x can be written as

$$x = \frac{\mu}{\lambda} y + \frac{\lambda - \mu}{\lambda} J_\lambda y.$$

Also from (3.16) we have

$$J_{\lambda, h_n} y = J_{\mu, h_n}\left(\frac{\mu}{\lambda} y + \frac{\lambda - \mu}{\lambda} J_{\lambda, h_n} y\right).$$

Hence

$$\|J_{\mu, h_n} x - J_{\lambda, h_n} y\| \leq \left\| x - \left(\frac{\mu}{\lambda} y + \frac{\lambda - \mu}{\lambda} J_{\lambda, h_n} y\right)\right\|$$

$$= \frac{\mu - \lambda}{\lambda} \|J_{\lambda, h_n} y - J_\lambda y\| \to 0 \qquad (n \to \infty).$$

Therefore, $\lim_{n \to \infty} J_{\mu, h_n} x = \lim_{n \to \infty} J_{\lambda, h_n} y = J_\lambda y$. □

THEOREM 3.11. *Let $\{T(t): t \geq 0\}$ be a semigroup of contractions on a closed convex set X_0 (of a Banach space X) and set $J_{\lambda, h} = (I - \lambda A_h)^{-1}$ $(\lambda > 0, h > 0)$. We assume that there exist two sequences $\{h_n\}$ and $\{\lambda_m\}$ of positive numbers which converge to 0 and satisfy condition (3.22):*

$$\lim_{n \to \infty} J_{\lambda_m, h_n} x \quad \text{exists for all } \lambda_m \text{ and all } x \in X_0. \tag{3.22}$$

Then, the (i) and (ii) hold:

(i) *For all $\lambda > 0$ and all $x \in X_0$, $\lim_{n \to \infty} J_{\lambda, h_n} x$ exists. If we set $J_\lambda x = \lim_{n \to \infty} J_{\lambda, h_n} x$ $(\lambda > 0, x \in X_0)$, then for every $\lambda > 0$, J_λ is a contraction operator with $X_0 \to X_0$.*

(ii) *For all $\lambda > 0$ and $x \in X_0$, $J_\lambda x \in \widehat{D} \equiv \{x \in X_0 : \liminf_{h \to 0+} \|A_h x\| < \infty\}$ $(= \{x \in X_0 ; T(t)x : [0, \infty) \to X_0$ is Lipschitz continuous\})$ [4], and also, for all $x \in X_0$, $\lim_{\lambda \to 0+} J_\lambda x = x$. Therefore, $\widehat{D}$ is dense in X_0.*

PROOF. (i) This can be obtained by Lemma 3.10.

(ii) Let $\lambda > 0$ and $x \in X_0$. From (3.17) we have

$$h_n^{-1}(T(h_n)J_{\lambda, h_n}x - J_{\lambda, h_n}x) = \lambda^{-1}(J_{\lambda, h_n}x - x).$$

Let $n \to \infty$; then from (i), the right-hand side $\to \lambda^{-1}(J_\lambda x - x)$. Hence, there is a constant $K \geq 0$ (independent of n) such that

$$h_n^{-1}\|T(h_n)J_{\lambda, h_n}x - J_{\lambda, h_n}x\| \leq K \qquad (n = 1, 2, \dots).$$

From this, for arbitrary $t \geq 0$

$$
\begin{aligned}
\|T(t)J_{\lambda, h_n}x - J_{\lambda, h_n}x\| &\leq \|T(t)J_{\lambda, h_n}x - T(h_n[t/h_n])J_{\lambda, h_n}x\| \\
&\quad + \|T(h_n[t/h_n])J_{\lambda, h_n}x - J_{\lambda, h_n}x\| \\
&\leq \|T(t - h_n[t/h_n])J_{\lambda, h_n}x - J_{\lambda, h_n}x\| \\
&\quad + [t/h_n]\|T(h_n)J_{\lambda, h_n}x - J_{\lambda, h_n}x\| \\
&\leq 2\|J_{\lambda, h_n}x - J_\lambda x\| + \|T(t - h_n[t/h_n])J_\lambda x - J_\lambda x\| + Kt.
\end{aligned}
$$

Now we let $n \to \infty$, and obtain

$$\|T(t)J_\lambda x - J_\lambda x\| \leq Kt \qquad (t \geq 0)$$

and hence $J_\lambda x \in \widehat{D}$.

Next, we prove $\lim_{\lambda \to 0+} J_\lambda x = x$ $(x \in X_0)$. Let $x \in X_0$ and let $\varepsilon > 0$ be arbitrary. By Lemma 3.9, we can select a $\delta > 0$ such that if $0 < \lambda < \delta$ and $0 < h_n < \delta$, then $\|J_{\lambda, h_n}x - x\| < \varepsilon$. Now let $n \to \infty$; then we obtain $\|J_\lambda x - x\| \leq \varepsilon$ if $0 < \lambda < \delta$, and we have $\lim_{\lambda \to 0+} J_\lambda x = x$. □

COROLLARY 3.12. *Let X be a Banach space of finite dimension and let $\{T(t): t \geq 0\}$ be a semigroup of contractions on a closed convex set X_0 $(\subset X)$. Then, there exist sequences $\{h_n\}$ and $\{\lambda_m\}$ of positive numbers which converge to 0 and satisfy (3.22). Therefore, $\widehat{D}$ is dense in X_0 (by the above theorem).*

PROOF. Fix $x_0 \in X_0$. By Lemma 3.9, there exists a proper constant $\delta_0 > 0$ such that if $0 < \lambda \leq \delta_0$ and $0 < h \leq \delta_0$, then $\|J_{\lambda, h}x_0 - x_0\| \leq 1$. From this, for every $x \in X_0$ we have

$$
\begin{aligned}
\|J_{\lambda, h}x\| &\leq \|J_{\lambda, h}x - J_{\lambda, h}x_0\| + \|J_{\lambda, h}x_0 - x_0\| + \|x_0\| \\
&\leq \|x - x_0\| + 1 + \|x_0\| \qquad (0 < \lambda, h \leq \delta_0).
\end{aligned}
$$

[4] See Definition 3.3 and (3.4).

Since X is of finite dimension, it is separable and hence so is X_0. Let $\{x_1, x_2, \ldots, x_k, \ldots\}$ be a dense subset of X_0 and set $\lambda_m = \delta_0/m$ ($m = 1, 2, \ldots$). Then from the above inequality, $\{J_{\lambda_m, h}x_k : 0 < h \le \delta_0\}$ is a bounded set for every m and k, and therefore, it is sequentially compact (because X is of finite dimension). Hence, applying the diagonal process we can select a sequence $\{h_n\}$ of positive numbers such that $\lim_{n\to\infty} h_n = 0$ and $\lim_{n\to\infty} J_{\lambda_m, h_n}x_k$ exists for all m and k. Thus, from the contractiveness of J_{λ_m, h_n} and $\{x_1, x_2, \ldots, x_k, \ldots\}$ being dense in X_0, we can see that $\lim_{n\to\infty} J_{\lambda_m, h_n}x$ exists for arbitrary $x \in X_0$ and λ_m. $\square$

REMARK. Let $D(A) = \bigcup_{\lambda>0} R(J_\lambda)$ in the above theorem and define Ax by

$$Ax = \bigcup_{\lambda \in \Lambda(x)} \lambda^{-1}(I - J_\lambda^{-1})x$$

for every $x \in D(A)$, where $\Lambda(x) = \{\lambda > 0 : x \in R(J_\lambda)\}$. Then, A is a dissipative operator with $\overline{D(A)} = X_0$ and

$$R(I - \lambda A) \supset X_0 \tag{3.23}$$

for all $\lambda > 0$.

Next, we prove this. First, note that

$$J_\lambda x = J_\mu \left(\frac{\mu}{\lambda}x + \frac{\lambda - \mu}{\lambda}J_\lambda x\right) \qquad (\lambda \ge \mu > 0, \, x \in X_0). \tag{3.24}$$

(In fact, since $J_\lambda : X_0 \to X_0$ for every $\lambda > 0$ and since X_0 is a convex set, we have $(\mu/\lambda)x + (1 - (\mu/\lambda))J_\lambda x \in X_0$ for $\lambda > \mu > 0$ and $x \in X_0$; thus we can define $J_{\mu, h_n}((\mu/\lambda)x + ((\lambda - \mu)/\lambda)J_\lambda x)$ (see (3.14)) and

$$\left\|J_{\lambda, h_n}x - J_{\mu, h_n}\left(\frac{\mu}{\lambda}x + \frac{\lambda - \mu}{\lambda}J_\lambda x\right)\right\|$$

$$= \left\|J_{\mu, h_n}\left(\frac{\mu}{\lambda}x + \frac{\lambda - \mu}{\lambda}J_{\lambda, h_n}x\right) - J_{\mu, h_n}\left(\frac{\mu}{\lambda}x + \frac{\lambda - \mu}{\lambda}J_\lambda x\right)\right\|$$

(see (3.16)) $\le ((\lambda - \mu)/\lambda)\|J_{\lambda, h_n}x - J_\lambda x\|$ which goes to 0 as $n \to \infty$. Thus we obtain (3.24).) From this, we can derive

$$\begin{cases} \text{If } \lambda \ge \mu, \quad \text{then } R(J_\lambda) \subset R(J_\mu) \text{ and} \\ \lambda^{-1}(I - J_\lambda^{-1})x \subset \mu^{-1}(I - J_\mu^{-1})x \qquad (x \in R(J_\lambda)). \end{cases} \tag{3.25}$$

In fact, if we let $x \in R(J_\lambda)$ and take an arbitrary y with $y \in \lambda^{-1}(I - J_\lambda^{-1})x$, then

$$x = J_\lambda(x - \lambda y) = J_\mu\left(\frac{\mu}{\lambda}(x - \lambda y) + \frac{\lambda - \mu}{\lambda}J_\lambda(x - \lambda y)\right) \quad \text{(by (3.24))}$$

$$= J_\mu\left(\frac{\mu}{\lambda}(x - \lambda y) + \frac{\lambda - \mu}{\lambda}x\right) = J_\mu(x - \mu y).$$

Hence $y \in \mu^{-1}(I - J_\mu^{-1})x$ and we have $\lambda^{-1}(I - J_\lambda^{-1})x \subset \mu^{-1}(I - J_\mu^{-1})x$.

Now we prove (3.23). For $x \in X_0$ and $\lambda > 0$,

$$\lambda^{-1}(J_\lambda x - x) \in \lambda^{-1}(I - J_\lambda^{-1})J_\lambda x \subset AJ_\lambda x,$$

that is, $x \in (I - \lambda A)J_\lambda x \subset R(I - \lambda A)$. Thus we obtain (3.23). Next, in order to show that A is a dissipative operator, we let $x, y \in D(A)$, $x' \in Ax$, and $y' \in Ay$. Then from the definition of A, there exist positive numbers $\lambda_1 \in \Lambda(x)$ and $\lambda_2 \in \Lambda(y)$ such that

$$x' \in \lambda_1^{-1}(I - J_{\lambda_1}^{-1})x, \qquad y' \in \lambda_2^{-1}(I - J_{\lambda_2}^{-1})y.$$

We set $\lambda_0 = \min\{\lambda_1, \lambda_2\}$. If $0 < \lambda \leq \lambda_0$, then $x' \in \lambda^{-1}(I - J_\lambda^{-1})x$ and $y' \in \lambda^{-1}(I - J_\lambda^{-1})y$ by (3.25); that is, we obtain

$$J_\lambda(x - \lambda x') = x, \qquad J_\lambda(y - \lambda y') = y.$$

Hence, from the contractiveness of J_λ we have

$$\|x - y\| \leq \|x - y - \lambda(x' - y')\| \qquad (0 < \lambda \leq \lambda_0).$$

Also, if $\lambda_0 < \lambda$, then from $(^5)$

$$0 \leq (\|x - y - \lambda_0(x' - y')\| - \|x - y\|)/\lambda_0 \leq (\|x - y - \lambda(x' - y')\| - \|x - y\|)/\lambda,$$

we have

$$\|x - y\| \leq \|x - y - \lambda(x' - y')\|.$$

Thus, Condition (D_2) of Theorem 2.9 holds and A is a dissipative operator. Finally, since $J_\lambda x \in D(A)$ $(\lambda > 0)$ and $\lim_{\lambda \to 0+} J_\lambda x = x$ for every $x \in X_0$, we have $\overline{D(A)} = X_0$.

THEOREM 3.13. *Let $\{T(t): t \geq 0\}$ be a semigroup of contractions on a closed convex set X_0 of a Hilbert space X, and set $J_{\lambda, h} = (I - \lambda A_h)^{-1}$ $(\lambda > 0, h > 0)$. Then,*

(i) *There exists $\lim_{h \to 0+} J_{\lambda, h}x$ for all $x \in X_0$ and all $\lambda > 0$.*

(ii) *The infinitesimal generator A_0 of $\{T(t): t \geq 0\}$ is a dissipative operator which has a dense domain in X_0 (that is, it satisfies $\overline{D(A_0)} = X_0$).*

PROOF. (i) Note that $\langle u, v \rangle_s = \mathrm{Re}(u, v)$ $(u, v \in X)$, because X is a Hilbert space, where $(\cdot, \cdot)$ denotes the inner product in X.

Let $x \in X_0$ and $\lambda > 0$, and set $y_\tau = J_{\lambda, \tau}x$ $(\tau > 0)$. By Lemma 3.8, for every $h > 0$ and natural number n, we have

$$\sum_{i=1}^{n}[\|y_{nh} - y_h\|^2 + \mathrm{Re}(y_h - x, y_{nh} - y_h) + \mathrm{Re}(y_h - x, y_h - T(ih)y_{nh})] \leq 0,$$

and from this we obtain

$$\|y_{nh} - y_h\|^2 \leq \mathrm{Re}(y_h - x, y_h - y_{nh}) + \frac{1}{n}\sum_{i=1}^{n}\mathrm{Re}(y_h - x, T(ih)y_{nh} - y_h)$$

$$= \mathrm{Re}(y_h - x, x - y_{nh}) + \frac{1}{n}\sum_{i=1}^{n}\mathrm{Re}(y_h - x, T(ih)y_{nh} - x).$$

$$(3.26)$$

$(^5)$ Note that for every $u, v \in X$, the function of $\lambda > 0$, $(\|u + \lambda v\| - \|u\|)/\lambda$, is an increasing function.

Now, for arbitrary $\varepsilon > 0$, there exists a positive number $\delta \leq 1$ such that if $0 < t \leq \delta$ then $\|T(t)x - x\| < \varepsilon$. Then, the following (3.27) holds:

$$\text{If } 0 < nh \leq \delta, \text{ then } \|y_h - y_{nh}\|^2 \leq 2\varepsilon\|y_h - x\|. \tag{3.27}$$

In fact, if we let $0 < nh \leq \delta$, then

$$\|T(ih)y_{nh} - x\| \leq \|T(ih)y_{nh} - T(ih)x\| + \|T(ih)x - x\|$$
$$\leq \|y_{nh} - x\| + \varepsilon \qquad (i = 1, 2, \dots, n).$$

Hence, from (3.26)

$$\|y_{nh} - y_h\|^2 \leq \operatorname{Re}(y_h - x, x - y_{nh}) + \|y_h - x\|(\|y_{nh} - x\| + \varepsilon)$$
$$\leq \operatorname{Re}(y_h - x, x - y_{nh}) + (\|y_h - x\|^2 + \|y_{nh} - x\|^2)/2 + \varepsilon\|y_h - x\|$$
$$= \|(y_h - x) + (x - y_{nh})\|^2/2 + \varepsilon\|y_h - x\|$$
$$= \|y_h - y_{nh}\|^2/2 + \varepsilon\|y_h - x\|.$$

Thus, (3.27) holds. Next, by the remark following Lemma 3.9, $\{\|y_h - x\| : 0 < h \leq 1\}$ is a bounded set. Then, if we set $M = \sup\{\|y_h - x\| : 0 < h \leq 1\}$ (M is dependent on $\lambda > 0$ and $x \in X_0$), then by (3.27)

$$\text{if } 0 < nh \leq \delta, \text{ then } \|y_h - y_{nh}\|^2 \leq 2M\varepsilon. \tag{3.28}$$

Let s and t be arbitrary rational numbers such that $0 < s, t \leq \delta$. Then since there exist an $h > 0$ and natural numbers n and m such that $s = nh$ and $t = mh$, it follows from (3.28) that

$$\|y_s - y_t\| \leq \|y_{nh} - y_h\| + \|y_h - y_{mh}\| \leq 2\sqrt{2M\varepsilon}.$$

But, since y_t $(= J_{\lambda,t}x)$ is continuous with respect to t (> 0) $(^6)$, the above inequality holds for every $s, t \in (0, \delta]$. That is, if $0 < s \leq \delta$ and $0 < t \leq \delta$, then $\|y_s - y_t\| \leq 2\sqrt{2M\varepsilon}$. From this, $\lim_{t \to 0+} y_t$ exists. Therefore, we have shown that $\lim_{t \to 0+} J_{\lambda,t}x$ exists for every $x \in X_0$ and $\lambda > 0$.

(ii) Since (i) holds, by (ii) of Theorem 3.11, $\widehat{D}$ is dense in X_0. Since a Hilbert space is a uniformly convex Banach space, by Corollary 3.6 we have $D(A_0) = \widehat{D}$, and hence $D(A_0)$ is dense in X_0. $\square$

REMARKS. (1) Part (i) of the theorem above does not hold even in a Banach space of finite-dimension (however, as we see from Corollary 3.12 and Theorem 3.11, we can select a sequence $\{h_n\}$ of positive numbers which converges to 0 such that $\lim_{n \to \infty} J_{\lambda,h_n}x$ exists for all $\lambda > 0$ and $x \in X_0$). In fact, there is known an example which does not hold in the three-dimensional real Banach space $l^\infty(3)$ (see [37] of the bibliography).

(2) In this section, we have considered a semigroup of contractions on a closed convex set, but the following is known for a Hilbert space. First,

$(^6)$ If $t_0 > 0$, $\|J_{\lambda,t}x - J_{\lambda,t_0}x\| = \|J_{\lambda,t}x - J_{\lambda,t}(I - \lambda A_t)J_{\lambda,t_0}x\| \leq \|x - (I - \lambda A_t)J_{\lambda,t_0}x\| = \lambda\|A_t J_{\lambda,t_0}x - A_{t_0}J_{\lambda,t_0}x\| \to 0$ as $t \to t_0$.

we call a semigroup of contractions a maximal semigroup of contractions if the domain cannot be extended (maintaining its contractiveness) any further. (Every semigroup of contractions has a maximal extension.) Then,

THEOREM. *The domain of a maximal semigroup of contractions in a Hilbert space is a closed convex set.*

(See [65] and [18] of the bibliography.)

§3. Approximation of semigroups of contractions and the exponential formula

Throughout this section, we let X_0 be a closed convex subset of a Banach space X.

LEMMA 3.14. *Let $\alpha > 0$ and let $C: X_0 \to X_0$ be a (single-valued) operator satisfying*

$$\|Cx - Cy\| \le \alpha \|x - y\| \qquad (x, y \in X_0). \tag{3.29}$$

Then, there exists a unique semigroup $\{T(t; C-I): t \ge 0\}$ on X_0 which has $(C - I)$ as the infinitesimal generator and the following inequality holds:

$$\|T(t; C-I)x - T(t; C-I)y\| \le e^{(\alpha-1)t}\|x-y\| \qquad (x, y \in X_0, t \ge 0). \tag{3.30}$$

Furthermore, if we set $u(t; x) = T(t; C - I)x \quad (x \in X_0, t \ge 0)$, then for every $x \in X_0$, $u(t; x): [0, \infty) \to X_0$ is a unique function that satisfies $u(\cdot; x) \in C^1([0, \infty); X_0)$ [7] and also

$$\begin{cases} (d/dt)u(t) = (C - I)u(t) & (t \ge 0), \\ u(0) = x. \end{cases} \tag{3.31}$$

REMARK. If $T(t): X_0 \to X_0$, where $t \ge 0$, satisfies

$$T(0)x = x, \quad \lim_{s \to t} T(s)x = T(t)x \qquad (x \in X_0, t \ge 0),$$

and condition (ii) of Definition 3.1; then we call $\{T(t): t \ge 0\}$ a semigroup on X_0 and define the infinitesimal generator of this semigroup in the same way as in the case of a semigroup of contractions.

PROOF OF LEMMA 3.14. Let $x \in X_0$. A necessary and sufficient condition for $u(t): [0, \infty) \to X_0$ to satisfy (3.31) is that $u(t): [0, \infty) \to X_0$ is continuous and

$$u(t) = e^{-t}x + \int_0^t e^{s-t}Cu(s)\,ds \qquad (t \ge 0). \tag{3.32}$$

Now we show that there exists a continuous function $u(t): [0, \infty) \to X_0$ that satisfies (3.32). First, note that if $v(t): [0, \infty) \to X_0$ is continuous, then

---

[7] If $u(t): [0, \infty) \to X_0$ is strongly continuously (once) differentiable (that is, $u(t)$ is strongly differentiable at every $t \ge 0$, and its strong derivative du/dt is continuous on $[0, \infty)$), then we denote by $u(\cdot) \in C^1([0, \infty); X_0)$.

for every $t \geq 0$

$$e^{-t}x + \int_0^t e^{s-t} Cv(s)\,ds$$

$$= \left(1 - \int_0^t e^{s-t}\,ds\right)x + \int_0^t e^{s-t}\,ds\left(\frac{1}{\int_0^t e^{s-t}\,ds}\int_0^t e^{s-t}Cv(s)\,ds\right) \in X_0$$

(here we apply the fact that X_0 is a closed convex set). Now we set $u_0(t) \equiv x$ and define $u_n(t)$ $(n \geq 1)$ by

$$u_n(t) = e^{-t}x + \int_0^t e^{s-t}Cu_{n-1}(s)\,ds \qquad (t \geq 0)$$

(here we apply the remark mentioned above). Then, every $u_n(t): [0, \infty) \to X_0$ is continuous and

$$\|u_{n+1}(t) - u_n(t)\| \leq K(\alpha t)^n/n! \qquad (t \geq 0, n = 0, 1, 2, \dots),$$

where $K = \|Cx - x\|$. (This can be shown by induction.) From this, $\{u_n(t)\}$ converges uniformly on every bounded interval of $[0, \infty)$. Thus, if we set

$$u(t; x) = \lim_{n \to \infty} u_n(t) \qquad (t \geq 0),$$

then $u(t; x)$ $[0, \infty) \to X_0$ is a continuous function satisfying

$$u(t; x) = e^{-t}x + \int_0^t e^{s-t}Cu(s; x)\,ds \qquad (t \geq 0). \tag{3.33}$$

That is, $u(t; x): [0, \infty) \to X_0$ is a continuous function that satisfies (3.32). Also, it is easy to see that there is a unique continuous function that satisfies (3.32). Hence, $u(t; x)$ is a unique function that is $u(\cdot; x) \in C^1([0, \infty); X_0)$ and satisfies (3.31).

Next, let $s \geq 0$. If we replace x by $u(s; x)$ in the above discussion, then $u(t; u(s; x))$ satisfies

$$\begin{cases} (d/dt)u(t; u(s; x)) = (C - I)u(t; u(s; x)) & (t \geq 0), \\ u(0; u(s; x)) = u(s; x). \end{cases}$$

Thus, if we define $v(t) = u(t; x)$ $(0 \leq t \leq s)$ and $v(t) = u(t - s; u(s; x))$ $(s \leq t)$, then $v(t): [0, \infty) \to X_0$ is a solution of (3.31). Then, from the uniqueness of the solution, we obtain $v(t + s) = u(t + s; x)$ $(t \geq 0)$; that is,

$$u(t; u(s; x)) = u(t + s; x) \qquad (t \geq 0).$$

Finally, from (3.33)

$$\|e^t u(t; x) - e^t u(t; y)\| \leq \|x - y\| + \alpha \int_0^t \|e^s u(s; x) - e^s u(s; y)\|\,ds,$$

where $x, y \in X_0$, $t \geq 0$. From this, for every natural number n, we have

$$\|e^t[u(t; x) - u(t; y)]\|$$

$$\leq \left[\sum_{k=0}^n \frac{(\alpha t)^k}{k!}\right]\|x - y\| + \frac{\alpha^{n+1}}{n!}\int_0^t (t - s)^n\|e^s[u(s; x) - u(s; y)]\|\,ds.$$

Let $n \to \infty$. Then we obtain $\|e^t[u(t;x) - u(t;y)]\| \le e^{\alpha t}\|x - y\|$; that is,

$$\|u(t;x) - u(t;y)\| \le e^{(\alpha-1)t}\|x - y\| \qquad (x, y \in X_0, t \ge 0).$$

Now, if we let $\{U(t): t \ge 0\}$ be a semigroup on X_0 which has $(C - I)$ as the infinitesimal generator, then for every $x \in X_0$, the strongly right derivative $D^+U(t)x = (C - I)U(t)x$ of $U(T)x$ is continuous on $[0, \infty)$. Hence, $U(t)x$ is strongly differentiable at every $t \ge 0$ (see Theorem 1.15), and is a solution of (3.31). Therefore, from the uniqueness of the solution of (3.31), $U(t)x = u(t;x)$ $(t \ge 0)$ for every $x \in X_0$. If we define an operator $T(t; C - I)$ $(t \ge 0)$ by

$$T(t; C - I)x = u(t;x) \qquad (x \in X_0),$$

then $\{T(t; C - I): t \ge 0\}$ is a unique semigroup on X_0 which has $(C - I)$ as the infinitesimal generator and satisfies the required properties. $\square$

LEMMA 3.15. *Let* $\alpha \ge 1$ *and let* $C: X_0 \to X_0$ *be an operator satisfying* (3.29). *If we let* $\{T(t; C - I): t \ge 0\}$ *be a semigroup on* X_0 *of Lemma 3.14, then for every* $x \in X_0$ *and nonnegative integer* m *and* $t \ge 0$, *we have*

$$\|T(t; C - I)x - C^m x\| \le \alpha^m e^{(\alpha-1)t}[(m - \alpha t)^2 + \alpha t]^{1/2}\|Cx - x\|. \qquad (3.34)$$

In particular, if $\alpha = 1$ *(that is, if* $C: X_0 \to X_0$ *is a contraction operator), then we have*

$$\|T(m; C - I)x - C^m x\| \le \sqrt{m}\|Cx - x\| \qquad (x \in X_0, m = 0, 1, 2, \ldots), \qquad (3.35)$$

where $C^0 = I$.

PROOF. Let $x \in X_0$ and let m be a nonnegative integer. Set

$$u(t;x) = T(t; C - I)x \qquad (t \ge 0).$$

From (3.30) and the fact that $u(t;x)$ is the solution of (3.31), we have

$$\|(d/dt)u(t;x)\| = \lim_{h \to 0+} \|h^{-1}[T(t + h; C - I)x - T(t; C - I)x]\|$$
$$\le e^{(\alpha-1)t} \lim_{h \to 0+} \|h^{-1}[T(h; C - I)x - x]\|$$
$$= e^{(\alpha-1)t}\|Cx - x\|,$$

$$\|u(t;x) - x\| \le \int_0^t \|(d/ds)u(s;x)\| ds \le e^{(\alpha-1)t}t\|Cx - x\| \qquad (t \ge 0).$$

From this, (3.34) holds if $m = 0$ or $Cx = x$ (in fact, because if $Cx = x$, then $u(t;x) = x = C^m x$). We now show that (3.34) holds when $m \ne 0$ and $\|Cx - x\| \ne 0$. From (3.33) and $C^n x = e^{-t}C^n x + \int_0^t e^{s-t}C^n x \, ds$, it follows that for every natural number n and $t \ge 0$, we have

$$u(t;x) - C^n x = e^{-t}(x - C^n x) + \int_0^t e^{s-t}(Cu(s;x) - C^n x) \, ds.$$

Since

$$\|x - C^n x\| \le \sum_{k=1}^{n} \|C^k x - C^{k-1} x\| \le \sum_{k=1}^{n} \alpha^{k-1} \|Cx - x\| \le n\alpha^n \|Cx - x\|,$$

we have

$$\|u(t; x) - C^n x\| \le e^{-t} n\alpha^n \|Cx - x\| + \alpha \int_0^t e^{s-t} \|u(s; x) - C^{n-1} x\| \, ds$$

$$(t \ge 0, \, n = 1, 2, \dots).$$

Therefore, if we set $\varphi_n(t) = \alpha^{-n} e^t \|u(t; x) - C^n x\| / \|Cx - x\|$ $(n = 0, 1, 2, \dots)$, we obtain

$$\varphi_n(t) \le n + \int_0^t \varphi_{n-1}(s) \, ds \qquad (t \ge 0, \, n = 1, 2, \dots). \tag{3.36}$$

Also from the inequality shown earlier, we have

$$\varphi_0(t) \le e^{\alpha t} t \le \alpha t e^{\alpha t} \qquad (t \ge 0).$$

From this, it follows that for every natural number n, we have (by induction)

$$\varphi_n(t) \le \sum_{k=0}^{n-1} \frac{(n-k)t^k}{k!} + \frac{1}{(n-1)!} \int_0^t (t-s)^{n-1} \varphi_0(s) \, ds$$

$$\le \sum_{k=0}^{n-1} \frac{(n-k)t^k}{k!} + \frac{\alpha}{(n-1)!} \int_0^t (t-s)^{n-1} s e^{\alpha s} \, ds \qquad (t \ge 0).$$

Since

$$\int_0^t (t-s)^{n-1} s e^{\alpha s} \, ds = \sum_{k=0}^{\infty} \frac{\alpha^k}{k!} \int_0^t (t-s)^{n-1} s^{k+1} \, ds$$

$$= (n-1)! \sum_{k=0}^{\infty} \frac{(k+1)\alpha^k t^{k+n+1}}{(k+n+1)!}$$

$$\le (n-1)! \sum_{k=n+1}^{\infty} \frac{(k-n)\alpha^{k-1} t^k}{k!},$$

where we use

$$\int_0^t (t-s)^{n-1} s^{k+1} \, ds = t^{k+n+1} \int_0^1 (1-s)^{n-1} s^{k+1} \, ds = t^{k+n+1} \frac{(k+1)!(n-1)!}{(k+n+1)!},$$

we obtain

$$\varphi_n(t) \le \sum_{k=0}^{n-1} \frac{(n-k)\alpha^k t^k}{k!} + \sum_{k=n+1}^{\infty} \frac{(k-n)\alpha^k t^k}{k!} = \sum_{k=0}^{\infty} \frac{|k-n|\alpha^k t^k}{k!}$$

$(t \ge 0, \, n = 1, 2, \dots)$. Next, from the Schwarz inequality we have

$$\sum_{k=0}^{\infty} \frac{|k-n|\alpha^k t^k}{k!} \le e^{\alpha t/2} \left(\sum_{k=0}^{\infty} \frac{(k-n)^2 \alpha^k t^k}{k!} \right)^{1/2} = e^{\alpha t}[(n - \alpha t)^2 + \alpha t]^{1/2}.$$

Thus, $\varphi_n(t) \le e^{\alpha t}[(n - \alpha t)^2 + \alpha t]$ and

$$\|u(t; x) - C^n x\| \le \alpha^n e^{(\alpha-1)t}[(n - \alpha t)^2 + \alpha t]^{1/2}\|Cx - x\|$$

$(t \ge 0, n = 1, 2, \ldots)$. Here, if we let $n = m$, then we obtain (3.34). Thus, when $m \ne 0$ and $\|Cx - x\| \ne 0$, (3.34) holds.

Finally, if we put $\alpha = 1$ and $t = m$ in (3.34), we obtain (3.35). $\square$

LEMMA 3.16. *Let* $C \in \text{Cont}(X_0)$ *and* $h > 0$, *and set* $C_h = h^{-1}(C - I)$. *Then, there exists a unique semigroup of contractions* $\{T(t; C_h): t \ge 0\}$ *on* X_0 *which has* C_h *as its infinitesimal generator, and for every* $x \in X_0$ *and* $t \ge 0$, *we have*

$$\|T(t; C_h)x - C^{[t/h]}x\| \le (\sqrt{th} + h)\|C_h x\|. \tag{3.37}$$

Furthermore, for an arbitrary $x \in X_0$, $T(\cdot; C_h)x \in C^1([0, \infty); X_0)$ *and* $(d/dt)T(t; C_h)x = C_h T(t; C_h)x$ $(t \ge 0)$.

PROOF. Let $\alpha = 1$. Then since (3.29) holds, from Lemmas 3.14 and 3.15, there exists a unique semigroup of contractions $\{T(t; C - I): t \ge 0\}$ on X_0 which has $(C - I)$ as the infinitesimal generator, and for every $x \in X_0$ we have $T(\cdot; C - I)x \in C^1([0, \infty); X_0)$ and $(d/dt)T(t; C - I)x = (C - I)T(t; C - I)x$ $(t \ge 0)$, and also

$$\|T([t/h]; C - I)x - C^{[t/h]}x\| \le \sqrt{\frac{t}{h}}\|Cx - x\| \qquad (t \ge 0). \tag{3.38}$$

Now, we define $T(t; C_h)$ by

$$T(t; C_h) = T(th^{-1}; C - I) \qquad (t \ge 0).$$

Then, $\{T(t; C_h): t \ge 0\}$ is a semigroup of contractions on X_0 and its infinitesimal generator is C_h. For every $x \in X_0$, we have $T(\cdot; C_h)x \in C^1([0, \infty); X_0)$ and $(d/dt)T(t; C_h)x = C_h T(t; C_h)x$ $(t \ge 0)$. Furthermore, we have from (3.38)

$$\|T([t/h]h; C_h)x - C^{[t/h]}x\| = \|T([t/h]; C - I)x - C^{[t/h]}x\|$$
$$\le \sqrt{th}\|C_h x\| \qquad (x \in X_0, t \ge 0).$$

From this and

$$\|T(t; C_h)x - T([t/h]h; C_h)x\| \le \int_{[t/h]h}^{t} \|(d/ds)T(s; C_h)x\| \, ds \le h\|C_h x\|$$

$(x \in X_0, t \ge 0)$, we obtain (3.37).

Finally, if we let $\{U(t): t \ge 0\}$ be a semigroup of contractions on X_0 which has C_h as infinitesimal generator, then $\{U(ht): t \ge 0\}$ is a semigroup of contractions on X_0 which has $(C - I)$ as infinitesimal generator. Then, since $\{T(t; C - I): t \ge 0\}$ is a unique semigroup of contractions on X_0, which has $(C - I)$ as infinitesimal generator, we have $U(ht) = T(t; C - I)$ $(t \geq 0)$. Thus, we obtain

$$U(t) = U(h \cdot th^{-1}) = T(th^{-1}; C - I) = T(t; C_h) \qquad (t \geq 0)$$

and $\{T(t; C_h): t \geq 0\}$ is a unique semigroup of contractions on X_0 that has C_h as infinitesimal generator. $\square$

Having prepared for the above, we now show that some of the approximations and exponential formulas that are known for semigroups of linear contractions hold also for semigroups of nonlinear contractions. In the following, we let $\{T(t): t \geq 0\}$ be a semigroup of contractions on X_0, and set

$$A_h = h^{-1}(T(h) - I), \quad J_{\lambda, h} = (I - \lambda A_h)^{-1} \qquad (h > 0, \lambda > 0)$$

as in the previous section. First,

THEOREM 3.17. *For each $h > 0$, there exists a unique semigroup of contractions $\{T(t; A_h): t \geq 0\}$ on X_0 that has A_h as infinitesimal generator, and $T(\cdot; A_h)x \in C^1([0, \infty); X_0)$ $(x \in X_0)$. And for each $x \in X_0$, we have*

$$T(t)x = \lim_{h \to 0+} T(t; A_h)x \qquad (t \geq 0), \tag{3.39}$$

and this convergence is uniform on every bounded interval of $[0, \infty)$.

PROOF. We set $C = T(h)$ and apply Lemma 3.16. Then, since $C_h = A_h$, there is a unique semigroup of contractions $\{T(t; A_h): t \geq 0\}$ on X_0 that has A_h as infinitesimal generator and $T(\cdot; A_h)z \in C^1([0, \infty); X_0)$ $(z \in X_0)$, and also

$$\|T(t; A_h)z - T([t/h]h)z\| = \|T(t; A_h)z - T(h)^{[t/h]}z\|$$
$$\leq (\sqrt{th} + h)\|A_h z\| \qquad (z \in X_0, t \geq 0).$$

To prove (3.39), we set $x \in X_0$ and $z = J_{\sqrt{h}, h}x$, and apply the above inequality to obtain

$$\|T(t; A_h)x - T([t/h]h)x\| \leq \|T(t; A_h)x - T(t; A_h)J_{\sqrt{h}, h}x\|$$
$$+ \|T(t; A_h)J_{\sqrt{h}, h}x - T([t/h]h)J_{\sqrt{h}, h}x\|$$
$$+ \|T([t/h]h)J_{\sqrt{h}, h}x - T([t/h]h)x\|$$
$$\leq 2\|J_{\sqrt{h}, h}x - x\| + (\sqrt{th} + h)\|A_h J_{\sqrt{h}, h}x\|$$
$$= (2 + \sqrt{t} + \sqrt{h})\|J_{\sqrt{h}, h}x - x\| \qquad (t \geq 0, h > 0).$$

From Lemma 3.9 we have $\lim_{h \to 0+} J_{\sqrt{h}, h}x = x$. Hence, if we let $h \to 0+$, $\|T(t; A_h)x - T([t/h]h)x\|$ converges to 0 uniformly on every bounded interval of $[0, \infty)$. Furthermore, if $h \to 0+$, the following holds uniformly on $[0, \infty)$:

$$\|T(t)x - T([t/h]h)x\| \leq \|T(t - [t/h]h)x - x\| \to 0.$$

Thus (3.39) holds and the convergence is uniform on every bounded interval in $[0, \infty)$. $\square$

THEOREM 3.18. *For each* $x \in X_0$

$$T(t)x = \lim_{n \to \infty} [(1-t)I + tT(1/n)]^n x \qquad (0 \le t \le 1), \qquad (3.40)$$

and the convergence is uniform on $[0, 1]$.

PROOF. Let $\xi \in [0, 1]$ and $h > 0$. Since $\xi T(h) + (1-\xi)I \in \mathrm{Cont}(X_0)$, we let $C = \xi T(h) + (1-\xi)I$ and apply Lemma 3.16. Then, since $C_h = h^{-1}(C-I) = \xi A_h$, we note that $T(t; C_h) = T(t; \xi A_h) = T(\xi t; A_h)$ $(t \ge 0)$. Thus, from (3.37) we have

$$\|T(\xi t; A_h)x - [\xi T(h) + (1-\xi)I]^{[t/h]}x\|$$
$$\le 2\|J_{\sqrt{h},h}x - x\| + (\sqrt{h} + h)\|A_h J_{\sqrt{h},h}x\|$$
$$= (3 + \sqrt{h})\|J_{\sqrt{h},h}x - x\| \qquad (x \in X_0, 0 \le t \le 1).$$

Here, if we set $t = 1$, then we obtain

$$\|T(\xi; A_h)x - [\xi T(h) + (1-\xi)I]^{[1/h]}x\| \le (3 + \sqrt{h})\|J_{\sqrt{h},h}x - x\|$$

$(x \in X_0)$, and from $\lim_{h \to 0+} J_{\sqrt{h},h}x = x$ (Lemma 3.9), if we let $h \to 0+$, the left-hand side of the above inequality converges to 0 uniformly with respect to $\xi \in [0, 1]$. From this and (3.39), (3.40) holds. □

REMARK. If $\{T(t): t \ge 0\}$ is a semigroup of linear contractions on X, since $A_h : X \to X$ is a bounded linear operator, $T(t; A_h) = \exp(tA_h)$ $(t \ge 0, h > 0)$, and it is well known that $T(t)x = \lim_{h \to 0+} \exp(tA_h)x$ $(x \in X)$ holds. (For example, see [49, Theorem 10.4.1] or [97, Theorem 18.2]); also, see [49, Theorem 10.4.3] (or [97, Theorem 18.2]) for (3.40).

For a semigroup of contractions $\{T(t): t \ge 0\}$ on X_0, the following exponential formula holds:

THEOREM 3.19. *For each* $x \in X_0$,

$$T(t)x = \lim_{(\lambda, h) \to (0,0)} (I - \lambda A_h)^{-[t/\lambda]}x \qquad (t \ge 0). \qquad (3.41)$$

Moreover, the convergence is uniform on every bounded interval of $[0, \infty)$.

PROOF. First, let $h > 0$ and $\lambda > 0$. Consider a semigroup of contractions $\{T(t; A_h): t \ge 0\}$ on X_0 which has A_h as the infinitesimal generator, and set

$$\varepsilon(t, \mu) = \mu^{-1}[T(t; A_h)x - T(t - \mu; A_h)x] - A_h T(t; A_h)x,$$

where $x \in X_0$ and $t \ge \mu > 0$. From the definition we have

$$T(t; A_h)x = J_{\mu,h}(T(t - \mu; A_h)x + \mu\varepsilon(t, \mu)). \qquad (3.42)$$

Also,

$$\varepsilon(t, \mu) = \mu^{-1} \int_{t-\mu}^{t} [A_h T(s; A_h)x - A_h T(t; A_h)x]\,ds,$$

$$\|A_h T(s; A_h)x - A_h T(t; A_h)x\| \le 2h^{-1}\|T(s; A_h)x - T(t; A_h)x\|.$$

From this,

$$\|\varepsilon(t,\mu)\| \le 2(\mu h)^{-1} \int_0^\mu \|T(s;A_h)x - x\| ds$$

$$\le 2(\mu h)^{-1} \int_0^\mu s\|A_h x\| ds \le (\mu/h)\|A_h x\|.$$

Then from (3.42), for every natural number n and $0 < \mu < \lambda$, we have

$$\|T(t;A_h)x - J_{\lambda,h}^n x\|$$

$$= \left\| J_{\mu,h}(T(t-\mu;A_h)x + \mu\varepsilon(t,\mu)) - J_{\mu,h}\left[\frac{\mu}{\lambda}J_{\lambda,h}^{n-1}x + \frac{\lambda-\mu}{\lambda}J_{\lambda,h}^n x\right] \right\|$$

$$\le \frac{\mu}{\lambda}\|T(t-\mu;A_h)x - J_{\lambda,h}^{n-1}x\| + \frac{\lambda-\mu}{\lambda}\|T(t-\mu;A_h)x - J_{\lambda,h}^n x\|$$

$$+ (\mu^2/h)\|A_h x\|. \quad \text{(Here we used (3.16).)}$$

Here we set $t = m\mu$ (m is a natural number), $\alpha = \mu/\lambda$, $\beta = 1 - \alpha$, $a_{i,j} = \|T(i\mu;A_h)x - J_{\lambda,h}^j x\|$ $(i, j = 0, 1, 2, \ldots)$. Then

$$a_{m,n} \le \alpha a_{m-1,n-1} + \beta a_{m-1,n} + (\mu^2/h)\|A_h x\|. \quad (3.43)$$

As we will show later, if $0 < \mu < \lambda$, then for arbitrary nonnegative integers m and n, the following inequality holds:

$$a_{m,n} \le ([(n\lambda - m\mu)^2 + n\lambda^2]^{1/2} + [(n\lambda - m\mu)^2 + \lambda\mu m]^{1/2})\|A_h x\| + (\mu^2 m/h)\|A_h x\|. \quad (3.44)$$

Let $t \ge 0$ and $0 < \mu < \lambda$, and set $n = [t/\lambda]$ and $m = [t/\mu]$ in the above inequality. Then we have

$$\|T([t/\mu]\mu;A_h)x - J_{\lambda,h}^{[t/\lambda]}x\| \le [2((\lambda+\mu)^2 + t\lambda)^{1/2} + \mu t/h]\|A_h x\|.$$

Now let $\mu \to 0+$; then for every $x \in X_0$, we have

$$\|T(t;A_h)x - J_{\lambda,h}^{[t/\lambda]}x\| \le 2(\lambda^2 + \lambda t)^{1/2}\|A_h x\|. \quad (3.45)$$

Next, we show that the exponential formula (3.41) holds. Let $x \in X_0$, $h > 0$, and $\lambda > 0$. Since $\|A_h J_{\sqrt{\lambda},h}x\| = \|J_{\sqrt{\lambda},h}x - x\|/\sqrt{\lambda}$, from (3.45) we have

$$\|T(t;A_h)J_{\sqrt{\lambda},h}x - J_{\lambda,h}^{[t/\lambda]}J_{\sqrt{\lambda},h}x\| \le 2(\lambda + t)^{1/2}\|J_{\sqrt{\lambda},h}x - x\|.$$

Hence, for every $t \ge 0$

$$\|T(t;A_h)x - J_{\lambda,h}^{[t/\lambda]}x\| \le [2 + 2(\lambda + t)^{1/2}]\|J_{\sqrt{\lambda},h}x - x\|.$$

Here we let $(\lambda, h) \to (0, 0)$ and apply Lemma 3.9. Then the right-hand side of the above inequality converges to 0 uniformly on every bounded interval of $[0, \infty)$. From this and Theorem 3.17, we obtain (3.41). $\square$

PROOF OF INEQUALITY (3.44). Since $a_{0,n} = \|x - J_{\lambda,h}^n x\| \le n\|J_{\lambda,h}x - x\| \le n\lambda\|A_h x\|$ $(n = 0, 1, 2, \ldots)$, and $a_{m,0} = \|T(m\mu;A_h)x - x\| \le m\mu\|A_h x\|$ $(m = 0, 1, 2, \ldots)$, (3.44) holds for either $m = 0$ or $n = 0$. We prove that if

(3.44) holds for $(m+1, n)$ when (3.44) is true for (m, n) and $(m, n-1)$, then (3.44) holds for all (m, n) by the inductive method.

We now assume that (3.44) holds for (m, n) and $(m, n-1)$. Then, from (3.43) and the Schwarz inequality we have

$$
\begin{aligned}
a_{m+1,n} &\leq \alpha a_{m,n-1} + \beta a_{m,n} + (\mu^2/h)\|A_h x\| \\
&\leq [\alpha\{[((n-1)\lambda - m\mu)^2 + (n-1)\lambda^2]^{1/2} \\
&\quad + [((n-1)\lambda - m\mu)^2 + \lambda\mu m]^{1/2}\} \\
&\quad + \beta\{[(n\lambda - m\mu)^2 + n\lambda^2]^{1/2} + [(n\lambda - m\mu)^2 + \lambda\mu m]^{1/2}\}]\|A_h x\| \\
&\quad + (\mu^2(m+1)/h)\|A_h x\| \\
&\leq [(\alpha+\beta)^{1/2}\{\alpha[((n-1)\lambda - m\mu)^2 + (n-1)\lambda^2] \\
&\qquad\qquad\qquad\qquad + \beta[(n\lambda - m\mu)^2 + n\lambda^2]\}^{1/2} \\
&\quad + (\alpha+\beta)^{1/2}\{\alpha[((n-1)\lambda - m\mu)^2 + \lambda\mu m] \\
&\qquad\qquad\qquad\qquad + \beta[(n\lambda - m\mu)^2 + \lambda\mu m]\}^{1/2}]\|A_h x\| \\
&\quad + (\mu^2(m+1)/h)\|A_h x\| \\
&\leq ([(n\lambda - (m+1)\mu)^2 + n\lambda^2]^{1/2} + [(n\lambda - (m+1)\mu)^2 + (m+1)\lambda\mu]^{1/2}) \\
&\quad \times \|A_h x\| + (\mu^2(m+1)/h)\|A_h x\|.
\end{aligned}
$$

Thus, (3.44) holds for $(m+1, n)$. $\square$

COROLLARY 3.20. *Let* $\{T(t): t \geq 0\}$ *be a semigroup of contractions on* X_0, *and assume that* $\{h_n\}$ *and* $\{\lambda_m\}$ *are sequences of positive numbers that converge to* 0 *and satisfy condition* (3.22) *of Theorem* 3.11. *Let* A *be the dissipative operator defined after the remark of Corollary* 3.12. *Then,*

(i) *For each* $x \in X_0$ *we have*

$$T(t)x = \lim_{\lambda \to 0+} (I - \lambda A)^{-[t/\lambda]}x \qquad (t \geq 0), \tag{3.46}$$

where the convergence is uniform on every bounded subinterval of $[0, \infty)$.

(ii) *For every* $\lambda > 0$, *if we set* $A_\lambda x = \lambda^{-1}[(I - \lambda A)^{-1}x - x]$ $(x \in X_0)$, *there exists a unique semigroup of contractions* $\{T(t; A_\lambda): t \geq 0\}$ *on* X_0 *which has* A_λ *as the infinitesimal generator, and for each* $x \in X_0$ *we have*

$$T(t)x = \lim_{\lambda \to 0+} T(t; A_\lambda)x \qquad (t \geq 0), \tag{3.47}$$

where convergence is uniform on every bounded subinterval of $[0, \infty)$.

PROOF. As we have shown in Theorem 3.11, $\lim_{n\to\infty} J_{\lambda,h_n}x$ exists for all $\lambda > 0$ and $x \in X_0$, and if we set

$$J_\lambda x = \lim_{n\to\infty} J_{\lambda, h_n} x \qquad (\lambda > 0, x \in X_0),$$

then for each $\lambda > 0$ J_λ is a contraction operator of $X_0 \to X_0$. Let $\lambda > 0$, and $x \in X_0$. Then from the definition of the operator A we have

$$\lambda^{-1}(J_\lambda x - x) \in \lambda^{-1}(I - J_\lambda^{-1})J_\lambda x \subset AJ_\lambda x,$$

that is, $x \in (I - \lambda A)J_\lambda x$. Since A is a dissipative operator, $(I - \lambda A)^{-1}$ is a single-valued (contraction) operator and from the above equation, we obtain

$$J_\lambda x = (I - \lambda A)^{-1} x \qquad (x \in X_0). \tag{3.48}$$

(i) Let $x \in X_0$ and choose $T > 0$ arbitrarily. Then from Theorem 3.19, given any $\varepsilon > 0$ there exists a $\delta > 0$ such that if $0 < h < \delta$ and $0 < \lambda < \delta$, then

$$\|T(t)x - J_{\lambda, h_n}^{[t/\lambda]} x\| < \varepsilon \qquad (t \in [0, T]).$$

Here, we let $n \to \infty$; then if $0 < \lambda < \delta$ we have

$$\|T(t)x - J_\lambda^{[t/\lambda]} x\| \le \varepsilon \qquad (t \in [0, T]).$$

From this and (3.48), we have $T(t)x = \lim_{\lambda \to 0+} (I - \lambda A)^{-[t/\lambda]} x$ (uniformly on $[0, T]$).

(ii) Let $\lambda > 0$. Since $A_\lambda = \lambda^{-1}(J_\lambda - I)$ and $J_\lambda \colon X_0 \to X_0$ is a contraction operator, by Lemma 3.16, there exists a unique semigroup of contractions $\{T(t; A_\lambda) \colon t \ge 0\}$ on X_0 which has A_λ as the infinitesimal generator. For every $x \in D(A)$ and $t \ge 0$

$$\|T(t; A_\lambda)x - (I - \lambda A)^{-[t/\lambda]} x\| = \|T(t; A_\lambda)x - J_\lambda^{[t/\lambda]} x\|$$
$$\le (\sqrt{t\lambda} + \lambda)\|A_\lambda x\| \le (\sqrt{t\lambda} + \lambda)\||Ax\||.$$

From this and (3.46), it follows that for each $x \in D(A)$ we have that $T(t)x = \lim_{\lambda \to 0+} T(t; A_\lambda)x$ (the convergence is uniform on every bounded subinterval of $[0, \infty)$). Since $T(t)$ and $T(t; A_\lambda)$ are contraction operators and $\overline{D(A)} = X_0$ (see the remark following Corollary 3.12), the equation above also holds for each $x \in X_0$. $\square$

COROLLARY 3.21. *Let* $\{T(t) \colon t \ge 0\}$ *be a semigroup of contractions on* X_0 *whose infinitesimal generator* A_0 *satisfies*

$$R(I - \lambda A_0) \supset X_0 \qquad (\lambda > 0). \tag{3.49}$$

Then:

(i) *For each* $x \in X_0$,

$$T(t)x = \lim_{\lambda \to 0+} (I - \lambda A_0)^{-[t/\lambda]} x \qquad (t \ge 0), \tag{3.50}$$

where the convergence is uniform on every bounded subinterval of $[0, \infty)$.

(ii) *For each* $\lambda > 0$, *if we set* $A_\lambda x = A_0(I - \lambda A_0)^{-1} x$ $(x \in X_0)$, *then there exists a unique semigroup of contractions* $\{T(t; A_\lambda) \colon t \ge 0\}$ *on* X_0 *that has* A_λ *as the infinitesimal generator, and for each* $x \in X_0$ *we have*

$$T(t)x = \lim_{\lambda \to 0+} T(t; A_\lambda)x \qquad (t \ge 0), \tag{3.51}$$

where the convergence is uniform on bounded subintervals of $[0, \infty)$.

PROOF. For every $\lambda > 0$ and $x \in X_0$, we have

$$\lim_{h \to 0+} J_{\lambda, h} x = (I - \lambda A_0)^{-1} x.$$

In fact, for a given $\lambda > 0$, from (3.49), $x \in X_0$ can be written in the form $x = y - \lambda A_0 y$. Since $\|J_{\lambda, h} x - y\| \leq \|x - (I - \lambda A_h)y\| = \lambda \|A_h y - A_0 y\| \to 0$ as $h \to 0+$ and $y = (I - \lambda A_0)^{-1} x$, $\lim_{h \to 0+} J_{\lambda, h} x = (I - \lambda A_0)^{-1} x$. Therefore, by the method used in the proof of Corollary 3.20, we obtain the required result from Theorem 3.19. (Further, note that

$$A_0 (I - \lambda A_0)^{-1} x = \lambda^{-1} [(I - \lambda A_0)^{-1} x - x] \qquad (x \in X_0).)$$

□

REMARK. Let $\{T(t): t \geq 0\}$ be a semigroup of linear contractions and let A_0 be its infinitesimal generator. Then, A_0 is an m-dissipative operator with dense domain in X. Therefore, from the above corollary, we can derive the well-known exponential formula ([8]) for semigroups of linear contractions:

$$T(t)x = \lim_{\lambda \to 0+} (I - \lambda A_0)^{-[t/\lambda]} x$$

$$= \lim_{\lambda \to 0+} \exp[t A_0 (I - \lambda A_0)^{-1}] x \qquad (t \geq 0, x \in X).$$

(We note that since $A_\lambda = A_0 (I - \lambda A_0)^{-1} : X \to X$ is a bounded linear operator, $T(t; A_\lambda) = \exp(t A_\lambda) = \exp[t A_0 (I - \lambda A_0)^{-1}]$.)

([8]) See, for example, [49, Theorems 11.6.5 and 11.6.6]; the remark on p. 248 of [132]; or [97, Theorems 18.3 and 18.4].

CHAPTER 4

Generation of Semigroups of Nonlinear Contractions

Let X be a real or complex Banach space. In the previous chapter, we examined the relation between semigroups and dissipative operators when a semigroup of contractions is given in X. In this chapter, we consider the problem of the generation of semigroups of contractions by dissipative operators.

§1. Generation of semigroups of contractions I; the case of a general Banach space

Let A be a dissipative operator and set $J_\lambda = (I - \lambda A)^{-1}$ $(\lambda > 0)$. Then, we note that J_λ is a contraction operator, and remind the reader of Lemma 2.11.

LEMMA 4.1. *Let A be a dissipative operator and let $\lambda \geq \mu > 0$. If $x \in D(J_\mu^k) \cap D(J_\lambda^k) \cap D(A)$ $(k = 1, 2, \ldots)$, then for all nonnegative integers m and n*

$$\|J_\mu^n x - J_\lambda^m x\| \leq ([(m\lambda - n\mu)^2 + m\lambda^2]^{1/2} + [(m\lambda - n\mu)^2 + \lambda\mu n]^{1/2}) \|\|Ax\|\|. \quad (4.1)$$

PROOF. Let $x \in D(J_\mu^k) \cap D(J_\lambda^k) \cap D(A)$ and set $a_{n,m} = \|J_\mu^n x - J_\lambda^m x\|$ $(n, m = 0, 1, 2, \ldots)$. From part (iii) of Lemma 2.11, we have

$$\|J_\mu^n x - J_\lambda^m x\| = \left\| J_\mu^n x - J_\mu \left[\frac{\mu}{\lambda} J_\lambda^{m-1} x + \frac{\lambda - \mu}{\lambda} J_\lambda^m x \right] \right\|$$

$$\leq \frac{\mu}{\lambda} \|J_\mu^{n-1} x - J_\lambda^{m-1} x\| + \frac{\lambda - \mu}{\lambda} \|J_\mu^{n-1} x - J_\lambda^m x\|.$$

Thus, if we set $\alpha = \mu/\lambda$, $\beta = 1 - \alpha$, then we obtain

$$a_{n,m} \leq \alpha a_{n-1,m-1} + \beta a_{n-1,m} \quad (n, m = 1, 2, \ldots). \quad (4.2)$$

Then, we can show (4.1) by the method used in the proof of inequality (3.44). $\square$

THEOREM 4.2. *Let A be a dissipative operator that satisfies the following condition $(\mathbf{R}_1)$:*

$$R(I - \lambda A) \supset D(A) \quad \text{for all } \lambda > 0. \quad (\mathbf{R}_1)$$

Then, there is a semigroup of contractions $\{T(t) : t \geq 0\}$ *on* $\overline{D(A)}$ *that satisfies:*

(i) *For every* $x \in R \cap \overline{D(A)}$, *where* $R = \bigcap_{\lambda > 0} R(I - \lambda A)$,

$$T(t)x = \lim_{\lambda \to 0+} (I - \lambda A)^{-[t/\lambda]}x \qquad (t \geq 0). \qquad (4.3)$$

where convergence is uniform on bounded subintervals of $[0, \infty)$.

(ii) $\|T(t)x - T(s)x\| \leq \||Ax\|| |t - s|$ $(x \in D(A); t, s \geq 0)$.

PROOF. Since we obtain $D(A) \subset D(J_\lambda^k)$ $(\lambda > 0; k = 1, 2, \ldots)$ from condition (R_1), (4.1) holds for $x \in D(A)$, $\lambda \geq \mu > 0$, and nonnegative integers m and n, by Lemma 4.1. In particular, when we set $m = [t/\lambda]$ and $n = [t/\mu]$, where $t \geq 0$, we have

$$\|J_\mu^{[t/\mu]}x - J_\lambda^{[t/\lambda]}x\| \leq 2(\lambda^2 + \lambda t)^{1/2}\||Ax\||. \qquad (4.4)$$

Now we let $x \in R \cap \overline{D(A)}$ and choose a sequence $\{x_k\}$ in $D(A)$ such that $\|x_k - x\| \to 0$ as $k \to \infty$. Then for $\lambda \geq \mu > 0$ and $t \geq 0$, we have from (4.4)

$$\|J_\mu^{[t/\mu]}x - J_\lambda^{[t/\lambda]}x\| \leq 2\|x - x_k\| + 2(\lambda^2 + \lambda t)^{1/2}\||Ax_k\||.$$

Therefore, for an arbitrary $T > 0$ we obtain

$$\limsup_{\lambda, \mu \to 0+} \left[\sup_{0 \leq t \leq T} \|J_\mu^{[t/\mu]}x - J_\lambda^{[t/\lambda]}x\| \right] \leq 2\|x - x_k\| \to 0 \qquad (k \to \infty).$$

As $\lambda \to 0+$, $J_\lambda^{[t/\lambda]}x$ $(= (I - \lambda A)^{-[t/\lambda]}x)$ converges uniformly on arbitrary bounded subintervals of $[0, \infty)$. We now define an operator $T(t)$ for each $t \geq 0$ by

$$T(t)x = \lim_{\lambda \to 0+} J_\lambda^{[t/\lambda]}x \qquad (x \in R \cap \overline{D(A)}).$$

Obviously, $T(t) : R \cap \overline{D(A)} \to \overline{D(A)}$ is a contraction operator and $T(0)x = x$ $(x \in R(A) \cap \overline{D(A)})$. Since $R \cap \overline{D(A)}$ $(\supset D(A))$ is dense in $\overline{D(A)}$, we can uniquely extend every $T(t)$ to a contraction operator on $\overline{D(A)}$. We denote this operator by $T(t)$ again.

For each $t > 0$, $T(t) \in \text{Cont}(\overline{D(A)})$ and $T(0)x = x$ $(x \in \overline{D(A)})$. Next, we prove that $T(t + s) = T(t)T(s)$ $(t, s \geq 0)$. For this, we need to show that

$$T(t + s)z = T(t)T(s)z \qquad (z \in D(A)) \qquad (4.5)$$

for each $t, s \geq 0$. Now, to prove (4.5) we let $z \in D(A)$. Let $\mu \to 0+$ in (4.4); then we have

$$\|T(t)x - J_\lambda^{[t/\lambda]}x\| \leq 2(\lambda^2 + \lambda_t)^{1/2}\||Ax\|| \qquad (x \in D(A)).$$

Since $J_\lambda^{[s/\lambda]}z \in D(A)$, if we set $x = J_\lambda^{[s/\lambda]}z$ in the above equation and note that $|||AJ_\lambda^k z||| \leq |||Az|||$ $(k = 1, 2, \ldots)$ ([1]), then

$$\|T(t)J_\lambda^{[s/\lambda]}z - J_\lambda^{[t/\lambda]}J_\lambda^{[s/\lambda]}z\| \leq 2(\lambda^2 + \lambda_t)^{1/2}|||Az|||. \tag{4.6}$$

Also, since the value of $[(t+s)/\lambda] - ([t/\lambda] + [s/\lambda])$ is 0 or 1, we have

$$\|J_\lambda^{[t/\lambda]}J_\lambda^{[s/\lambda]}z - J_\lambda^{[(t+s)/\lambda]}z\| \leq \|J_\lambda z - z\| = \leq \lambda|||Az|||. \tag{4.7}$$

From (4.6) and (4.7)

$$
\begin{aligned}
\|T(t)&T(s)z - T(t+s)z\| \\
&\leq \|T(t)T(s)z - T(t)J_\lambda^{[s/\lambda]}z\| + \|T(t)J_\lambda^{[s/\lambda]}z - J_\lambda^{[t/\lambda]}J_\lambda^{s/\lambda}z\| \\
&\quad + \|J_\lambda^{[t/\lambda]}J_\lambda^{[s/\lambda]}z - J_\lambda^{[(t+s)/\lambda]}z\| + \|J_\lambda^{[(t+s)/\lambda]}z - T(t+s)z\| \\
&\leq \|T(s)z - J_\lambda^{[s/\lambda]}z\| + 2(\lambda^2 + \lambda t)^{1/2}|||Az||| + \lambda|||Az||| \\
&\quad + \|J_\lambda^{[(t+s)/\lambda]}z - T(t+s)z\| \to 0 \qquad (\lambda \to 0+).
\end{aligned}
$$

Therefore (4.5) holds.

Finally, if we let $\lambda \to 0+$ for

$$\|J_\lambda^{[t/\lambda]}x - x\| \leq [t/\lambda]\|J_\lambda x - x\| \leq t|||Ax||| \qquad (t \geq 0, \ \lambda > 0, \ x \in D(A)),$$

we have

$$\|T(t)x - x\| \leq t|||Ax||| \qquad (x \in D(A), \ t \geq 0) \tag{4.8}$$

From this we obtain $\lim_{t \to 0+} T(t)x = x$ $(x \in \overline{D(A)})$, and $\{T(t) : t \geq 0\}$ is a semigroup of contractions on $\overline{D(A)}$. Also, from (4.8) we obtain

$$\|T(t)x - T(s)x\| \leq \|T(t-s)x - x\| \leq |||Ax|||(t-s) \qquad (x \in D(A), \ t \geq s \geq 0)$$

and thus (ii) holds. $\square$

Let A be a dissipative operator satisfying the condition (R_2):

$$R(I - \lambda A) \supset \operatorname{co} D(A) \quad \text{for all } \lambda > 0. \tag{R_2}$$

We set $X_0 = \overline{\operatorname{co}}\, D(A)$. The restriction $J_\lambda|_{\operatorname{co} D(A)}$ of J_λ $(\lambda > 0)$ to $\operatorname{co} D(A)$ can be extended uniquely to a contraction operator on X_0. We denote this extension by $\widetilde{J}_\lambda$, and set $\widetilde{A}_\lambda = \lambda^{-1}(\widetilde{J}_\lambda - I)$. Since $\widetilde{J}_\lambda \in \operatorname{Cont}(X_0)$, from Lemma 3.16, there is only one semigroup of contractions $\{T(t; \widetilde{A}_\lambda) : t \geq 0\}$ on X_0 which has $\widetilde{A}_\lambda$ as the infinitesimal generator. And for each $x \in X_0$, $T(\cdot; \widetilde{A}_\lambda)x \in C^1([0, \infty); X_0)$, and we obtain

$$\|T(t; \widetilde{A}_\lambda)x - \widetilde{J}_\lambda^{[t/\lambda]}x\| \leq (\sqrt{t\lambda} + \lambda)\|\widetilde{A}_\lambda x\| \qquad (t \geq 0). \tag{4.9}$$

We have the following theorem.

([1])For an arbitrary $z \in D(A)$, $|||AJ_\lambda^k z||| \leq |||Az|||$ $(k = 1, 2, \ldots)$. In fact, by parts (i) and (ii) of Lemma 2.11, $|||AJ_\lambda z||| \leq |||A_\lambda z||| \leq |||Az|||$. Assuming this is true for k, then $|||AJ_\lambda^{k+1}z||| = |||AJ_\lambda^k(J_\lambda z)||| \leq |||AJ_\lambda z||| \leq |||Az|||$ and so the inequality also holds for $(k+1)$.

THEOREM 4.3. *Let A be a dissipative operator satisfying condition* (R_2) *and let $\{T(t) : t \geq 0\}$ be a semigroup of contractions on $\overline{D(A)}$ of Theorem 4.2. Then, for each $x \in \overline{D(A)}$ we have*

$$T(t)x = \lim_{\lambda \to 0+} T(t; \widetilde{A}_\lambda)x \qquad (t \geq 0), \qquad (4.10)$$

where the convergence is uniform on bounded subintervals of $[0, \infty)$. ($T(t; \widetilde{A}_\lambda)$ is the so-called Yosida approximation of $T(t)$.)

PROOF. Let $x \in D(A)$. Since $\widetilde{J}_\lambda x = J_\lambda x \in D(A)$, $\widetilde{J}_\lambda^{[t/\lambda]} x = J_\lambda^{[t/\lambda]} x$ $(t \geq 0, \lambda > 0)$. Therefore, from (4.9)

$$\|T(t; \widetilde{A}_\lambda)x - J_\lambda^{[t/\lambda]} x\| \leq (\sqrt{t\lambda} + \lambda)\|A_\lambda x\| \leq (\sqrt{t\lambda} + \lambda)\||Ax|\|$$

$(t \geq 0, \lambda > 0)$. From this and (4.3), (4.10) holds for each $x \in D(A)$ and the convergence is uniform on bounded subintervals of $[0, \infty)$. Since $T(t)$ and $T(t; \widetilde{A}_\lambda)$ are contraction operators, we obtain the same result for each $x \in \overline{D(A)}$. $\square$

From the two theorems above, we have the following corollary:

COROLLARY 4.4. *Let A be an m-dissipative operator and set $A_\lambda = \lambda^{-1}(J_\lambda - I)$ $(\lambda > 0)$. Then, there exists a semigroup of contractions $\{T(t) : t \geq 0\}$ on $\overline{D(A)}$, and for each $x \in \overline{D(A)}$*

$$T(t)x = \lim_{\lambda \to 0+} (I - \lambda A)^{-[t/\lambda]} x = \lim_{\lambda \to 0+} T(t; A_\lambda)x \qquad (t \geq 0),$$

where the convergence is uniform on bounded subintervals of $[0, \infty)$. Here, $\{T(t; A_\lambda) : t \geq 0\}$ is a semigroup of contractions on X that has A_λ $(\lambda > 0)$ as infinitesimal generator.

PROOF. This is because when we let $\{T(t; \widetilde{A}_\lambda) : t \geq 0\}$ be a semigroup of contractions on $\overline{\text{co}}\, D(A)$ of Theorem 4.3,

$$T(t; A_\lambda)x = T(t; \widetilde{A}_\lambda)x \qquad (x \in \overline{\text{co}}\, D(A)) \qquad (4.11)$$

holds for arbitrary $\lambda > 0$ and $t \geq 0$. $\square$

REMARK. Equation (4.11), for example, can be shown as follows. Since $\widetilde{J}_\lambda x = J_\lambda x$ $(x \in \overline{\text{co}}\, D(A))$, $\widetilde{A}_\lambda x = A_\lambda x$. Therefore, we obtain $(I - \mu A_\lambda)^{-1} x = (I - \mu \widetilde{A}_\lambda)^{-1} x \in \overline{\text{co}}\, D(A)$ $(x \in \overline{\text{co}}\, D(A))$. Thus

$$(I - \mu A_\lambda)^{-[t/\mu]} x = (I - \mu \widetilde{A}_\lambda)^{-[t/\mu]} x \qquad (x \in \overline{\text{co}}\, D(A), \ \mu > 0). \qquad (4.12)$$

Then, since $R(I - \mu A_\lambda) = X$ (that is, A_λ is an m-dissipative operator), $R(I - \mu \widetilde{A}_\lambda) \supset \overline{\text{co}}\, D(A)$ (see Example 2.4), from (i) of Corollary 3.21 we have

$$T(t; A_\lambda)x = \lim_{\mu \to 0+} (I - \mu A_\lambda)^{-[t/\mu]} x \qquad (x \in X),$$

$$T(t; \widetilde{A}_\lambda)x = \lim_{\mu \to 0+} (I - \mu \widetilde{A}_\lambda)^{-[t/\mu]} x \qquad (x \in \overline{\text{co}}\, D(A)).$$

From this and (4.12), we obtain (4.11).

Here, we naturally ask whether the semigroup of contractions formulated with Theorem 4.2 is a solution of the Cauchy problem for the operator A:

$$(d/dt)u(t) \in Au(t) \quad (t > 0), \qquad u(0) = x \quad (\in \overline{D(A)}),$$

but we will investigate this later and state the following theorem.

THEOREM 4.5. *Let A be a closed dissipative operator satisfying condition (R_1) and let $\{T(t) : t \geq 0\}$ be a semigroup of contractions on $\overline{D(A)}$ formulated with Theorem 4.2 and let $x \in \overline{D(A)}$. If $T(t)x$ is strongly differentiable at $t_0 > 0$, then*

$$T(t_0)x \in D(A), \qquad [(d/dt)T(t)x]_{t=t_0} \in AT(t_0)x; \qquad (4.13)$$

and furthermore, we have

$$T(t_0)x \in D(A^0), \qquad [(d/dt)T(t)x]_{t=t_0} \in A^0 T(t_0)x. \qquad (4.14)$$

We will use the following lemma in the proof of Theorem 4.5.

LEMMA 4.6. *Let A be a dissipative operator satisfying condition (R_1) and let $\{T(t) : t \geq 0\}$ be a semigroup of contractions on $\overline{D(A)}$ formulated with Theorem 4.2. Then, for every $x \in \overline{D(A)}$, $x_0 \in D(A)$, and $y_0 \in Ax_0$,*

$$\|T(t)x - x_0\|^2 - \|T(r)x - x_0\|^2$$
$$\leq 2 \int_r^t \langle y_0, T(\tau)x - x_0 \rangle_s \, d\tau \qquad (0 \leq r < t < \infty), \qquad (4.15)$$

and

$$\begin{cases} \text{for every } f \in F(x - x_0) \\ \limsup_{t \to 0+} \mathrm{Re}\left(\frac{T(t)x - x}{t}, f\right) \leq \langle y_0, x - x_0 \rangle_s. \end{cases} \qquad (4.16)$$

PROOF. Let $x_0 \in D(A)$ and $y_0 \in Ax_0$. To prove (4.15), we first let $x \in D(A)$. From Lemma 2.11, for each $\lambda > 0$ and k $(= 1, 2, \ldots)$

$$y_{\lambda, k} \equiv \lambda^{-1}(J_\lambda^k x - J_\lambda^{k-1} x) = A_\lambda J_\lambda^{k-1} x \in A J_\lambda^k x.$$

Since A is a dissipative operator, there exists $g \in F(J_\lambda^k x - x_0)$ with $\mathrm{Re}(y_{\lambda, k} - y_0, g) \leq 0$. From

$$\mathrm{Re}(y_{\lambda, k}, g) = \lambda^{-1} \mathrm{Re}(J_\lambda^k x - x_0 - [J_\lambda^{k-1} x - x_0], g)$$
$$\geq \lambda^{-1}(\|J_\lambda^k x - x_0\|^2 - \|J_\lambda^{k-1} x - x_0\| \|J_\lambda^k x - x_0\|)$$
$$\geq (2\lambda)^{-1}(\|J_\lambda^k x - x_0\|^2 - \|J_\lambda^{k-1} x - x_0\|^2),$$

we have

$$\|J_\lambda^k x - x_0\|^2 - \|J_\lambda^{k-1} x - x_0\|^2 \leq 2\lambda \mathrm{Re}(y_{\lambda, k}, g)$$
$$\leq 2\lambda \mathrm{Re}(y_0, g) \leq 2\lambda \langle y_0, J_\lambda^k x - x_0 \rangle_s.$$

Since $J_\lambda^{[\tau/\lambda]}x = J_\lambda^k x$ $(k\lambda \le \tau < (k+1)\lambda)$, the above inequality can be written as

$$\|J_\lambda^k x - x_0\|^2 - \|J_\lambda^{k-1}x - x_0\|^2 \le 2\int_{k\lambda}^{(k+1)\lambda} \langle y_0, J_\lambda^{[\tau/\lambda]}x - x_0\rangle_s \, d\tau.$$

Now, we let $t > r \ge \lambda$ and add respective sides of the above inequality for $k = [r/\lambda]+1, \ldots, [t/\lambda]$, and obtain

$$\|J_\lambda^{[t/\lambda]}x - x_0\|^2 - \|J_\lambda^{[\tau/\lambda]}x - x_0\|^2 \le 2\int_{([r/\lambda]+1)\lambda}^{([t/\lambda]+1)\lambda} \langle y_0, J_\lambda^{[\tau/\lambda]}x - x_0\rangle_s \, d\tau.$$

Here,(^{2}) letting $\lambda \to 0+$, by the Lebesgue convergence theorem (since $\|J_\lambda^{[\tau/\lambda]}x - x\| \le \tau\|\|Ax\|\|$, note $|\langle y_0, J_\lambda^{[\tau/\lambda]}x - x_0\rangle_s| \le \|y_0\|(\tau\|\|Ax\|\| + \|x - x_0\|)$) and from the upper semicontinuity of $\langle \cdot, \cdot\rangle_s : X \times X \to (-\infty, \infty)$ (see part (iv) of Corollary 2.6), we obtain

$$\|T(t)x - x_0\|^2 - \|T(r)x - x_0\|^2 \le 2\int_r^t \langle y_0, T(\tau)x - x_0\rangle_s \, d\tau$$

$(0 \le r \le t < \infty)$. Next, let $x \in \overline{D(A)}$; then there exists a sequence $\{x_k\}$ in $D(A)$ such that $\lim_{k\to\infty} x_k = x$. From what we have shown above, we have

$$\|T(t)x_k - x_0\|^2 - \|T(r)x_k - x_0\|^2 \le 2\int_r^t \langle y_0, T(\tau)x_k - x_0\rangle_s \, d\tau.$$

Then we obtain (4.15) by letting $k \to \infty$ and applying again the Lebesgue convergence theorem and the upper semicontinuity of $\langle \cdot, \cdot\rangle_s$.

Next we prove (4.16). For any given $f \in F(x - x_0)$, from $\|T(t)x - x_0\|^2 - \|x - x_0\|^2 \ge 2\,\mathrm{Re}\langle T(t)x - x, f\rangle$ and (4.15) we have

$$\mathrm{Re}\langle T(t)x - x, f\rangle \le \int_0^t \langle y_0, T(\tau)x - x_0\rangle_s \, d\tau \qquad (t > 0).$$

Since $\langle y_0, T(\tau)x - x_0\rangle_s : [0, \infty) \to (-\infty, \infty)$ is upper semicontinuous, we obtain (4.16) from the inequality above. $\square$

PROOF OF THEOREM 4.5. From the assumption that A is a closed (dissipative) operator, $R(I - \lambda A)$ is a closed set for each $\lambda > 0$. (In fact, we let $z_n \in R(I - \lambda A)$ and $z_n \to z$. Then we can write $z_n = x_n - \lambda y_n$, where $x_n \in D(A)$, $y_n \in Ax_n$, and $\|x_n - x_m\| \le \|z_n - z_m\| \to 0$ as n and $m \to \infty$. Thus, there exists an $x \in X$ such that $\lim_{n\to\infty} x_n = x$, and $y_n = \lambda^{-1}(x_n - z_n) \to \lambda^{-1}(x - z)$. Then, since A is a closed operator, $x \in D(A)$ and $\lambda^{-1}(x - z) \in Ax$. Hence, we obtain $z \in (I - A)x \subset R(I - \lambda A)$ and $R(I - \lambda A)$ is a closed set.) Thus from condition (R_1) we obtain

$$R(I - \lambda A) \supset \overline{D(A)} \quad \text{for all } \lambda > 0. \tag{R_1'}$$

Now, if we set $[(d/dt)T(t)x]_{t=t_0} = y$, then

$$T(t_0 - \lambda)x = T(t_0)x - \lambda y + o(\lambda)(^3) \qquad (\lambda \to 0+).$$

_____

(^{2})We consider $\limsup_{\lambda\to 0+}$.

(^{3}) $o(\lambda)$ means that $\|o(\lambda)\|/\lambda \to 0$ as $\lambda \to 0+$

Since $T(t_0 - \lambda)x \in \overline{D(A)}$, from (R_1'), there exist $x_\lambda \in D(A)$ and $y_\lambda \in Ax_\lambda$ such that $T(t_0 - \lambda)x = x_\lambda - \lambda y_\lambda$. Thus,

$$x_\lambda - \lambda y_\lambda = T(t_0)x - \lambda y + o(\lambda) \qquad (\lambda \to 0+). \qquad (4.17)$$

If we let $x_0 = x_\lambda$, $y_0 = y_\lambda$, and $x = T(t_0)x$, and apply (4.16), then for every $f \in F(T(t_0)x - x_\lambda)$, we have

$$\mathrm{Re}(y, f) \le \langle y_\lambda, T(t_0)x - x_\lambda \rangle_s.$$

Since there exists $g \in F(T(t_0)x - x_\lambda)$ such that $\langle y_\lambda, T(t_0)x - x_\lambda \rangle_s = \mathrm{Re}(y_\lambda, g)$, we have[4]

$$\mathrm{Re}(y - y_\lambda, g) \le 0.$$

From (4.17) we have $\lambda(y - y_\lambda) = T(t_0)x - x_\lambda + o(\lambda)$. Hence

$$\mathrm{Re}(T(t_0)x - x_\lambda + o(\lambda), g) \le 0.$$

From this we obtain $\|T(t_0)x - x_\lambda\|^2 \le \|o(\lambda)\| \|T(t_0)x - x_\lambda\|$ and

$$\lim_{\lambda \to 0+} \frac{T(t_0)x - x_\lambda}{\lambda} = 0.$$

From this and $\lambda(y - y_\lambda) = T(t_0)x - x_\lambda + o(\lambda)$, we have

$$x_\lambda \to T(t_0)x, \qquad y_\lambda \to y \quad (\lambda \to 0+).$$

Then, from the assumption of A being a closed operator, we obtain

$$T(t_0)x \in D(A), \qquad y \in AT(t_0)x. \qquad (4.18)$$

Therefore, (4.13) is proved.

Next, from part (ii) of Theorem 4.2, $\|T(h)z - z\| \le \|\|Az\|\|h$ ($z \in D(A)$, $h \ge 0$). In particular, if we consider $z = T(t_0)x$ ($\in D(A)$), then

$$\|T(t_0 + h)x - T(t_0)x\| \le \|\|AT(t_0)x\|\|h \qquad (h > 0).$$

Thus,

$$\|y\| \le \|\|AT(t_0)x\|\|.$$

From this and (4.18) we obtain

$$T(t_0)x \in D(A^0), \qquad y \in A^0 T(t_0)x,$$

and (4.14) is proved. □

THEOREM 4.7. *Let A be a dissipative operator satisfying the condition:*

$$R(I - \lambda A) \supset \overline{D(A)} \quad \textit{for all } \lambda > 0, \qquad (R_1')$$

and let $\{T(t) : t \ge 0\}$ be the semigroup of contractions on $\overline{D(A)}$ of Theorem 4.2. Then:

(i) For every $x \in \overline{D(A)}$

$$\lim_{\lambda \to 0+} \|A_\lambda x\| \left(= \lim_{\lambda \to 0+} \lambda^{-1}\|J_\lambda x - x\|\right) = \lim_{h \to 0+} \inf h^{-1}\|T(h)x - x\|.$$

[4] See Theorem 2.5 and Equation (2.12) of the definition of $\langle \cdot, \cdot \rangle_s$.

(ii) *Let* $x \in \overline{D(A)}$. *The following three conditions are equivalent:*

(ii$_1$) $\liminf_{h \to 0+} h^{-1} \|T(h)x - x\| < \infty$ ($\rightleftarrows T(t)x : [0, \infty) \to \overline{D(A)}$ *is Lipschitz continuous* (*see Lemma* 3.1),

(ii$_2$) $\lim_{\lambda \to 0+} \|A_\lambda x\| < \infty$,

(ii$_3$) *there is a sequence* $\{x_n\}$ *in* $D(A)$ *such that* $\lim_{n \to \infty} x_n = x$ *and* $\sup_n \||Ax_n\|| < \infty$.

PROOF. (i) Let $x \in \overline{D(A)}$. Note that, by part (iv) of Lemma 2.11, $\lim_{\lambda \to 0+} \|A_\lambda x\|$ (including $+\infty$) exists. From (R$'_1$), since $\|J_\lambda^{[t/\lambda]} x - x\| \leq t\|A_\lambda x\|$ and $T(t)x = \lim_{\lambda \to 0+} J_\lambda^{[t/\lambda]} x$, we have $\|T(t)x - x\| \leq t \lim_{\lambda \to 0+} \|A_\lambda x\|$. Thus,

$$\liminf_{t \to 0+} t^{-1} \|T(t)x - x\| \leq \lim_{\lambda \to 0+} \|A_\lambda x\|.$$

Next, we show the reverse direction of the inequality. From (4.16)

$$\limsup_{t \to 0+} \left(-\frac{\|T(t)x - x\|}{t} \|x - x_0\| \right) \leq \langle y_0, x - x_0 \rangle_s \quad (x_0 \in D(A), \ y_0 \in Ax_0).$$

Here we set $x_0 = J_\lambda x$ and $y_0 = A_\lambda x$ ($\in AJ_\lambda x$). Since $x - x_0 = -\lambda A_\lambda x$,

$$\limsup_{t \to 0+} (-t^{-1} \|T(t)x - x\|) \lambda \|A_\lambda x\| \leq -\lambda \|A_\lambda x\|^2;$$

that is, we obtain

$$\liminf_{t \to 0+} t^{-1} \|T(t)x - x\| \geq \|A_\lambda x\| \quad (\lambda > 0).$$

From this we have $\liminf_{t \to 0+} t^{-1} \|T(t)x - x\| \geq \lim_{\lambda \to 0+} \|A_\lambda x\|$.

(ii) From (i), (ii$_1$) and (ii$_2$) are equivalent. Assume (ii$_2$), then $\|J_\lambda x - x\| = \lambda \|A_\lambda x\| \to 0$ as $\lambda \to 0+$. Thus, if we set $x_n = J_{1/n} x$ ($n = 1, 2, \ldots$), then $\lim_{n \to \infty} x_n = x$ and $\sup_n \||Ax_n\|| \leq \sup_n \|A_{1/n} x\| < \infty$ and we obtain (ii$_3$). Next, we assume (ii$_3$). Then from (ii) of Theorem 4.2 we have

$$\|T(t)x_n - x_n\| \leq Mt \quad (t \geq 0, n = 1, 2, \ldots), \quad \text{where } M = \sup_n \||Ax_n\||.$$

If we let $n \to \infty$, then $\|T(t)x - x\| \leq Mt$ ($t \geq 0$). Thus, we obtain (ii$_1$). $\square$

REMARK. Let A be a dissipative operator satisfying condition (R$_1$), and let $\{T(t) : t \geq 0\}$ be a semigroup of contractions on $\overline{D(A)}$ formulated in Theorem 4.2. Consider the closure $\overline{A}$ of A (see the remark following Lemma 2.16). Then :

$1°$. Let $x \in \overline{D(A)}$. If $T(t)x$ is strongly differentiable at $t_0 > 0$, then

$$T(t_0)x \in D((\overline{A})^0), \quad [(d/dt)T(t)x]_{t=t_0} \in (\overline{A})^0 T(t_0)x.$$

$2°$. If we set $\overline{J}_\lambda = (I - \lambda\overline{A})^{-1}$ and $\overline{A}_\lambda = \lambda^{-1}(\overline{J}_\lambda - I)$ ($\lambda > 0$), then for $x \in \overline{D(A)}$

$$\lim_{\lambda \to 0+} \|\overline{A}_\lambda x\| = \liminf_{h \to 0+} h^{-1} \|T(h)x - x\|.$$

Also, if we replace A and A_λ by $\overline{A}$ and $\overline{A}_\lambda$, respectively, then part (ii) of Theorem 4.7 holds as is.

In fact, $\overline{A}$ is a closed dissipative operator satisfying

$$R(I - \lambda\overline{A}) \supset \overline{D(\overline{A})} = \overline{D(A)} \quad \text{for all } \lambda > 0. \tag{$*$}$$

Further $T(t)x = \lim_{\lambda \to 0+}(I - \lambda\overline{A})^{-[t/\lambda]}x(t \geq 0, \ x \in \overline{D(A)})$ holds. So the results are obtained from Theorems 4.5 and 4.7.

We will show in Example 4.7 of §4 that in a general Banach space, a semigroup of contractions formulated in Theorem 4.2 from a dissipative operator A does not necessarily have the (weak) infinitesimal generator. However, we can show the following:

THEOREM 4.8. *Let A be a single-valued dissipative operator whose domain $D(A)$ is a closed set and satisfies condition (R_1), and let $\{T(t) : t \geq 0\}$ be a semigroup of contractions on $D(A)$ formulated in Theorem 4.2.*

(i) If A is a demicontinuous operator (that is, is a continuous operator from $D(A)$ with the strong topology into X with the weak topology), then A is the weak infinitesimal generator of $\{T(t) : t \geq 0\}$.

(ii) If A is a continuous operator (that is, A is a continuous operator from $D(A)$ with the strong topology into X with the strong topology), then A is the infinitesimal generator of $\{T(t) : t \geq 0\}$.

PROOF. Let $x \in D(A)$. Then

$$T(t)x = \lim_{\lambda \to 0+}(I - \lambda A)^{-[t/\lambda]}x = \lim_{n \to \infty}\left(I - \frac{t}{n}A\right)^{-n}x = \lim_{n \to \infty}J^n_{t/n}x \quad (t \geq 0),$$

$$J^n_{t/n}x - x = \sum_{i=0}^{n-1}[J_{t/n}(J^i_{t/n}x) - J^i_{t/n}x] = \frac{t}{n}\sum_{i=0}^{n-1}A(J^{i+1}_{t/n}x),$$

where we use $J_\lambda z - z = \lambda A J_\lambda z \ (z \in D(J_\lambda))$. Thus, for every $t > 0$

$$t^{-1}[T(t)x - x] = \lim_{n \to \infty}t^{-1}[J^n_{t/n}x - x] = \lim_{n \to \infty}\frac{1}{n}\sum_{i=1}^{n}A(J^i_{t/n}x). \tag{4.19}$$

Also, for every natural number n and $t > 0$,

$$\|J^i_{t/n}x - x\| \leq i\|J_{t/n}x - x\| \leq n\|J_{t/n}x - x\| \leq t\|Ax\| \quad (1 \leq i \leq n). \tag{4.20}$$

(i) Let $x^* \in X^*$, and for every $\varepsilon > 0$ set

$$U(\varepsilon) = \{z \in X : |x^*(z)| < \varepsilon\}.$$

Since A is a demicontinuous operator, we can choose a $\delta > 0$ such that if $\|y - x\| < \delta$ and $y \in D(A)$, then $Ay - Ax \in U(\varepsilon)$. We take a $\delta' > 0$ such that $\delta'\|Ax\| < \delta$. Then, from (4.20), if $0 < t < \delta'$, then for every n, $AJ^i_{t/n}x - Ax \in U(\varepsilon)$, that is, $|x^*(AJ^i_{t/n}x - Ax)| < \varepsilon$ $(1 \leq i \leq n)$. Thus,

$$\left|x^*\left(\frac{1}{n}\sum_{i=1}^{n}A(J^i_{t/n}x) - Ax\right)\right| \leq \frac{1}{n}\sum_{i=1}^{n}|x^*(AJ^i_{t/n}x - Ax)| < \varepsilon.$$

Therefore, from (4.19) we obtain

$$|x^*(t^{-1}[T(t)x - x]) - x^*(Ax)| \le \varepsilon \qquad (0 < t < \delta'),$$

and hence

$$\lim_{t \to 0+} x^*(t^{-1}[T(t)x - x]) = x^*(Ax).$$

(ii) From (4.19)

$$\|t^{-1}[T(t)x - x] - Ax\| \le \limsup_{n \to \infty} \frac{1}{n} \sum_{i=1}^{n} \|AJ_{t/n}^i x - Ax\| \qquad (t \ge 0).$$

Then, from the continuity of A and (4.20) we obtain

$$\lim_{t \to 0+} t^{-1}[T(t)x - x] = Ax. \quad \square$$

As we have considered above, we can construct a semigroup of contractions on $\overline{D(A)}$ from a dissipative operator A that satisfies condition (R_1). In fact, it can be shown that we can construct a semigroup of contractions on $\overline{D(A)}$ even when the dissipative operator A satisfies the following condition weaker than (R_1):

"For all $x \in \overline{D(A)}$, $\displaystyle\liminf_{\lambda \to 0+} \lambda^{-1} d(R(I - \lambda A), x) = 0.$"[5]

We will examine this in the next chapter (see Theorem 5.12). (Suppose $R(I - \lambda A) \supset D(A)$; then from $\overline{R(I - \lambda A)} \supset \overline{D(A)}$ $d(R(I - \lambda A), x) = d(\overline{R(I - \lambda A)}, x) = 0$ $(x \in \overline{D(A)})$. Therefore, if A satisfies condition (R_1), then for all $x \in \overline{D(A)}$, $\lim_{\lambda \to 0+} \lambda^{-1} d(R(I - \lambda A), x) = 0$.)

REMARK. Applying the theory in this section, let us derive a theorem on the generation of semigroups of linear contractions. Let A be an m-dissipative linear operator whose domain is dense in X. Then from Theorem 4.2, $\{T(t) : t \ge 0\}$, defined by $T(t)x = \lim_{n \to \infty} J_{t/n}^n x$ $(t \ge 0, x \in X)$, forms a semigroup of contractions on X. From the linearity of A, we have that $J_{t/n} = (I - \frac{t}{n}A)^{-1}$ is a linear operator and $\{T(t) : t \ge 0\}$ is a semigroup of linear contrations on X. Let $x \in D(A)$. Then again from the linearity of A, we obtain $AJ_{t/n}x = J_{t/n}Ax$ and $AJ_{t/n}^n x = J_{t/n}^n Ax$ $(t \ge 0; n = 1, 2, \ldots)$. Therefore, from (4.19)

$$\frac{1}{t}[T(t)x - x] = \lim_{n \to \infty} \frac{1}{n} \sum_{i=1}^{n} J_{t/n}^i Ax \qquad (t \ge 0). \tag{4.21}$$

Since $\|J_{t/n}^i z - z\| \le t\|Az\|$ $(1 \le i \le n; z \in D(A))$, for every $t \ge 0$ and

---

[5] $d(R(I - \lambda A), x)$ denotes the distance between a set $R(I - \lambda A)$ and a point x.

natural number n we have

$$\left\| \frac{1}{n} \sum_{i=1}^{n} J_{t/n}^{i} Ax - Ax \right\|$$

$$\leq \left\| \frac{1}{n} \sum_{i=1}^{n} J_{t/n}^{i} (Ax - z_k) \right\| + \frac{1}{n} \sum_{i=1}^{n} \| J_{t/n}^{i} z_k - z_k \| + \| z_k - Ax \|$$

$$\leq 2\| z_k - Ax \| + t\| Az_k \|,$$

where $\{z_k\}$ is a sequence in $D(A)$ satisfying $\| z_k - Ax \| \to 0$ as $k \to \infty$. From this and (4.21) we obtain

$$\lim_{t \to 0+} t^{-1}[T(t)x - x] = Ax.$$

Hence, if we let A_0 be the infinitesimal generator of $\{T(t) : t \geq 0\}$, then A_0 is the dissipative extension of A. But, since A is a maximal dissipative operator (because it is an m-dissipative operator), we have $A_0 = A$ and hence we have shown that A is the infinitesimal generator of $\{T(t) : t \geq 0\}$.

From the above discussion, we obtain the following generation theorem.

THEOREM. *An m-dissipative linear operator whose domain is dense in X is always the infinitesimal generator of (a unique) semigroup of linear contractions on X.*

§2. Cauchy's problems

Let A be an operator (not necessarily single-valued) with $D(A) \subset X$ and $R(A) \subset X$ and let T be a positive number. Consider the following Cauchy problem (for A):

$$(\mathrm{CP}; x) \qquad \begin{cases} (d/dt)u(t) \in Au(t) & (t \in [0, T]), \\ u(0) = x, \end{cases}$$

where x is a given element (of X).

DEFINITION 4.1. $u(t) : [0, T] \to X$ is called a solution of the Cauchy problem $(\mathrm{CP}; x)$, if

(i) $u(0) = x$.

(ii) $u(t)$ is Lipschitz continuous on $[0, T]$.

(iii) $u(t)$ is strongly differentiable a.e. $t \in [0, T]$ and satisfies

$$(d/dt)u(t) \in Au(t) \quad (\text{a.e. } t \in [0, T]). \tag{4.22}$$

$u(t) : [0, \infty) \to X$ is called a solution of the Cauchy problem $(\mathrm{CP}; x)_\infty$, if for every $T > 0$, the restriction $u_T(t)$ of $u(t)$ to $[0, T]$ (i.e., $u_T(t) = u(t)$, $0 \leq t \leq T$) is a solution of the Cauchy problem $(\mathrm{CP}; x)$ on $[0, T]$.

REMARK. Note that if $u(t) : [0, T] \to X$ is a solution of the Cauchy problem $(\mathrm{CP}; x)$, then its initial value x always satisfies $x \in \overline{D(A)}$.

LEMMA 4.9. *If A is a dissipative operator, then the Cauchy problems $(CP; x)$ and $(CP; x)_\infty$ have at most one solution, respectively.*

PROOF. Let $u(t)$ and $v(t)$ be solutions of the Cauchy problem $(CP; x)$ on $[0, T]$. Then, since the real-valued function $\|u(t) - v(t)\|$ is Lipschitz continuous on $[0, T]$, it is differentiable a.e. $t \in [0, T]$. Therefore, by Lemma 2.8, $(d/dt)\|u(t) - v(t)\|^2 = 2\langle u'(t) - v'(t), u(t) - v(t)\rangle_i$ (a.e. $t \in [0, T]$). Since $u'(t) = (d/dt)u(t) \in Au(t)$ and $v'(t) = (d/dt)v(t) \in Av(t)$ (a.e. $t \in [0, T]$), from the dissipativity of A we obtain that $(d/dt)\|u(t) - v(t)\|^2 \le 0$ (a.e. $t \in [0, T]$). From this we obtain

$$\|u(t) - v(t)\|^2 = \int_0^t (d/ds)\|u(s) - v(s)\|^2 \, ds \le 0 \qquad (0 \le t \le T),$$

and hence $u(t) = v(t)$ $(0 \le t \le T)$. Similarly, we can show the uniqueness of the solution of $(CP; x)_\infty$. □

From Theorems 4.5, 4.8, and the lemma above we obtain the following theorem:

THEOREM 4.10. *Let A be a closed dissipative operator satisfying condition (R_1) and let $\{T(t) : t \ge 0\}$ be a semigroup of contractions on $\overline{D(A)}$ formulated in Theorem 4.2 and let $x \in D(A)$.*

(i) If $T(t)x$ is strongly differentiable a.e. $t \ge 0$, then $T(t)x : [0, \infty) \to X$ is a unique solution of the Cauchy problem $(CP; x)_\infty$.

(ii) If X is reflexive, then $T(t)x : [0, \infty) \to X$ is a unique solution of the Cauchy problem $(CP; x)_\infty$.

PROOF. We obtain (i) by (ii) of Theorem 4.2, Theorem 4.5, and Lemma 4.9. Next, if X is reflexive, then $T(t)x : [0, \infty) \to X$ is strongly differentiable a.e. $t \ge 0$, and we obtain (ii) from (i). □

THEOREM 4.11. *Let A be a single-valued dissipative operator satisfying condition (R_1). Assume that the domain $D(A)$ is a closed set and let $\{T(t) : t \ge 0\}$ be a semigroup of contractions on $D(A)$ formulated in Theorem 4.2. If A is a demicontinuous operator, then for each $x \in D(A)$ $T(t)x$ is a unique solution of the Cauchy problem $(CP; x)_\infty$.*

PROOF. Let $x \in D(A)$. Since $T(t)x \in D(A)$ for every $t \ge 0$, and since A is the weak infinitesimal generator of $\{T(t) : t \ge 0\}$ (see Theorem 4.8)[6],

$$\text{w-lim}_{h \to 0+} h^{-1}[T(t + h)x - T(t)x] = \text{w-lim}_{h \to 0+} A_h T(t)x$$
$$= AT(t)x \qquad (t \ge 0).$$

Therefore, $T(t)x$ has a weak right derivative $AT(t)x$ at every $t \ge 0$, and also $AT(t)x : [0, \infty) \to X$ is weakly continuous. From this, $T(t)x$ is weakly differentiable at every $t \ge 0$ and its weak derivative is $AT(t)x$. Also, since

---

[6] $A_h = h^{-1}(T(h) - I)$.

$AT(s)x$ is Bochner integrable on $[0, t]$ for every $t > 0$ (see Theorem 1.14), we obtain for every $x^* \in X^*$ and $t \geq 0$

$$x^*[T(t)x - x] = \int_0^t x^*(AT(s)x)\,ds = x^* \left(\int_0^t AT(s)x\,ds \right).$$

Thus

$$T(t)x - x = \int_0^t AT(s)x\,ds \qquad (t \geq 0).$$

Hence $T(t)x$ is strongly differentiable a.e. $t \geq 0$ and $(d/dt)T(t)x = AT(t)x$ (a.e. $t \geq 0$) (see Theorem 1.16). Also $T(t)x : [0, \infty) \to D(A)$ is Lipschitz continuous. Therefore $T(t)x$ is a unique solution of $(CP; x)_\infty$. $\quad\square$

We now consider a semigroup of contractions $\{T(t); t \geq 0\}$ on $\overline{D(A)}$ defined by $T(t)x = \lim_{\lambda \to 0+}(I - \lambda A)^{-[t/\lambda]}x$ $(t \geq 0, \; x \in \overline{D(A)})$, where A is a closed dissipative operator that satisfies condition (R_1). We mentioned above that, when $x \in D(A)$, if there is an appended condition such as whether $T(t)x$ is strongly differentiable a.e. $t \geq 0$ or X is reflexive, then $T(t)x : [0, \infty) \to X$ is a unique solution of the Cauchy problem $(CP; x)_\infty$. However, if no such a condition is attached, then this does not hold in general. In fact, as we will show later in Example 4.7, we can construct a semigroup of contractions $\{T(t): t \geq 0\}$ on $\overline{D(A)}$ such that for every $x \in \overline{D(A)}$, $T(t)x$ is not weakly differentiable at all $t \geq 0$. Moreover, even for such a case, we can show that $T(t)x$ $(= \lim_{\lambda \to 0+}(I - \lambda A)^{-[t/\lambda]}x, \; x \in \overline{D(A)})$, which is constructed above, is "a unique solution of $(CP; x)_\infty$ for A in a generalilzed sense." (From this, $\{T(t) : t \geq 0\}$ as formulated in Theorem 4.2 may be called "a semigroup of contractions on $\overline{D(A)}$ generated by A.") We will examine this in the next chapter.

§3. Generation of semigroups of contractions II; the case where X^* is a uniformly convex Banach space

In this section we consider first the characterizations of the (g)-operator of a semigroup of contractions, which is defined in §3.1, in the case where the conjugate space X^* of a Banach space X is uniformly convex. Then, applying this, we characterize the infinitesimal generator of a semigroup of contractions, when both X and X^* are uniformly convex Banach spaces or X is a Hilbert space. Throughout this section, X^* is always a uniformly convex Banach space.

DEFINITION 4.2. Let A be a not necessarily single-valued operator with $D(A) \subset X$, and $R(A) \subset X$. We say that A has property $(\mathfrak{G})$ if it satisfies (I) and (II).

(I) A is a maximal dissipative operator on $D(A)$.

(II) For every $x \in D(A)$ and $\varepsilon > 0$, there is a $\delta = \delta(\varepsilon, x) > 0$ with $\delta \leq \varepsilon$ and there exist three functions, $f_\delta(t) : [0, \delta] \to X$, $g_\delta(t) : [0, \delta) \to D(A)$, and $h_\delta(t) : [0, \delta) \to X$, that satisfy (II_1) through (II_3):

(II_1) $f_\delta(t)$ is (strongly) absolutely continuous on $[0, \delta]$, $f_\delta(0) = x$, $f_\delta(\delta) \in D(A)$, and $\||Af_\delta(\delta)\|| \le e^{\varepsilon\delta}\||Ax\||$,

(II_2) $\|g_\delta(t) - f_\delta(t)\| \le \varepsilon e^{\varepsilon\delta}(\||Ax\|| + \varepsilon)$ $(t \in [0, \delta))$,

(II_3) $h_\delta(t) \in Ag_\delta(t)$, $\|h_\delta(t)\| \le e^{\varepsilon\delta}\||Ax\||$ $(t \in [0, \delta))$, and

$$\|(d/dt)f_\delta(t) - h_\delta(t)\| \le \varepsilon \quad (\text{a.e. } t \in [0, \delta)).$$

(Note that since $f_\delta(t)$ is (strongly) absolutely continuous and X is reflexive, it follows that $f_\delta(t)$ is strongly differentiable a.e. $t \in [0, \delta]$.)

Next we give examples which have property $(\mathfrak{G})$.

EXAMPLE 4.1. If A is a maximal dissipative operator on $D(A)$ and satisfies:

(a) for every $x \in D(A)$ and $\varepsilon > 0$, there exist $\delta = \delta(\varepsilon, x) > 0$ with $\delta \le \varepsilon$, $x_\delta \in D(A)$ and $y_\delta \in Ax_\delta$ satisfying the following (a_1) and (a_2):

$$\|x_\delta - \delta y_\delta - x\| \le \delta\varepsilon, \tag{a_1}$$

$$\|y_\delta\| \le e^{\varepsilon\delta}\||Ax\||, \tag{a_2}$$

then A has property $(\mathfrak{G})$.

PROOF. Let $x \in D(A)$ and let $\varepsilon > 0$. Define $f_\delta(t): [0, \delta] \to X$, $g_\delta(t): [0, \delta) \to D(A)$, and $h_\delta(t): [0, \delta) \to X$ as follows:

$$f_\delta(t) = \delta^{-1}[(\delta - t)x + tx_\delta] \quad (t \in [0, \delta]),$$
$$g_\delta(t) = x_\delta, \quad h_\delta(t) = y_\delta \quad (t \in [0, \delta)).$$

Then, from (a_2) we obtain $\||Af_\delta(\delta)\|| = \||Ax_\delta\|| \le \|y_\delta\| \le e^{\varepsilon\delta}\||Ax\||$, and (II_1) holds. Next, from (a_1) and (a_2), for every $t \in [0, \delta)$ we have

$$\|g_\delta(t) - f_\delta(t)\| \le \|x_\delta - x\| \le \|x_\delta - \delta y_\delta - x\| + \delta\|y_\delta\|$$
$$\le \delta\varepsilon + \delta e^{\varepsilon\delta}\||Ax\|| \le \varepsilon e^{\varepsilon\delta}(\||Ax\|| + \varepsilon).$$

Thus, (II_2) holds. Finally, since $(d/dt)f_\delta(t) = \delta^{-1}(x_\delta - x)$ $(t \in (0, \delta))$, from (a_1) we obtain

$$\|(d/dt)f_\delta(t) - h_\delta(t)\| = \|\delta^{-1}(x_\delta - x) - y_\delta s\| \le \varepsilon \quad (t \in (0, \delta)),$$

and (II_3) holds. $\square$

EXAMPLE 4.2. (i) If A is a maximal dissipative operator on $D(A)$, and for every $x \in D(A)$, there is a sequence $\{\lambda_n\}$ of positive numbers such that $\lambda_n \to 0$ as $n \to \infty$ and

$$R(I - \lambda_n A) \supset \{x\} \quad (n = 1, 2, \ldots),$$

then A has property $(\mathfrak{G})$.

(ii) If A is a maximal dissipative operator on $D(A)$ and A satisfies condition (R_1) (i.e., $R(I - \lambda A) \supset D(A)$ for all $\lambda > 0$), then A has property $(\mathfrak{G})$.

(iii) Every m-dissipative operator has property $(\mathfrak{G})$.

PROOF. (i) Let $x \in D(A)$. From the assumption, for every natural number n there exist $x_n \in D(A)$ and $y_n \in Ax_n$ such that $x_n - \lambda_n y_n = x$. From this, $\|y_n\| = \lambda_n^{-1}\|x_n - x\| = \lambda_n^{-1}\|J_{\lambda_n}x - x\| \leq \||Ax\||$ (see Lemma 2.11). Thus, for a given $\varepsilon > 0$, we select an n such that $\lambda_n \leq \varepsilon$ and set $\delta = \lambda_n$, $x_\delta = x_n$, and $y_\delta = y_n$; then conditions (a_1) and (a_2) of Example 4.1 are satisfied.

(ii) This is obvious from (i). Also, if A is an m-dissipative operator, then $R(I - \lambda A) = X$ $(\lambda > 0)$, and since A is a maximal dissipative operator on $D(A)$ (see Lemma 2.12), (iii) is obtained from (ii). $\square$

EXAMPLE 4.3. If A is a maximal dissipative operator on $D(A)$ and satisfies the following condition (b), then A has property $(\mathfrak{G})$:

(b) For every $x \in D(A)$ and $\varepsilon > 0$, there exist $\delta = \delta(\varepsilon, x) > 0$ with $\delta \leq \varepsilon$ and $x_\delta \in D(A)$ that satisfies the following conditions (b_1) and (b_2) $(^7)$:

$$d(x + \delta A^0 x, x_\delta) \leq \varepsilon\delta, \qquad (b_1)$$

$$\||Ax_\delta\|| \leq e^{\varepsilon\delta}\||Ax\||. \qquad (b_2)$$

PROOF. Let $x \in D(A)$ and let $\varepsilon > 0$. Then, from the assumption, there exist $\delta > 0$ and $x_\delta \in D(A)$ such that $d(x + \delta A^0 x, x_\delta) \leq \delta\varepsilon/2$, $\||Ax_\delta\|| \leq e^{\varepsilon\delta/2}\||Ax\||$, and $\delta \leq \varepsilon/2$. Now, choose $y_\delta \in A^0 x$ such that $\|x + \delta y_\delta - x_\delta\| < d(x + \delta A^0 x, x_\delta) + \delta\varepsilon/2$. Then

$$\|x + \delta y_\delta - x_\delta\| \leq \delta\varepsilon, \quad \||Ax_\delta\|| \leq e^{\varepsilon\delta}\||Ax\||, \qquad 0 < \delta \leq \varepsilon.$$

We define functions $f_\delta(t) : [0, \delta] \to X$, $g_\delta(t) : [0, \delta) \to D(A)$, and $h_\delta(t) : [0, \delta) \to X$ as follows:

$$f_\delta(t) = \delta^{-1}[(\delta - t)x + tx_\delta] \qquad (t \in [0, \delta]),$$
$$g_\delta(t) = x, \quad h_\delta(t) = y_\delta \qquad (t \in [0, \delta)).$$

Obviously, $f_\delta(t)$, $g_\delta(t)$, and $h_\delta(t)$ satisfy (II_1) through (II_3) of Definition 4.2. $\square$

REMARK. In particular, consider the case where X is a uniformly convex Banach space. Then, since from Lemma 2.19, the canonical restriction A^0 of a maximal dissipative operator A on $D(A)$ is a single-valued operator and $D(A^0) = D(A)$, we can write condition (b) of Example 4.3 as follows:

(b') For $x \in D(A)$ and $\varepsilon > 0$, there exist $\delta = \delta(\varepsilon, x) > 0$ with $\delta \leq \varepsilon$ and $x_\delta \in D(A)$ that satisfies the following (b_1') and (b_2'):

$$\|x + \delta A^0 x - x_\delta\| \leq \varepsilon\delta, \qquad (b_1')$$

$$\|A^0 x_\delta\| \leq e^{\varepsilon\delta}\|A^0 x\|. \qquad (b_2')$$

$(^7)$Note that $A^0 x \neq \varnothing$ (see Lemma 2.19).

Furthermore, we see that the above (b$'$) is equivalent with the following condition (a$'$):

(a$'$) For every $x \in D(A)$ and $\varepsilon > 0$, there exist $\delta = \delta(\varepsilon, x) > 0$ with $\delta \leq \varepsilon$ and $x_\delta \in D(A)$, that satisfies the following (a$'_1$) and (a$'_2$):

$$\|x_\delta - \delta A^0 x_\delta - x\| \leq \varepsilon \delta, \qquad (a'_1)$$

$$\|A^0 x_\delta\| \leq e^{\varepsilon\delta}\|A^0 x\|. \qquad (a'_2)$$

To show (a$'$) and (b$'$) are equivalent, we select and fix $x \in D(A)$. We assume either condition (a$'$) or condition (b$'$), and write $\delta(\varepsilon)$ and $x_{\delta(\varepsilon)}$ for $\delta (= \delta(\varepsilon, x))$ and x_δ, respectively, which are determined by $\varepsilon > 0$. Then we have

$$\lim_{\varepsilon \to 0+} \|A^0 x_{\delta(\varepsilon)} - A^0 x\| = 0. \qquad (*)$$

In fact, $\|x_{\delta(\varepsilon)} - x\| \leq \varepsilon(e^{\varepsilon^2}\|A^0 x\| + \varepsilon) \to 0$ as $\varepsilon \to 0+$. Because $\|A^0 x_{\delta(\varepsilon)}\| = O(1)$ as $\varepsilon \to 0+$, and also X is reflexive, there exist $y \in X$ and a sequence $\{\varepsilon_k\}$ of positive numbers with $\lim_{k\to\infty} \varepsilon_k = 0$, with w-$\lim_{k\to\infty} A^0 x_{\delta(\varepsilon_k)} = y$. Since A is a dissipative operator, for every $u \in D(A)$ and $v \in Au$ we have

$$\mathrm{Re}(A^0 x_{\delta(\varepsilon_k)} - v, F(x_{\delta(\varepsilon_k)} - u)) \leq 0.$$

If we let $k \to \infty$, then for every $u \in D(A)$ and $v \in Au$ we have

$$\mathrm{Re}(y - v, F(x - u)) \leq 0.$$

Since $x \in D(A)$ and A is a maximal dissipative operator on $D(A)$, we obtain $y \in Ax$ from the above inequality. Hence

$$\|A^0 x\| \leq \|y\| \leq \liminf_{k\to\infty}\|A^0 x_{\delta(\varepsilon_k)}\| \leq \limsup_{k\to\infty}\|A^0 x_{\delta(\varepsilon_k)}\| \leq \|A^0 x\|.$$

Thus we obtain $y = A^0 x$ and $\lim_{k\to\infty}\|A^0 x_{\delta(\varepsilon_k)}\| = \|A^0 x\|$, and also $A^0 x = y = $ w-$\lim_{k\to\infty} A^0 x_{\delta(\varepsilon_k)}$. Therefore, from the uniform convexity of X, we obtain $\lim_{k\to\infty}\|A^0 x_{\delta(\varepsilon_k)} - A^0 x\| = 0$. Since the same conclusion holds when we start with a sequence $\{\varepsilon^{(n)}\}$ of positive numbers with $\varepsilon^{(n)} \to 0$ (instead of $\varepsilon > 0$), $(*)$ is satisfied. Hence, from $(*)$ we can derive easily that conditions (a$'$) and (b$'$) are equivalent.

Now we show that if A is almost demiclosed and has property $(\mathfrak{G})$, then we can generate a semigroup of contractions on $\overline{D(A)}$ from A. First,

LEMMA 4.12. *Let A be an operator which is almost demiclosed and has property $(\mathfrak{G})$, and let $x \in D(A)$, $T > 0$, and $0 < \varepsilon < T$. Then, there exist three functions, $u(t): [0, T] \to X$, $g(t): [0, T) \to D(A)$, and $h(t): [0, T) \to X$, that satisfy the following properties* (i) *through* (v):

    (i) $u(0) = x$,

    (ii) $\|u(t) - u(s)\| \leq |t - s|[e^{\varepsilon T}\|\,\|Ax\|\| + \varepsilon]$ $(t, s \in [0, T])$,

(iii) $u(T) \in D(A)$, $\||Au(T)\|| \leq e^{\varepsilon T}\||Ax\||$,

(iv) $\|u(t) - g(t)\| \leq \varepsilon e^{\varepsilon T}(\||Ax\|| + \varepsilon)$ $(t \in [0, T))$,

(v) $h(t) \in Ag(t)$, $\|h(t)\| \leq e^{\varepsilon T}\||Ax\||$ $(t \in [0, T))$,
$\|(d/dt)u(t) - h(t)\| \leq \varepsilon$ $(a.e.\ t \in [0, T])$.

PROOF. We substitute c $(0 < c \leq T)$, $v(t)$, $p(t)$, and $q(t)$ for T, $u(t)$, $g(t)$, and $h(t)$, respectively, and denote by $\mathfrak{P}$ the class of all sets (v, p, q) of the three functions $v(t) : [0, c] \to X$, $p(t) : [0, c) \to D(A)$, and $q(t) : [0, c) \to X$ that satisfy (i) through (v) above. When $(v_i, p_i, q_i) \in \mathfrak{P}$ $(i = 1, 2)$ satisfy that $c_1 \leq c_2$, $v_1(t) = v_2(t)$ $(t \in [0, c_1])$, $p_1(t) = p_2(t)$ $(t \in [0, c_1))$, and $q_1(t) = q_2(t)$ $(t \in [0, c_1))$, where $v_i(t)$ is defined on $[0, c_i]$; $p_i(t)$ and $q_i(t)$ are defined on $[0, c_i)$, we write this

$$(v_1, p_1, q_1) \preceq (v_2, p_2, q_2).$$

From the assumption that A has property $(\mathfrak{G})$, there are functions $f_\delta(t)$, $g_\delta(t)$, and $h_\delta(t)$ that satisfy (II_1) through (II_3) of Definition 4.2. Obviously, $(f_\delta, g_\delta, h_\delta) \in \mathfrak{P}$ and the set $\mathfrak{P}$ is partially ordered by the relation "$\preceq$" introduced above. In order to show that $\mathfrak{P}$ has a maximal element, we let $Q = \{(v_\alpha, p_\alpha, q_\alpha) : \alpha \in \Lambda\}$ be a totally ordered subset of $\mathfrak{P}$, where v_α is defined on $[0, c_\alpha]$, and p_α and q_α are defined on $[0, c_\alpha)$. Let $\bar{c} = \sup\{c_\alpha : \alpha \in \Lambda\}$. If $\bar{c} = c_\alpha$ for some $\alpha \in \Lambda$, then $(v_\alpha, p_\alpha, q_\alpha)$ is an upper bound for Q. When it is not, that is, if $c_\alpha < \bar{c}$ for all $\alpha \in \Lambda$, for every $t \in [0, \bar{c})$ we select $\alpha \in \Lambda$ such that $t < c_\alpha$, and define $v(t) = v_\alpha(t)$, $p(t) = p_\alpha(t)$, and $q(t) = q_\alpha(t)$ (recall that Q is a totally ordered set so this definition makes sense).

Then, it is self-evident that these three functions (on $[0, \bar{c})$) $v(t)$, $p(t)$, and $q(t)$ satisfy (i),(ii), (iv), and (v) on $[0, \bar{c})$. From $\|v(t) - v(s)\| \leq |t - s|[e^{\varepsilon \bar{c}}\||Ax\|| + \varepsilon]$ $(t, s \in [0, \bar{c}))$, $\lim_{t \to \bar{c}-0} v(t)$ exists. We let $v(\bar{c}) = \lim_{t \to \bar{c}-0} v(t)$. Obviously, $v(t) : [0, \bar{c}] \to X$, $p(t) : [0, \bar{c}) \to D(A)$, and $q(t) : [0, \bar{c}) \to X$ satisfy (i), (ii), (iv), and (v). To show that $v(\bar{c}) \in D(A)$ and $\||Av(\bar{c})\|| \leq e^{\varepsilon \bar{c}}\||Ax\||$ (i.e., (iii)) hold, we select a sequence $\{(v_n, p_n, q_n)\} \subset Q$ such that $\lim_{n \to \infty} c_n = \bar{c}$. Here, v_n is a function that can be defined on $[0, c_n]$. Then $v(\bar{c}) = \lim_{n \to \infty} v(c_n) = \lim_{n \to \infty} v_n(c_n)$ and $\||Av_n(c_n)\|| \leq e^{\varepsilon \bar{c}}\||Ax\||$. Since A is a maximal dissipative operator on $D(A)$, and almost demiclosed, A is demiclosed (see Lemma 2.17); therefore, we obtain $v(\bar{c}) \in D(A)$ and $\||Av(\bar{c})\|| \leq e^{\varepsilon \bar{c}}\||Ax\||$ from Lemma 2.18. Thus, $(v, p, q) \in \mathfrak{P}$, and obviously (v, p, q) is an upper bound of Q. After all, we have shown that every totally ordered subset of $\mathfrak{P}$ has an upper bound, and so from Zorn's lemma it follows that $\mathfrak{P}$ has a maximal element (u, g, h).

Now, if we show that the function u is defined on $[0, T]$, then u, g, and h are the required three functions and the proof of the lemma is complete. We will assume that the function u is defined on $[0, c]$, with $c < T$, and derive a contradiction. Choose $\eta > 0$ such that $c + \eta \leq T$. Since $u(c) \in D(A)$ and A has property $(\mathfrak{G})$, there are $\delta > 0$ with $\delta \leq \min\{\varepsilon, n\}$ and three functions $u_\delta(t) : [0, \delta] \to X$, $g_\delta(t) : [0, \delta) \to D(A)$, and $h_\delta(t) : [0, \delta) \to X$ satisfying the following (4.23) through (4.25) hold:

$$\begin{cases} u_\delta(t) \text{ is (strongly) absolutely continuous on } [0, \delta], \text{ and} \\ u_\delta(0) = u(c), \ u_\delta(\delta) \in D(A), \text{ and } |||Au_\delta(\delta)||| \leq e^{\varepsilon\delta}|||Au(c)|||, \end{cases} \quad (4.23)$$

$$\|g_\delta(t) - u_\delta(t)\| \leq \varepsilon e^{\varepsilon\delta}(|||Au(c)||| + \varepsilon) \qquad (t \in [0, \delta)), \qquad (2.24)$$

$$\begin{cases} h_\delta(t) \in Ag_\delta(t), \quad \|h_\delta(t)\| \leq e^{\varepsilon\delta}|||Au(c)||| \qquad (t \in [0, \delta)), \\ \|(d/dt)u_\delta(t) - h_\delta(t)\| \leq \varepsilon \quad \text{(a.e. } t \in [0, \delta)). \end{cases} \quad (4.25)$$

We set $d = c + \delta$ $(\leq T)$, and define functions $u_0(t) : [0, d] \to X$, $g_0(t) : [0, d) \to D(A)$, and $h_0(t) : [0, d) \to X$ by the following equations:

$$u_0(t) = \begin{cases} u(t) & (0 \leq t \leq c), \\ u_\delta(t - c) & (c \leq t \leq d); \end{cases}$$

$$g_0(t) = \begin{cases} g(t) & (0 \leq t < c), \\ g_\delta(t - c) & (c \leq t < d); \end{cases}$$

$$h_0(t) = \begin{cases} h(t) & (0 \leq t < c), \\ h_\delta(t - c) & (c \leq t < d). \end{cases}$$

Then, $(u_0, g_0, h_0) \in \mathfrak{P}$. In fact (i) holds since $u_0(0) = u(0) = x$. Next, from (4.23), $u_0(d) = u_\delta(\delta) \in D(A)$ and

$$|||Au_0(d)||| = |||Au_\delta(\delta)||| \leq e^{\varepsilon\delta}|||Au(c)||| \leq e^{\varepsilon d}|||Ax|||.$$

Thus, (iii) holds. Also, from (4.24), for $t \in [c, d)$

$$\|u_0(t) - g_0(t)\| = \|u_\delta(t - c) - g_\delta(t - c)\| \leq \varepsilon e^{\varepsilon\delta}(|||Au(c)||| + \varepsilon)$$
$$\leq \varepsilon e^{\varepsilon\delta}(e^{\varepsilon c}|||Ax||| + \varepsilon) \leq \varepsilon e^{\varepsilon d}(|||Ax||| + \varepsilon).$$

From this, (iv) holds. Further, from (4.25) we see that u_0, g_0, and h_0 satisfy (v). Finally, from (4.25), for $0 \leq s_1 \leq s_2 \leq \delta$

$$\|u_\delta(s_2) - u_\delta(s_1)\| \leq \int_{s_1}^{s_2} \|(d/ds)u_\delta(s)\| \, ds \leq (s_2 - s_1)(e^{\varepsilon\delta}|||Au(c)||| + \varepsilon).$$

From this, we obtain

$$\|u_0(t) - u_0(s)\| = \|u_\delta(t - c) - u_\delta(s - c)\|$$
$$\leq |t - s|(e^{\varepsilon\delta}|||Au(c)||| + \varepsilon)$$
$$\leq |t - s|(e^{\varepsilon d}|||Ax||| + \varepsilon) \qquad (s, t \in [c, d]),$$

and u_0 satisfies (ii). Hence, $(u_0, g_0, h_0) \in \mathfrak{P}$. Obviously,

$$(u, g, h) \preceq (u_0, g_0, h_0) \quad \text{and} \quad (u, g, h) \neq (u_0, g_0, h_0).$$

This contradicts the property that (u, g, h) is a maximal element of $\mathfrak{P}$. Thus we have shown that $u(t)$ is defined on $[0, T]$. $\square$

THEOREM 4.13. *Let A be an almost demiclosed operator satisfying property $(\mathfrak{G})$, and $T > 0$. Then, for each $x \in D(A)$, the Cauchy problem*

$$(\text{CP}; x) \qquad \begin{cases} (d/dt)u(t) \in Au(t) & (t \in [0, T]), \\ u(0) = x \end{cases}$$

has a unique solution $u(t) = u(t; x)$. Furthermore, this solution $u(t; x)$ satisfies

(i) $\|u(t; x) - u(s; x)\| \leq \||Ax|\| \, |t - s| \ (x \in D(A); t, s \in [0, T])$,

(ii) $u(t; x) \in D(A) \ (x \in D(A), \ t \in [0, T])$.

PROOF. Let $x \in D(A)$ and $\{\varepsilon_n\}$ be a sequence of numbers such that $0 < \varepsilon_n \leq 1$ and $\lim_{n \to \infty} \varepsilon_n = 0$. Then, by Lemma 4.12, for each natural number n, there exist functions $u_n(t; x): [0, T] \to X$, $g_n(t): [0, T) \to D(A)$, and $h_n(t): [0, T) \to X$ that satisfy (4.26) through (4.31):

$$u_n(0; x) = x, \tag{4.26}$$

$$\|u_n(t; x) - u_n(s; x)\| \leq |t - s|[e^{\varepsilon_n T}\||Ax|\| + \varepsilon_n] \tag{4.27}$$
$$(t, s \in [0, T]),$$

$$u_n(T; x) \in D(A), \qquad \||Au_n(T; x)|\| \leq e^{\varepsilon_n T}\||Ax|\|, \tag{4.28}$$

$$\|u_n(t; x) - g_n(t)\| \leq \varepsilon_n e^{\varepsilon_n T}(\||Ax|\| + \varepsilon_n) \qquad (t \in [0, T)), \tag{4.29}$$

$$h_n(t) \in Ag_n(t), \quad \|h_n(t)\| \leq e^{\delta_n T}\||Ax|\| \qquad (t \in [0, T)), \tag{4.30}$$

$$\|(d/dt)u_n(t; x) - h_n(t)\| \leq \varepsilon_n \quad (\text{a.e. } t \in [0, T)). \tag{4.31}$$

Now, from (4.30) and (4.31) and the dissipativity of A, we have for a.e. $t \in [0, T)$ [8]

$$
\begin{aligned}
(d/dt)&\|u_n(t; x) - u_m(t; x)\|^2 \\
&= 2\,\text{Re}((d/dt)u_n(t; x) - (d/dt)u_m(t; x), F(u_n(t; x) - u_m(t; x))) \\
&= 2\,\text{Re}([(d/dt)u_n(t; x) - h_n(t)] \\
&\quad - [(d/dt)u_m(t; x) - h_m(t)], F(u_n(t; x) - u_m(t; x))) \\
&\quad + 2\,\text{Re}(h_n(t) - h_m(t), F(u_n(t; x) - u_m(t; x)) - F(g_n(t) - g_m(t))) \\
&= +2\,\text{Re}(h_n(t) - h_m(t), F(g_n(t) - g_m(t))) \\
&\leq 2(\varepsilon_n + \varepsilon_m)\|u_n(t; x) - u_m(t; x)\| \\
&\quad + 4e^T\||Ax|\| \, \|F(u_n(t; x) - u_m(t; x)) - F(g_n(t) - g_m(t))\|.
\end{aligned}
$$

From (4.26), (4.27), and (4.29), the set

$$B = \{u_n(t; x) - u_m(t; x), g_n(t) - g_m(t): n, m = 1, 2, \ldots, 0 \leq t < T\}$$

is a bounded set. Since F is uniformly continuous on B (see Theorem 2.3), for every $\varepsilon > 0$, we can select a $\delta > 0$ such that if $y, z \in B$ and

---

[8] We use Lemma 2.8 (note that the duality mapping F is single valued, because X^* is uniformly convex).

$\|y - z\| < \delta$, then $\|F(y) - F(z)\| \, \|\|Ax\|\| < \varepsilon/(8e^T)$. Now, we let $M = \sup\{\|z\| : z \in B\}$ and select a natural number n_0 such that if $n \geq n_0$, then $\varepsilon_n e^T(\|\|Ax\|\| + 1) < \delta/2$ and $M\varepsilon_n < \varepsilon/8$. Let $n, m \geq n_0$. Then from (4.29)

$$\|(u_n(t; x) - u_m(t; x)) - (g_n(t) - g_m(t))\| \leq (\varepsilon_m + \varepsilon_n)e^T(\|\|Ax\|\| + 1) < \delta$$

$(t \in [0, T))$, and thus

$$4e^T \|\|Ax\|\| \, \|F(u_n(t; x) - u_m(t; x)) - F(g_n(t) - g_m(t))\| < \varepsilon/2$$

$(t \in [0, T))$ Further,

$$2(\varepsilon_n + \varepsilon_m)\|u_n(t; x) - u_m(t; x)\| \leq 2M(\varepsilon_n + \varepsilon_m) < \varepsilon/2 \qquad (t \in [0, T]).$$

Hence, we obtain $(d/dt)\|u_n(t; x) - u_m(t; x)\|^2 < \varepsilon$ (a.e. $t \in [0, T)$), and

$$\|u_n(t; x) - u_m(t; x)\|^2 = \int_0^t (d/d\tau)\|u_n(\tau; x) - u_m(\tau; x)\|^2 \, d\tau$$
$$\leq \varepsilon T \qquad (t \in [0, T]).$$

It follows that when $n \to \infty$ $\{u_n(t; x)\}$ converges uniformly on $[0, T]$. Thus we set

$$u(t; x) = \lim_{n \to \infty} u_n(t; x) \qquad (t \in [0, T]).$$

Then, from (4.26) and (4.27), we obtain $u(0; x) = x$ and

$$\|u(t; x) - u(s; x)\| \leq \|\|Ax\|\| \, |t - s| \qquad (t, s \in [0, T]).$$

Next, we show (ii), that is,

$$u(t; x) \in D(A) \qquad (t \in [0, T]). \tag{4.32}$$

In fact, we obtain $u(T; x) \in D(A)$ from (4.28) and the assumption that A is almost demiclosed. Let $0 \leq t < T$. From (4.29)

$$\|g_n(t) - u(t; x)\| \leq \varepsilon_n e^{\varepsilon_n T}(\|\|Ax\|\| + \varepsilon_n) + \|u_n(t; x) - u(t; x)\| \to 0$$

as $n \to \infty$. From this and (4.30) and the assumption that A is almost demiclosed, we obtain $u(t; x) \in D(A)$.

Next we show that $u(t; x)$ is a solution of $(CP; x)$. Let $x_0 \in D(A)$ and $y_0 \in Ax_0$. Then from (4.30) and (4.31), and the dissipativity of A, we have for a.e. $t \in [0, T)$

$$(d/dt)\|u_n(t; x) - x_0\|^2 = 2\operatorname{Re}((d/dt)u_n(t; x), F(u_n(t; x) - x_0))$$
$$= 2\operatorname{Re}((d/dt)u_n(t; x) - h_n(t), F(u_n(t; x) - x_0))$$
$$\quad + 2\operatorname{Re}(h_n(t), F(u_n(t; x) - x_0) - F(g_n(t) - x_0))$$
$$\quad + 2\operatorname{Re}(h_n(t), F(g_n(t) - x_0))$$
$$\leq 2\varepsilon_n\|u_n(t; x) - x_0\| + 2e^{\varepsilon_n T}\|F(u_n(t; x) - x_0) - F(g_n(t) - x_0)\| \, \|\|Ax\|\|$$
$$\quad + 2\operatorname{Re}(y_0, F(g_n(t) - x_0)).$$

Also, from (4.29) $\lim_{n\to\infty} \|F(u_n(t\,;x) - x_0) - F(g_n(t) - x_0)\| = 0$ (uniformly on $[0, T)$, and also from $\lim_{n\to\infty} g_n(t) = u(t\,;x)$ (uniformly on $[0, T)$), we have $\lim_{n\to\infty} \|F(g_n(t) - x_0) - F(u(t\,;x) - x_0)\| = 0$ converges uniformly on $[0, T)$. Hence, for an arbitrary $\varepsilon > 0$, we can select a proper natural number n_0 such that if $n \geq n_0$

$$(d/dt)\|u_n(t\,;x) - x_0\|^2 \leq 2\,\mathrm{Re}(y_0, F(u(t\,;x) - x_0)) + \varepsilon \quad (\text{a.e. } t \in [0, T)).$$

When we integrate this over $[s, t]$ and let $n \to \infty$, we obtain for $0 \leq s < t \leq T$

$$\|u(t\,;x) - x_0\|^2 - \|u(s\,;x) - x_0\|^2 \leq 2 \int_s^t \mathrm{Re}(y_0, F(u(\tau\,;x) - x_0))\,d\tau.$$

From this and

$$2\,\mathrm{Re}(u(t\,;x) - u(s\,;x), F(u(s\,;x) - x_0)) \leq \|u(t\,;x) - x_0\|^2 - \|u(s\,;x) - x_0\|^2,$$

we obtain

$$\mathrm{Re}(u(t\,;x) - u(s\,;x), F(u\,;x) - x_0)$$
$$\leq \int_s^t \mathrm{Re}(y_0, F(u(\tau\,;x) - x_0))d\tau \quad (0 \leq s \leq t \leq T). \tag{4.33}$$

From this, at a point t where $u(t\,;x)$ is strongly differentiable, we have

$$\mathrm{Re}((d/dt)u(t\,;x), F(u(t\,;x) - x_0)) \leq \mathrm{Re}(y_0, F(u(t\,;x) - x_0))$$

for all $x_0 \in D(A)$ and $y_0 \in Ax_0$. Then, from (4.32) and the assumption that A is a maximal dissipative operator on $D(A)$, we obtain

$$(d/dt)u(t\,;x) \in Au(t\,;x).$$

Since $u(t\,;x)$ is Lipschitz continuous on $[0, T]$, it is strongly differentiable a.e. $t \in [0, T]$. Thus

$$(d/dt)u(t\,;x) \in Au(t\,;x) \quad (\text{a.e. } t \in [0, T]).$$

By the above, we have shown that $u(t\,;x)$ is a solution of $(CP\,;x)$. The uniqueness of the solution follows from Lemma 4.9. $\square$

From this, we obtain the following theorem on generation.

THEOREM 4.14. *Let A be an almost demiclosed operator with property $(\mathfrak{G})$. Then, for each $x \in D(A)$, there exists a unique semigroup of contractions $\{T(t) : t \geq 0\}$ on $\overline{D(A)}$ such that $T(t)x$ is a solution of the Cauchy problem $(CP\,;x)_\infty$ (for the operator A), and the following* (i) *through* (iii) *hold:*

(i) $\|T(t)x - T(s)x\| \leq \|\|Ax\|\|\,|t - s| \quad (x \in D(A)\,;t, s \geq 0).$

(ii) *If $x \in D(A)$, then $T(t)x \in D(A)$ $(t \geq 0)$ and $(d/dt)T(t)x \in A^0T(t)x$ (a.e. $t \geq 0$).*

(iii) *For arbitrary $x_0 \in D(A)$, $y_0 \in Ax_0$, and $x \in \overline{D(A)}$,*

$$\limsup_{t\to 0+} \mathrm{Re}(t^{-1}[T(t)x - x], F(x - x_0)) \leq \mathrm{Re}(y_0, F(x - x_0)).$$

PROOF. For each $T > 0$ and $x \in D(A)$, we let $u_T(t; x)$ be the (unique) solution of $(CP; x)$. Then we easily see that

$$\text{If } t, s \geq 0 \text{ and } t + s \leq T, \text{ then } u_T(t + s; x) = u_T(t; u_T(s; x)) \quad (4.34)$$

and

$$\text{if } 0 < T_1 \leq T_2, \text{ then } u_{T_2}(t; x) = u_{T_1}(t; x) \ (t \in [0, T_1]) \quad (4.35)$$

hold.

We now define an operator $T(t)$ $(t \geq 0)$ with domain $D(A)$ by setting

$$T(t)x = u_T(t; x) \qquad (x \in D(A))$$

for $0 \leq t \leq T$. From (4.35), for each $t \geq 0$ and $x \in D(A)$, $T(t)x$ is determined uniquely and this definition makes sense. Obviously, for each $x \in D(A)$, $T(t)x$ is the unique solution of $(CP; x)_\infty$, and (i) and $T(t)x \in D(A)$ $(t \geq 0)$ hold. Also, from $u_T(0; x) = x$ and (4.34), we have

$$T(0)x = x \ (x \in D(A)) \quad \text{and} \quad T(t + s) = T(t)T(s) \ (t, s \geq 0).$$

From this and (i), we have

$$\|T(t + h)x - T(t)x\| = \|T(h)T(t)x - T(t)x\| \leq h\||AT(t)x||| \qquad (t, h > 0),$$

and hence

$$\|(d/dt)T(t)x\| \leq \||AT(t)x||| \quad (\text{a.e. } t \geq 0).$$

From this inequality and $(d/dt)T(t)x \in AT(t)x$ (a.e. $t \geq 0$), we obtain

$$(d/dt)T(t)x0 \in A^0 T(t)x \quad (\text{a.e. } t \geq 0).$$

Next, from $(d/ds)T(s)x \in AT(s)x$ (a.e. $s \geq 0$) and the dissipativity of A, we have $(d/ds)\|T(s)x - T(s)y\|^2 = 2\operatorname{Re}((d/ds)T(s)x - (d/ds)T(s)y)$, and $F(T(s)x - T(s)y) \leq 0$ (a.e. $s \geq 0$). Hence, we obtain

$$\|T(t)x - T(t)y\|^2 - \|x - y\|^2 = \int_0^t (d/ds)\|T(s)x - T(s)y\|^2 \, ds \leq 0;$$

that is,

$$\|T(t)x - T(t)y\| \leq \|x - y\| \qquad (t \geq 0; x, y \in D(A)).$$

Since a semigroup of contractions $\{T(t) : t \geq 0\}$ on $D(A)$ can be extended uniquely to a semigroup of contractions on $\overline{D(A)}$, if we again denote the semigroup of contractions by $\{T(t) : t \geq 0\}$, then this is the required semigroup of contractions. Further, we can show (iii) as follows: Let $x \in \overline{D(A)}$ and take a sequence $\{x_n\} \subset D(A)$ such that $\lim_{n \to \infty} x_n = x$, and let $x_0 \in D(A)$ and $y_0 \in Ax_0$. Then, from (4.33), for $t \geq 0$ and $n = 1, 2, \dots$,

$$\operatorname{Re}(T(t)x_n - x_n, F(x_n - x_0)) \leq \int_0^t \operatorname{Re}(y_0, F(T(\tau)x_n - x_0)) \, d\tau.$$

If we let $n \to \infty$, then

$$\text{Re}(T(t)x - x, F(x - x_0)) \le \int_0^t \text{Re}(y_0, F(T(\tau)x - x_0)) \, d\tau \qquad (t \ge 0).$$

Hence, (iii) holds. □

REMARK. When A is either an m-dissipative operator or satisfies the condition in (ii) of Example 4.2, the semigroup of contractions $\{T(t) : t \ge 0\}$ may, of course, be constructed through the approach of Theorem 4.2 directly without making use of the above method. Also, when A is a dissipative operator satisfying condition (R_2) $(R(I - \lambda A) \supset \text{co} \, D(A)$ for all $\lambda > 0)$, the semigroup of contractions $\{T(t) : t \ge 0\}$ can also be constructed as follows: Let $X_0 = \overline{\text{co}} \, D(A)$ and consider a semigroup of contractions $\{T(t; \tilde{A}_\lambda) : t \ge 0\}$ on X_0 described before Theorem 4.3. We set $T_\lambda(t) = T(t; \tilde{A}_\lambda)$ and let $x \in D(A)$, $t > 0$, and $\lambda, \mu > 0$. Then,

$$\|T_\lambda(t)x - T_\mu(t)x\|^2 = \int_0^t (d/ds)\|T_\lambda(s)x - T_\mu(s)x\|^2 \, ds$$

$$= 2 \int_0^t \text{Re}(\tilde{A}_\lambda T_\lambda(s)x - \tilde{A}_\mu T_\mu(s)x, F(T_\lambda(s)x - T_\mu(s)x)) \, ds,$$

$$\text{Re}(\tilde{A}_\lambda T_\lambda(s)x - \tilde{A}_\mu T_\mu(s)x, F(T_\lambda(s)x - T_\mu(s)x))$$

$$\le \text{Re}(\tilde{A}_\lambda T_\lambda(s)x - \tilde{A}_\mu T_\mu(s)x, F(T_\lambda(s)x - T_\mu(s)x)$$

$$- F(\tilde{J}_\lambda T_\lambda(s)x - \tilde{J}_\mu T_\mu(s)x))$$

$$\le 2\||Ax\|| \, \|F(T_\lambda(s)x - T_\mu(s)x) - F(\tilde{J}_\lambda T_\lambda(s)x - \tilde{J}_\mu T_\mu(s)x)\|,$$

where we use the dissipativity of the closure $\overline{A}$ of A, $\tilde{J}_\lambda = (I - \lambda\overline{A})^{-1}$, $\tilde{A}_\lambda z \in \overline{A}\tilde{J}_\lambda z$ $(z \in X_0)$, and $\|\tilde{A}_\lambda T_\lambda(s)x\| \le \|\tilde{A}_\lambda x\| \le \||\overline{A}x\|| \le \||Ax\||$. From

$$\|(T_\lambda(s)x - T_\mu(s)x) - (\tilde{J}_\lambda T_\lambda(s)x - \tilde{J}_\mu T_\mu(s)x)\| \le (\lambda + \mu)\||Ax\||$$

and F is uniformly continuous on bounded sets, if we let $\lambda \to 0+$, then we see that $T_\lambda(t)x$ converges uniformly on bounded intervals of $[0, \infty)$. We let $T(t)x = \lim_{\lambda \to 0+} T_\lambda(t)x$ and use the same notation $T(t)$ for the extension of the operator to $\overline{D(A)}$; then $\{T(t) : t \ge 0\}$ is a semigroup of contractions on $\overline{D(A)}$.

Since we used the uniform convergence of F on bounded sets, the above proof does not apply for a general Banach space. But, we see $\lim_{\lambda \to 0+} T_\lambda(t)x$ $(x \in D(A), t \ge 0)$ exists by using inequality (3.37) of Lemma 3.16 and the convergence property (4.3) in Theorem 4.2 (as we have done in Theorem 4.3). However, a direct proof of the existence of this limit appears to be unknown.

Here, in order to characterize the (g)-operator of a semigroup of contractions on X_0, we introduce the following set $\mathfrak{G}(X_0)$: let X_0 be a subset of X, and denote by $\mathfrak{G}(X_0)$ the set of all operators A (not necessarily single-valued) that are almost demiclosed and have property $(\mathfrak{G})$ with $D(A) \subset X_0$.

If $A_i \in \mathfrak{G}(X_0)$ $(i = 1, 2)$ satisfy $A_1 \subset A_2$, then we write $A_1 \preceq A_2$. Then, $\mathfrak{G}(X_0)$ is a partially ordered set with the relation " $\preceq$ ".

LEMMA 4.15. *If* A *is the* (g)-*operator of a semigroup of contractions* $\{T(t) : t \geq 0\}$ *on a closed set* X_0, *then* A *is a maximal dissipative operator on* $\overline{D(A)}$ *and has property* $(\mathfrak{G})$.

PROOF. From Theorem 3.5, A is a maximal dissipative operator on $\overline{D(A)}$ and $A_0 \subset A^0$ and $D(A^0) = D(A) = \widehat{D}$ $(\equiv \{x \in X_0 ; T(t)x : [0, \infty) \to X_0$ is Lipschitz continuous $\})$.

Next, we show that A has property $(\mathfrak{G})$. Since A is a maximal dissipative operator on $\overline{D(A)}$, A is a maximal dissipative operator on $D(A)$. Let $x \in D(A)$ $(= \widehat{D})$. Then $T(t)x \in \widehat{D} = D(A)$ $(t \geq 0)$ and there exists a set N of measure 0, such that

$$(d/dt)T(t)x = A_0 T(t)x \in A^0 T(t)x \subset AT(t)x \qquad (t \in [0, \infty)\backslash N).$$

Further, from (iii) of Theorem 3.5 we have

$$\||AT(t)x\|| = \lim_{h \to 0+} h^{-1}\|T(t+h)x - T(t)x\|$$
$$\leq \lim_{h \to 0+} h^{-1}\|T(h)x - x\| = \||Ax\|| \qquad (t \geq 0).$$

For $\varepsilon > 0$, we let $\delta = \varepsilon$, $f_\delta(t) = g_\delta(t) = T(t)x$ $(0 \leq t \leq \delta)$ and define $h_\delta(t)$ by

$$h_\delta(t) = A_0 T(t)x \qquad (t \in [0, \delta)\backslash N)$$
$$= \text{"one element belonging to } A^0 T(t)x\text{"} \qquad (t \in 0, \delta) \cap N).$$

Then, $f_\delta(t)$, $g_\delta(t)$, and $h_\delta(t)$ satisfy (II_1) through (II_3) of Definition 4.2. $\square$

THEOREM 4.16. *Let* X_0 *be a closed set, and let* A *be an operator with* $\overline{D(A)} = X_0$ *and* $R(A) \subset X$ *(not necessarily single-valued). Then, the following* (i), (ii), *and* (iii) *are equivalent:*
    (i) A *is the* (g)-*operator of a semigroup of contractions on* X_0.
    (ii) A *is a maximal dissipative operator on* X_0 *and has property* $(\mathfrak{G})$.
    (iii) A *is a maximal element of the ordered set* $\mathfrak{G}(X_0)$ *(introduced above).*

PROOF. By Lemma 4.15, we obtain "(i) $\Rightarrow$ (ii)." To show "(ii) $\Rightarrow$ (iii)," we assume (ii). Then, since A is demiclosed (see Lemma 2.17), $A \in \mathfrak{G}(X_0)$. Now, suppose $\widetilde{A} \in \mathfrak{G}(X_0)$ satisfies $A \preceq \widetilde{A}$; then $\widetilde{A}$ is a dissipative extension of A, and $D(\widetilde{A}) \subset X_0$. Since A is a maximal dissipative operator on X_0, it follows that $\widetilde{A} = A$. Therefore, A is a maximal element of $\mathfrak{G}(X_0)$.

Finally, to show "(iii) $\Rightarrow$ (i)," we let A be a maximal element of $\mathfrak{G}(X_0)$. Since A is almost demiclosed and has property $(\mathfrak{G})$, from Theorem 4.14, there exists a semigroup of contractions $\{T(t) : t \geq 0\}$ on X_0 $(= \overline{D(A)})$ that satisfies

$$\|T(t)x - T(s)x\| \leq \||Ax\|| \, |t - s| \qquad (x \in D(A); \, t, s \geq 0), \qquad (4.36)$$

$$\limsup_{t \to 0+} \mathrm{Re}(t^{-1}[T(t)x - x], F(x - x_0)) \leq \mathrm{Re}(y_0, F(x - x_0)) \qquad (4.37)$$

$$(x_0 \in D(A), \, y_0 \in Ax_0, \, x \in \overline{D(A)}).$$

Now let A_0 be the infinitesimal generator of this semigroup of contractions. Then, from (4.37) we obtain

$$\mathrm{Re}(A_0 x - y_0, F(x - x_0)) \leq 0 \quad (x_0 \in D(A), \, y_0 \in Ax_0) \quad \text{if } x \in D(A_0). \quad (4.38)$$

We define an operator A_1 by $D(A_1) = D(A) \cup D(A_0)$, and

$$
\begin{aligned}
A_1 x &= Ax & (x \in D(A)\backslash D(A_0)), \\
A_1 x &= A_0 x & (x \in D(A_0)\backslash D(A)), \\
A_1 x &= Ax \cup \{A_0 x\} & (x \in D(A) \cap D(A_0)).
\end{aligned}
$$

Then from (4.38), we see that A_1 is a dissipative operator. Further, from (4.36), $D(A_1) \subset \hat{D} \equiv \{x \in X_0; \, T(t)x : [0, \infty) \to X_0 \text{ is Lipschitz continuous}\}$. Consider a maximal dissipative operator $\tilde{A}_1$ on $\hat{D}$ with $\tilde{A}_1 \supset A_1$ and $\hat{D} \supset D(\tilde{A}_1)$. Since $\tilde{A}_1 \supset A_0$, $\tilde{A}_1$ is the (g)-operator of the semigroup of contractions $\{T(t) : t \geq 0\}$ on X_0. Thus from Lemma 4.15, we obtain $\tilde{A}_1 \in \mathfrak{G}(X_0)$. Clearly, $\tilde{A}_1 \succeq A$. Then, since A is the maximal element of $\mathfrak{G}(X_0)$, $A = \tilde{A}_1$ and A is the (g)-operator of the semigroup of contractions $\{T(t) : t \geq 0\}$ on X_0. $\square$

If A is an m-dissipative operator, obviously A is a maximal dissipative operator on $\overline{D(A)}$ and has property $(\mathfrak{G})$ (see Example 4.2). Thus from Theorem 4.16, every m-dissipative operator A is always the (g)-operator of a semigroup of contractions on $\overline{D(A)}$. Now, we denote this semigroup of contractions on $\overline{D(A)}$ by $\{T_A(t) : t \geq 0\}$ (note that, from the definition of (g)-operator, there is a unique semigroup of contractions on $\overline{D(A)}$) which has A as the (g)-operator. It is self-evident that if both A and B are M-dissipative operators and $A = B$, then $T_A(t) = T_B(t) \ (t \geq 0)$. Conversely,

COROLLARY 4.17. *Let A and B be m-dissipative operators. If $T_A(t) = T_B(t) \ (t \geq 0)$, then $A = B$.*

PROOF. Let $T(t) = T_A(t) \ (= T_B(t)) \ (t \geq 0)$. Since both A and B are (g)-operators of a semigroup of contractions $\{T(t) : t \geq 0\}$ on $\overline{D(A)}$ $(= \overline{D(B)})$, from (iii) of Theorem 3.5 we obtain

$$D(A) = \hat{D} = D(B),$$

where $\hat{D} = \{x \in \overline{D(A)}; \, T(t)x : [0, \infty) \to \overline{D(A)} \text{ is Lipschitz continuous}\}$.

Let $x \in \hat{D}$. Since $t^{-1}\|T(t)x - x\| = O(1)$ as $t \to 0+$, there exists a sequence $\{t_n\}$ of positive numbers which converges to 0 such that w-$\lim_{n \to \infty} t_n^{-1}(T(t_n)x - x) = x'$ exists. Then, from (i) of Theorem 3.5, we obtain $x' \in A^0 x$, $x' \in B^0 x$, and $A^0 x \cap B^0 x \neq \varnothing$. Here we apply Corollary 2.24 and obtain $A = B$. $\square$

Using Theorem 4.16, we can obtain a characterization of the infinitesimal generator of semigroups of contractions, when (both X^* and) X is a uniformly convex Banach space, as follows.

THEOREM 4.18. *Let X_0 be a closed subset of a uniformly convex Banach space X, and let A_0 be a single-valued operator with $\overline{D(A_0)} = X_0$ and $R(A_0) \subset X$. Then the following four propositions are equivalent:*

(i) A_0 *is the infinitesimal generator of a semigroup of contractions on X_0.*

(ii) *There exists a maximal dissipative operator A on X_0 that satisfies condition (b') (described in the remark following Example 4.3) with $D(A) \subset X_0$, and $A_0 = A^0$.*

(iii) *There exists a maximal dissipative operator A on X_0 that satisfies condition (a') (described in the remark following Example 4.3) with $D(A) \subset X_0$, and $A_0 = A^0$.*

(iv) *There exists a maximal dissipative operator A on X_0, and $D(A) \subset X_0$, which has property (𝕊), and $A_0 = A^0$.*

PROOF. (i) $\Rightarrow$ (ii). Let A_0 be the infinitesimal generator of a semigroup of contractions $\{T(t) : t \geq 0\}$ on X_0, and let A be the (g)-operator of this semigroup of contractions. Then $(D(A_0) \subset) D(A) \subset X_0$, and A is a maximal dissipative operator on X_0 $(= \overline{D(A)})$ (see (v) of Theorem 3.5), and $A_0 = A^0$ (see (ii) of Corollary 3.6). Note also that $D(A^0) = D(A) = \hat{D}$. To prove condition (b') we choose $x \in D(A)$ arbitrarily. Then from Corollary 3.7

$$\lim_{h \to 0+} h^{-1}(T(h)x - x) = A_0 x = A^0 x, \quad \|A^0 T(t)x\| \leq \|A^0 x\| \quad (t \geq 0).$$

Hence, for any $\varepsilon > 0$, there exists $\delta > 0$ with $\delta < \varepsilon$ such that

$$\|\delta^{-1}(T(\delta)x - x) - A^0 x0\| < \varepsilon.$$

Now set $x_\delta = T(\delta)x$. Then we obtain

$$\|x + \delta A^0 x - x_\delta\| \leq \delta \varepsilon, \quad \|A^0 x_\delta\| \leq \|A^0 x\|$$

with $x_\delta \in D(A)$. Therefore (b') holds.

Next, as we showed in the previous remark, since conditions (a') and (b') are equivalent for a maximal dissipative operator A on $D(A)$, we have that (ii) and (iii) are equivalent.

(iii) $\Rightarrow$ (iv). Since Condition (a') satisfies condition (a) of Example 4.1, A has property (𝕊) and we obtain (iv) from (iii).

(iv) $\Rightarrow$ (i). Assume (iv). Then since $\overline{D(A)} = X_0$, we have from Theorem 4.16 that A is the (g)-operator of a semigroup of contractions $\{T(t) : t \geq 0\}$

on X_0. Thus from Corollary 3.6, A^0 is the infinitesimal generator of this semigroup of contractions. Since $A_0 = A^0$, A_0 is the infinitesimal generator of the semigroup of contractions $\{T(t) : t \geq 0\}$ on X_0. □

COROLLARY 4.19. *Let X be uniformly convex Banach space. If A is an m-dissipative operator, then $\overline{D(A)}$ is a (closed) convex set and A^0 is the infinitesimal generator of a semigroup of contractions on $\overline{D(A)}$.*

PROOF. From Theorem 2.20, $\overline{D(A)}$ is a (closed) convex set. Next, from Example 4.2, A is a maximal dissipative operator and has property $(\mathfrak{G})$. Also, since $D(A^0) = D(A)$ and A^0 is a single-valued operator, we consider A^0 as A_0 and apply Theorem 4.18 to finish the proof. □

Finally, let us recall the theorem of Hille-Yosida on linear semigroups. Let X be a general Banach space. Then the infinitesimal generator of a semigroup of linear contractions is an m-dissipative operator whose domain is dense in X. Also conversely, an m-dissipative operator whose domain is dense in X is the infinitesimal generator of a (unique) semigroup of linear contractions on X. This is the so-called Hille-Yosida theorem. When X is a Hilbert space, we can obtain the same result for semigroups of nonlinear contractions. That is, the following theorem holds:

THEOREM 4.20. *Let X be a Hilbert space. Then,*

(i) *The infinitesimal generator A_0 of a semigroup of contractions $\{T(t) : t \geq 0\}$ on a closed convex set X_0 $(\subset X)$ has a dense domain in X_0 and there exists a unique maximal dissipative operator A such that $A^0 = A_0$. Conversely,*

(ii) *If A is the maximal dissipative operator, then $\overline{D(A)}$ is a closed convex set and A^0 is the infinitesimal generator of a semigroup of contractions on $\overline{D(A)}$.*

PROOF. (ii) Since the maximal dissipative operator in a Hilbert space is an m-dissipative operator (see Corollary 2.27), we use Corollary 4.19 to prove this.

(i) We showed in Theorem 3.13 that $\overline{D(A_0)} = X_0$. Next, since A_0 is a dissipative operator with $D(A_0) \subset X_0$, by Theorem 2.26 there exists an m-dissipative operator with $D(A) \subset X_0$ and $A_0 \subset A$. Then from Corollary 4.19, there exists a semigroup of contractions $\{U(t) : t \geq 0\}$ on X_0 $(= \overline{D(A_0)} = \overline{D(A)})$ that has A^0 as the infinitesimal generator. In fact, $U(t) = T(t)$ $(t \geq 0)$. Because, for each $x \in D(A_0)$ $(\subset D(A) = D(A^0)$ should be noted), both $T(t)x$ and $U(t)x$ are solutions of the Cauchy problem, $(d/dt)u(t) \in Au(t)$ $(0 \leq t < \infty)$ with $u(0) = x$, and thus from the uniqueness of the solution (see Lemma 4.9), $U(t)x = T(t)x$ $(t \geq 0)$. Hence $U(t) = T(t)$ $(t \geq 0)$. From this we obtain $A^0 = A_0$, and thus we have shown that there exists an m-dissipative operator A with $A_0 = A^0$. Also, we see from (ii) of Corollary 2.24 that there is no m-dissipative operator ($=$ the maximal dissipative operator) other than A $\square$

§4. Examples

In this section, we describe examples of semigroups of contractions $\{T(t) : t \geq 0\}$ on X which, are constructed (by the method of Theorem 4.2) from an m-dissipative operator with dense domain in X, and have the property that for each $x \in X$, $T(t)x$ is not weakly differentiable at every $t \geq 0$. We also give examples of concrete applications to partial differential equations of the generation theory of semigroups of (nonlinear) contractions.

EXAMPLE 4.4. Let $g : R^1 \to R^1$ be a monotone decreasing function, and consider the m-dissipative operator $\tilde{g}$ (in R^1) given in Example 2.1. Since the canonical restriction of $\tilde{g}$ can be given as follows:

$$g^0(x) = \begin{cases} g_+(x) & \text{if } g_+(x) > 0, \\ 0 & \text{if } g_+(x) \leq 0 \leq g_-(x), \\ g_-(x) & \text{if } g_-(x) < 0, \end{cases}$$

from part (ii) of Theorem 4.20, there exists a semigroup of contractions $\{S(t) : t \geq 0\}$ on R^1 that has g^0 as infinitesimal generator, and the following holds (see Corollary 3.7):

$$D^+ S(t)x = g^0(S(t)x) \qquad (t \geq 0, \ x \in R^1)(^9), \tag{4.39}$$

$$|S(t+h)x - S(t)x| \leq h|D^+ S(t)x| \leq h|g^0(x)| \qquad (t, h \geq 0, \ x \in R^1). \tag{4.40}$$

EXAMPLE 4.5. Let g^0 and $\{S(t) : t \geq 0\}$ be the function and semigroup of contractions on R^1 given in Example 4.4. Let $X = C[-1, 1]$ and define an operator $T(t) : X \to X$ for each $t \geq 0$ by

$$(T(t)f)(x) = S(t)f(x) \qquad (f \in C[-1, 1]). \tag{4.41}$$

Then, $\{T(t) : t \geq 0\}$ is a semigroup of contractions on X and (4.42) through (4.44) hold:

(4.42) $\|T(t)f - f\| \leq M_f t \ (t \geq 0; f \in X)$, where

$$M_f = \sup\{|g^0(x)| : -\|f\| \leq z \leq \|f\|\}(< \infty),$$

(4.43) $(A'f)(x) = g^0(f(x)) \ (f \in D(A'))$,

(4.44) {all constant functions on $[-1, 1]$} $\subset D(A_0) \subset D(A')$, where A_0 and A' are the infinitesimal generator and the weak infinitesimal generator of $\{T(t) : t \geq 0\}$, respectively.

In fact, it is easy to see that $\{T(t) : t \geq 0\}$ is a semigroup of contractions on X from the fact that $\{S(t) : t \geq 0\}$ is a semigroup of contractions (on

---

(^{9}) Also, for each $x \in R^1$, $S(t)x$ is differentiable (with respect to t) except for at most a countable number of points t (see Corollary 3.7).

$\mathbf{R}^1$). Next, from (4.40) we obtain

$$|(T(t)f)(x) - f(x)| = |S(t)f(x) - f(x)| \le t|g^0(f(x))| \le tM_f \qquad (f \in X),$$

and thus (4.42) holds.

Also, considering a constant function $f(x) \equiv c$,

$$\lim_{t \to 0+} \|t^{-1}[T(t)f - f] - g^0(f(\cdot))\| = \lim_{t \to 0+} |t^{-1}[S(t)c - c] - g^0(c)| = 0$$

(see (4.39)) hence we obtain (4.44). To show (4.43) we let $f \in D(A')$. Then for every function of bounded variation $\alpha(s)$ on $[-1, 1]$, we have

$$\lim_{h \to 0+} h^{-1} \int_{-1}^1 [(T(h)f)(s) - f(s)] d\alpha(s) = \int_{-1}^1 (A'f)(s) d\alpha(s).$$

Now we let $x \in [-1, 1]$ and consider the characteristic function of $(x, 1]$ (the characteristic function of the set $\{1\}$ if $x = 1$) as $\alpha(s)$. Then from the equation above we obtain

$$\lim_{h \to 0+} h^{-1}[(T(h)f)(x) - f(x)] = (A'f)(x).$$

On the other hand, from (4.39)

$$\lim_{h \to 0+} h^{-1}[(T(h)f)(x) - f(x)] = \lim_{h \to 0+} h^{-1}[S(h)f(x) - f(x)] = g^0(f(x)).$$

Hence, (4.43) holds.

Furthermore, for this semigroup of contractions $\{T(t) : t \ge 0\}$ the following hold:

(i) There exists an m-dissipative operator A such that $\overline{D(A)} = X$ and

$$T(t)f = \lim_{\lambda \to 0+} (I - \lambda A)^{-[t/\lambda]} f \qquad (f \in X, \ t \ge 0).$$

(ii) If the set of all points of discontinuity of g is dense in $\mathbf{R}^1$, then

$$D(A') = D(A_0) = \{\text{all constant functions on } [-1, 1]\}.$$

PROOF OF (i). For $h > 0$ and $\lambda > 0$, we set $A_h = h^1(T(h) - I)$ and $J_{\lambda, h} = (I - \lambda A_h)^{-1}$. Then from the remark following Corollary 3.12 and from Corollary 3.20 we need to show that (4.45) holds:

$$\text{There exists } \lim_{h \to 0+} J_{\lambda, h} f \text{ for every } f \in X \text{ and } \lambda > 0. \qquad (4.45)$$

To show this, let $f \in X$ and $\lambda > 0$, and set $J_{\lambda, h} f = \varphi_h$. Then, for $x \in [-1, 1]$

$$f(x) = \varphi_h(x) - \lambda \frac{(T(h)\varphi_h)(x) - \varphi_h(x)}{h}$$
$$= \varphi_h(x) - \lambda \frac{S(h)\varphi_h(x) - \varphi_h(x)}{h}$$
$$= \left(i - \lambda \frac{S(h) - i}{h}\right) \varphi_h(x).$$

where i is the identity operator on R^1, that is, $iz = z$ $(z \in R^1)$. From this we have

$$\varphi_h(x) = (i - \lambda(S(h) - i)/h)^{-1}f(x),$$

and

$$(J_{\lambda,h}f)(x) = \left(i - \lambda\frac{S(h) - i}{h}\right)^{-1}f(x) \qquad (x \in [-1, 1]).$$

From Theorem 3.13,

$$\lim_{h\to 0+}(i - \lambda(S(h) - i)/h)^{-1}f(x) \qquad (x \in [-1, 1])$$

exists and we see that the convergence is uniform with respect to x. To show this, let us recall the proof of Theorem 3.13.

First, note that

$$|(i - \lambda(S(h) - i)/h)^{-1}f(x) - f(x)| \leq \lambda|S(h)f(x) - f(x)|/h$$
$$\leq \lambda\|T(h)f - f\|/h \leq \lambda M_f,$$

where we use the contractiveness of the operator $(i - \lambda(S(h)-i)/h)^{-1} : R^1 \to R^1$ and (4.42). Now, given $\varepsilon > 0$, we choose a positive $\delta < 1$ such that $\delta M_f < \varepsilon$; then $|S(t)f(x) - f(x)|$ $(\leq \|T(t)f - f\| \leq tM_f) < \varepsilon$ if $0 < t \leq \delta$. By the same reasoning as in the proof of Theorem 3.13, if $0 < s, t \leq \delta$, then for every $x \in [-1, 1]$ we have

$$\left|\left(i - \lambda\frac{S(s) - i}{s}\right)^{-1}f(x) - \left(i - \lambda\frac{S(t) - i}{t}\right)^{-1}f(x)\right| \leq 2\sqrt{2\varepsilon\lambda M_f}$$

(where the M in the proof of Theorem 3.13 is replaced by λM_f). Therefore, if $0 < s, t \leq \delta$, then $\|J_{\lambda,s}f - J_{\lambda,t}f\| \leq 2\sqrt{2\varepsilon\lambda M_f}$, and (4.45) is proved.

PROOF OF (ii). Since g^0 is also discontinuous at the points of discontinuity of g, the set of all points of discontinuity of g^0 is dense in R^1. From this we note that if $g^0(K)$ $(= \{g^0(x) : x \in K\})$ is connected then K is a set that consists of a single point.

Let $f \in D(A')$. Then from (4.43), $g^0(f(\cdot)) = A'f \in X = C[-1, 1]$; i.e., $g^0(f(\cdot))$ is continuous on $[-1, 1]$. Hence $g^0(f([-1, 1]))$ is connected and $f([-1, 1])$ $(= \{f(x) : x \in [-1, 1]\})$ must be a set that consists of a single point. Hence f is a constant function. From this and (4.44), (ii) is proved.

From the above, "if $g : R^1 \to R^1$ is monotone decreasing and the set of all points of discontinuity of g is dense in R^1, then the semigroup of contractions $\{T(t) : t \geq 0\}$ on $C[-1, 1]$ constructed above can be formulated by the exponential formula $T(t)f = \lim_{\lambda\to 0+}(I - \lambda A)^{-[t/\lambda]}f$ $(f \in C[-1, 1]$, $t \geq 0)$ with an m-dissipative operator A that has domain dense in $C[-1, 1]$, and the domain of the weak infinitesimal generator coincides with the set of all constant functions on $[-1, 1]$ (consequently, the domain of the weak infinitesimal generator is not dense in $C[-1, 1]$). Also from (4.42), $\hat{D} = C[-1, 1]$, where $\hat{D} = \{f \in C[-1, 1] : \liminf_{h \to 0+} h^{-1}\|T(h)f - f\| < \infty\}$."

REMARK. Let $\{x_1, x_2, \ldots, x_n, \ldots\} \subset R^1$, $x_i \neq x_j$ $(i \neq j)$, and let $c_n(\cdot)$ be the characteristic function of $(-\infty, x_n]$. If we set $g(x) = \sum_{n=1}^{\infty} 2^{-n} c_n(x)$ $(x \in R^1)$, then $g : R^1 \to R^1$ is a monotone decreasing function and becomes discontinuous only at x_n $(n = 1, 2, \ldots)$. Hence, if $\{x_1, x_2, \ldots, x_n, \ldots\}$ is dense in R^1, then g is a monotone decreasing function such that the set of all points of discontinuity is dense in R^1.

EXAMPLE 4.6. Let $g : R^1 \to R^1$ be a monotone decreasing function which is continuous at 0, and let g^0 and $\{S(t) : t \geq 0\}$ be the function and the semigroup of contractions on R^1, given in Example 4.4, respectively.

Now for each $x \in [-1, 1]$ we define a function $g_x : R^1 \to R^1$ by

$$g_x(y) = g(x^2 y + x) \qquad (y \in R^1). \tag{4.46}$$

Then, $g_x : R^1 \to R^1$ is a monotone decreasing function. Here we define $\tilde{g}_x$ in the same way as in Example 2.1 as follows:

$$\tilde{g}_x(y) = \{z : g_x^+(y) \leq z \leq g_x^-(y)\} \qquad (y \in R^1).$$

Then $\tilde{g}_x$ is an m-dissipative operator (on R^1) and its canonical restriction is as follows:

$$g_x^0(y) = \begin{cases} g_x^+(y) & \text{if } g_x^+(y) > 0, \\ 0 & \text{if } g_x^+(y) \leq 0 \leq g_x^-(y), \\ g_x^-(y) & \text{if } g_x^-(y) < 0, \end{cases}$$

where $g_x^+(y) = \lim_{h \to 0+} g_x(y+h)$ and $g_x^-(y) = \lim_{h \to 0+} g_x(y-h)$ $(y \in R^1)$. Note that from the definition (4.46) of g_x,

$$g_x^0(y) = g^0(x^2 y + x) \qquad (y \in R^1). \tag{4.47}$$

Thus, from (ii) of Theorem 4.20, there exists a semigroup of contractions $\{S_x(t) : t \geq 0\}$ on R^1 which has g_x^0 as its infinitesimal generator and satisfies the following properties:

$$D^+ S_x(t)y = g_x^0(S_x(t)y) \qquad (t \geq 0, \ y \in R^1), \tag{4.48}$$

$$|S_x(t+h)y - S_x(t)y| \leq h|D^+ S_x(t)y| \leq h|g_x^0(y)| \qquad (t, h \geq 0, \ y \in R^1), \tag{4.49}$$

and for each $t \geq 0$ and $y \in R^1$ we have

$$S_x(t) = \begin{cases} x^{-2}[S(tx^2)(x^2 y + x) - x] & \text{if } x \neq 0, \\ y + tg(0) & \text{if } x \neq 0. \end{cases} \tag{4.50}$$

In fact, we obtain (4.48) and (4.49) from Corollary 3.7 (also, for each $y \in R^1$, $S_x(t)y$ is differentiable (with respect to t) except for at most a countable number of points t). To show (4.50), we define $\widetilde{S}_x(t)y$ by the right-hand side of (4.50) and fix $y \in R^1$. Then from (4.40) we obtain

$$|\widetilde{S}_x(t_1)y - \widetilde{S}_x(t_2)y| \leq |t_1 - t_2| \, |g^0(x^2 y + x)|$$

(note that when $x = 0$, $g^0(0) = g(0)$ since g is continuous at 0), and $\widetilde{S}_x(t)y$ is Lipschitz continuous as a function of t. Also from (4.39) and (4.47), $D^+ \widetilde{S}_x(t)y = g_x^0(\widetilde{S}_x(t)y)\,(^{10})$ ($t \geq 0$, $y \in R^1$). Hence, $\widetilde{S}_x(t)y$ is a solution of the Cauchy problem $(d/dt)u(t) = g_x^0(u(t))$ with $u(0) = y$. Also, since $S_x(t)y$ is a solution of the same Cauchy problem, from the uniqueness of the solution we obtain $S_x(t)y = \widetilde{S}_x(t)y$ ($t \geq 0$) and therefore (4.50) holds. (Note that g_x^0 is a dissipative operator (on R^1) and also there is at most one solution of Cauchy's problem for the dissipative operator.)

EXAMPLE 4.7. Let $g : R^1 \to R^1$ be a monotone decreasing function which is continuous at 0, and let g^0 and $\{S(t) : t \geq 0\}$ be the function and the semigroup of contractions on R^1 given in Example 4.4 respectively. Also, for each $x \in [-1, 1]$, g_x, g_x^0, and $\{S_x(t) : t \geq 0\}$ are the functions and the semigroup of contractions on R^1 described in Example 4.6.

Let $X = C[-1, 1]$ and define an operator $T(t) : X \to X$ for each $t \geq 0$ by

$$(T(t)f)(x) = S_x(t)f(x) \qquad (f \in C[-1, 1]). \tag{4.51}$$

Then, $\{T(t) : t \geq 0\}$ is a semigroup of contractions on X and the following (4.52) and (4.53) hold:

$$\|T(t)f - f\| \leq tK_f \qquad (t \geq 0, \; f \in X), \tag{4.52}$$

where $K_f = \sup\{|g^0(z)| : -(\|f\| + 1) \leq z \leq \|f\| + 1\}$, and

$$(A'f)(x) = g^0(x^2 f(x) + x) \qquad (f \in D(A')), \tag{4.53}$$

where A' is the weak infinitesimal generator of $\{T(t) : t \geq 0\}$.

First, we show $T(t)f \in X$ ($t > 0$, $f \in X$). From (4.50) we have

$$(T(t)f)(x) = x^{-2}[S(tx^2)(x^2 f(x) + x) - x] \qquad (x \neq 0).$$

From this and since $\{S(t) : t \geq 0\}$ is a semigroup of contractions on R^1, we see that $(T(t)f)(x)$ is continuous at $x \neq 0$. Also

$(T(t)f)(x) - (T(t)f)(0)$
$$= x^{-2}[S(tx^2)(x^2 f(x) + x) - (x^2 f(x) + x)] + (f(x) - f(0)) - tg^0(0)$$
$$= x^{-2}\int_0^{tx^2} [g^0(S(s)(x^2 f(x) + x)) - g^0(0)]ds + (f(x) - f(0)) \quad (x \neq 0)$$

(we consider $x^2 f(x) + x$ instead of x in (4.39), and integrate it over $[0, tx^2]$; also note that $g(0) = g^0(0)$), and from g and hence g^0 being continuous

---

$(^{10})\, D^+$ denotes the right derivative with respect to t.

at 0, we obtain $(T(t)f)(x) \to (T(t)f)(0)$ as $x \to 0$, and thus $(T(t)f)(x)$ is also continuous at $x \in 0$. Therefore, $T(t)f \in C[-1, 1] = X$. Next, we see easily that $\{T(t) : t \geq 0\}$ is a semigroup of contractions on X. From (4.49) and (4.47)

$$|(T(t)f)(x) - f(x)| = |S_x(t)f(x) - f(x)| \leq t|g^0(x^2 f(x) + x)|,$$
$$\leq tK_f \qquad (-1 \leq x \leq 1, \ f \in X, \ t \geq 0),$$

and hence (4.52) holds. Finally, by the same reaoning as in the proof of (4.43), we have $\lim_{h \to 0+} h^{-1}[(T(h)f)(x) - f(x)] = (A'f)(x)$ for $f \in D(A')$. On the other hand, from (4.47) and (4.48) we have

$$\lim_{h \to 0+} h^{-1}[(T(h)f)(x) - f(x)] = \lim_{h \to 0+} h^{-1}[S_x(h)f(x) - f(x)] = g_x^0(f(x))$$
$$= g^0(x^2 f(x) + x) \qquad (f \in X).$$

Therefore, we obtain (4.53).

Furthermore, for this semigroup of contractions $\{T(t) : t \geq 0\}$, the following hold:

(i) There is an m-dissipative operator A such that $\overline{D(A)} = X$ and

$$T(t)f = \lim_{\lambda \to 0+} (I - \lambda A)^{-[t/\lambda]} f \qquad (f \in X, \ t \geq 0).$$

(ii) If the set of all points of discontinuity of g is dense in R^1, then $D(A') = \varnothing$, and hence, for arbitrary $f \in X$, $T(t)f$ is not weakly right differentiable at all $t \geq 0$.

PROOF OF (i). By the same reasoning used to prove (i) of Example 4.5, by letting $A_h = h^{-1}(T(h) - I)$, and $J_{\lambda, h} = (I - \lambda A_h)^{-1}$ $(h, \lambda > 0)$, we show (4.45). Here, we choose and fix $f \in X$ and $\lambda > 0$ arbitrarily, and set $J_{\lambda, h} f = \varphi_h$. Then

$$(J_{\lambda, h} f)(x) = \left(i - \lambda \frac{S_x(h) - i}{h}\right)^{-1} f(x) \qquad (x \in [-1, 1]),$$

where i is the identity operator on R^1. Now we see that $(J_{\lambda, h} f)(x)$ converges uniformly with respect to x as $h \to 0+$, by the same reasoning as in the proof of (i) of Example 4.5. In fact, since

$$|(i - \lambda (S_x(h) - i)/h)^{-1} f(x) - f(x)| \leq \lambda \|T(h)f - f\|/h \leq \lambda K_f$$

(see (4.52), for every $\varepsilon > 0$ we select a positive number δ (< 1) such that $\delta K_f < \varepsilon$; then $|S_x(t)f(x) - f(x)| < \varepsilon$ $(0 < t \leq \delta)$. From this, if $0 < s, t \leq \delta$,

$$\left|\left(i - \lambda \frac{S_x(s) - i}{s}\right)^{-1} f(x) - \left(i - \lambda \frac{S_x(t) - i}{t}\right)^{-1} f(x)\right|$$
$$\leq 2\sqrt{2\varepsilon\lambda K_f} \qquad (x \in [-1, 1]).$$

That is, we obtain $\|J_{\lambda,s}f - J_{\lambda,t}f\| \leq 2\sqrt{2\varepsilon\lambda K_f}$, and thus (4.45) holds.

PROOF OF (ii). We derive a contradiction to $D(A') \neq \varnothing$. In fact, suppose that there is a function f with $f \in D(A')$. Since $g^0(x^2 f(x) + x) = (A'f)(x)$ (see (4.53)), if we set $\psi(x) = x^2 f(x) + x$ $(x \in [-1, 1])$, then $g^0(\psi(\cdot)) = A'f \in X = C[-1, 1]$. Thus, by the same reasoning as in the proof of (ii) in Example 4.5, $\psi([-1, 1])$ is a set with only one point; that is, ψ must be a constant function. On the other hand, $\psi(x) = x^2 f(x) + x$ cannot be a constant function for any $f \in C[-1, 1]$, and this is a contradiction.

From the above, "If $g : R^1 \to R^1$ is continuous at 0, and a monotone decreasing function such that the set of all points of discontinuity is dense in R^1, then the semigroup of contractions $\{T(t) : t \geq 0\}$ on $C[-1, 1]$ constructed from this g by the method described above, can be formulated by the exponential formula $T(t)f = \lim_{\lambda \to 0+} (I - \lambda A)^{[t/\lambda]} f$ $(f \in C[-1, 1]$, $t \geq 0)$ from an m-dissipative operator A with dense domain in $C[-1, 1]$, and moreover, for every $f \in C[-1, 1]$, $T(t)f$ is not weakly right differentiable at all $t \geq 0$. Furthermore from (4.52), $\widehat{D} = C[-1, 1]$."

EXAMPLE 4.8 (equation of quasihyperbolic type). Consider the following problem:

$$\begin{cases} u_t = (\phi(u))_x & (t > 0, \ 0 < x < 1), \\ u(0, x) = u_0(x) & (0 < x < 1), \\ u(t, 0) = 0 & (t > 0) \end{cases} \tag{4.54}$$

where u_0 is a given function, and $\phi : R^1 \to R^1$ is a (given) continuous, strictly monotone decreasing function satisfying $\phi(0) = 0$ and $\phi(R^1) = R^1$.

Let $X = L^1(0, 1)$ and define an operator A by

$$D(A) = \{u \in C[0, 1] : u(0) = 0, \ \phi(u) \text{ is absolutely continuous on } [0, 1]\}$$

$$\text{and } Au = (\phi(u))' \quad (u \in D(A))(^{11}). \tag{4.55}$$

Then, A is an m-dissipative operator on $L^1(0, 1)$ (thus we can apply Theorem 4.2). We show this next.

I (The dissipativity of A). Let $u, v \in D(A)$ and $\lambda > 0$. First, note that for every monotone decreasing Lipschitz continuous function $p : R^1 \to R^1$ satisfying

$$|p| \leq 1 \quad \text{with } p(0) = 0, \tag{4.56}$$

we have

$$\int_0^1 |u - v - \lambda(Au - Av)| \, dx \geq \int_0^1 (u - v)p(\phi(u) - \phi(v)) \, dx. \tag{4.57}$$

In fact, if we set $q(s) = \int_0^s p(\xi) \, d\xi$ $(s \in R^1)$, then since $q(s) \leq 0$ and

---

$(^{11})\, v'$ denotes the derivative of function v.

$\phi(u(0)) = \phi(0) = \phi(v(0))$, we have

$$\int_0^1 (Au - Av)p(\phi(u) - \phi(v))\, dx = \int_0^1 (\phi(u) - \phi(v))'p(\phi(u) - \phi(v))\, dx$$

$$= \int_0^1 [q(\phi(u) - \phi(v))]'\, dx$$

$$= q(\phi(u(1)) - \phi(v(1))) \leq 0.$$

From this and $|p| \leq 1$,

$$\int_0^1 |u - v - \lambda(Au - Av)|\, dx \geq \int_0^1 (u - v)p(\phi(u) - \phi(v))\, dx$$

$$- \lambda \int_0^1 (Au - Av)p(\phi(u) - \phi(v))\, dx$$

$$\geq \int_0^1 (u - v)p(\phi(u) - \phi(v))\, dx.$$

In particular, we consider functions p_n ($n = 1, 2, \ldots$) such that

$$p_n(s) = -ns \quad (|s| \leq 1/n), \qquad p_n(s) = -\operatorname{sign} s \quad (|s| > 1/n), \tag{4.58}$$

and let $p = p_n$ and apply (4.57). Then we have

$$\int_0^1 |u - v - \lambda(Au - Av)|\, dx \geq \int_0^1 (u - v)p_n(\phi(u) - \phi(v))\, dx.$$

Since $(u - v)p_n(\phi(u) - \phi(v)) \to |u - v|$ as $n \to \infty$, if we let $n \to \infty$ in the above inequality, then we have $\|u - v - \lambda(Au - Av)\|_1 \geq \|u - v\|_1$ (where $\|\cdot\|_1$ denotes the norm in the space $L^1(0, 1)$). Therefore, A is a dissipative operator.

II $(R(I - A) = L^1(0, 1))$. Since A is a dissipative operator, in order to prove that A is m-dissipative, we need to $R(I - A) = X = L^1(0, 1)$ (see Lemma 2.13). Hence, when $g \in L^1(0, 1)$ is given arbitrarily, we need to obtain $u \in D(A)$ that satisfies $u - (\phi(u))' = g$. Now if we set $\psi = \phi^{-1}$ and $v = \phi(u)$, then this (when $g \in L^1(0, 1)$ is given arbitrarily) results in the problem of finding an absolutely continuous function v on $[0, 1]$ that satisfies

$$\psi(v) - v' = g \quad \text{with } v(0) = 0. \tag{4.59}$$

Then, from an existence theorem of solutions of ordinary differential equations, (4.59) has a local solution; that is, for some $a > 0$ there exists an absolutely continuous function v on $[0, a]$ satisfying

$$\psi(v(x)) - v'(x) = g(x) \quad (\text{a.e. } x \in [0, a]),\, v(0) = 0. \tag{4.60}$$

(For example, by the Carathéodory theorem of existence of solutions, we see that for the function

$$f(x, y) = \begin{cases} \psi(y) - g(x) & \text{if } 0 \leq x \leq 1, \\ \psi(y) & \text{if } -1 \leq x \leq 0 \end{cases}$$

defined on the rectangular $D = \{(x, y) : |x| \leq 1 \text{ and } |y| \leq 2\|g\|_1 + 1\}$,
the differential equation $v' = f(x, v)$ with $v(0) = 0$ has a (absolutely
continuous) solution in the neighborhood of $x = 0$.) But, for each absolutely
continuous function v on $[0, a]$ that satisfies (4.60),

$$
\begin{aligned}
|v(c)| &\leq \int_0^c |v'(x)|\, dx = \int_0^c |\psi(v(x)) - g(x)|\, dx \\
&\leq \int_0^c |\psi(v(x))|\, dx + \int_0^c |g(x)|\, dx \\
&\leq 2 \int_0^c |g(x)|\, dx \leq 2\|g\|_1
\end{aligned}
$$

$(0 \leq c \leq a)$ holds. (If we consider $L'(0, c)$ instead of $L^1(0, 1)$, then as in
the proof of I above, A becomes a dissipative operator on $L^1(0, c)$. From
this, if we set $u = \psi(v)$, we obtain

$$
\int_0^c |u|\, dx \leq \int_0^c |u - 0 - (Au - A0)|\, dx = \int_0^c |g|\, dx,
$$

where the notation 0 denotes the constant function that has the value 0.
That is, $\int_0^c |\psi(v)|\, dx \leq \int_0^c |g|\, dx$. We used this inequality above.) Hence if
$a < 1$, we can always extend the solution to the right from a. Therefore, for
an arbitrary $g \in L^1(0, 1)$, there is always an absolutely continuous function
v on $[0, 1]$ that satisfies (4.59).

EXAMPLE 4.9 (equation of quasiparabolic type). Consider the following
problem:

$$
\begin{cases}
u_t = (\phi(u))_{xx} & (t > 0, \; 0 < x < 1), \\
u(0, x) = u_0(x) & (0 < x < 1), \\
u(t, 0) = u(t, 1) = 0 & (t > 0).
\end{cases}
\tag{4.61}
$$

Here, u_0 is a given function, and $\phi : R^1 \to R^1$ is a (given) continuous,
strictly monotone increasing function satisfying $\phi(0) = 0$ and $\phi(R^1) = R^1$.

Let $X = L^1(0, 1)$ and define an operator A by

$$
\begin{cases}
D(A) = \quad \{u \in C[0, 1] : u(0) = u(1) = 0, \text{ and } \phi(u) \text{ and} \\
\qquad\qquad (\phi(u))' \text{ are absolutely continuous on } [0, 1]\}, \text{ and} \\
\qquad Au = (\phi(u))'' \quad (u \in D(A)).
\end{cases}
\tag{4.62}
$$

We show next that A is an m-dissipative operator on $L^1(0, 1)$.

I (The dissipativity of A). Let $u, v \in D(A)$ and $\lambda > 0$. Then, for every
monotone increasing Lipschitz continuous function $p : R^1 \to R^1$ satisfying

(4.56), we have

$$\int_0^1 (Au - Av)p(\phi(u) - \phi(v))\, dx$$

$$= \int_0^1 (\phi(u) - \phi(v))'' p(\phi(\mu) - \phi(v))\, dx$$

$$= -\int_0^1 [(\phi(u) - \phi(v))']^2 p'(\phi(u) - \phi(v))\, dx \leq 0.$$

From this and $|p| \leq 1$, we obtain

$$\int_0^1 |u - v - \lambda(Au - Av)|\, dx \geq \int_0^1 (u - v)p(\phi(u) - \phi(v))\, dx$$

$$- \lambda \int_0^1 (Au - Av)p(\phi(u) - \phi(v))\, dx$$

$$\geq \int_0^1 (u - v)p(\phi(u) - \phi(v))\, dx.$$

Here as p, consider $p_n(s) = ns$ (if $|s| \leq 1/n$), $p_n(s) = \text{sign}\, s$ (if $|s| > 1/n$) and let $n \to \infty$. Then since $(u - v)p_n(\phi(u) - \phi(v)) \to |u - v|$, we obtain

$$\int_0^1 |u - v - \lambda(Au - Av)|\, dx \geq \int_0^1 |u - v|\, dx.$$

That is, $\|u - v - \lambda(Au - Av)\|_1 \geq \|u - v\|_1$, and hence A is a dissipative operator.

II $(R(I - A) = L^1(0, 1))$. Given an arbitrary $g \in L^1(0, 1)$. In order to solve $u - Au = g$, we set $\psi = \phi^{-1}$ and seek a function v such that v satisfies

$$\psi(v) - v'' = g \quad \text{with } v(0) = v(1) = 0, \tag{4.63}$$

and v and v' are absolutely continuous on $[0, 1]$ (in fact, if we find such a v, then $u = \psi(v)$ is a solution of the equation $u - Au = g$). Below, we solve this.

First, note the following: suppose that we have found an absolutely continuous function v on $[0, 1]$ (v' is also absolutely continuous on $[0, 1]$) satisfying (4.63). Then by the dissipativity of A, we have

$$\|\psi(v)\|_1 \ (= \|u\|_1 \leq \|u - 0 - (Au - A0)\|_1) = \|g\|_1.$$

From $v(0) = v(1) = 0$, there exists $c \in (0, 1)$ such that $v'(c) = 0$. Then

$$|v'(x)| \leq \left| \int_c^x |v''(s)|\, ds \right| \leq \int_0^1 |v''(s)|\, ds$$

$$= \int_0^1 |\psi(v) - g|\, ds \leq \|\psi(v)\|_1 + \|g\|_1 \leq 2\|g\|_1.$$

Therefore, we obtain $|v(x)| \leq \int_0^x |v'(s)| \, ds \leq \int_0^1 |v'(s)| \, ds \leq 2\|g\|_1$ $(x \in [0, 1])$. Then we consider the following function

$$\tilde{\psi}(s) = \begin{cases} 2\|g\|_1 & \text{if } \psi(s) \geq 2\|g\|_1, \\ \psi(s) & \text{if } |\psi(s)| < 2\|g\|_1, \\ -2\|g\|_1 & \text{if } \psi(s) \leq -2\|g\|_1. \end{cases}$$

$\tilde{\psi} : R^1 \to R^1$ is a continuous monotone increasing function, and $\tilde{\psi}(0) = 0$ and $|\tilde{\psi}(s)| \leq 2\|g\|_1$ $(s \in R^1)$. Now, consider the following ordinary differential equation instead of (4.63):

$$\tilde{\psi}(v) - v'' = g \quad \text{with } v(0) = v(1) = 0. \tag{4.64}$$

We solve this by the fixed point theorem $(^{12})$ of Schauder. For this, we introduce the following operator $T : X \to X$:

$$(Tv)(x) = \int_0^1 K(x, y)(\tilde{\psi}(v(y)) - g(y)) \, dy \qquad (v \in X = L^1(0, 1)),$$

where $K(x, y) = y(x - 1)$ $(0 \leq y \leq x \leq 1)$ and $K(x, y) = x(y - 1)$ $(0 \leq x \leq y \leq 1)$. Since

$$(Tv)(x) = (x-1) \int_0^x y(\tilde{\psi}(v(y)) - g(y)) \, dy + x \int_x^1 (y-1)(\tilde{\psi}(v(y)) - g(y)) \, dy,$$

we have $(Tv)(0) = (Tv)(1) = 0$, and

$$(Tv)'(x) = \int_0^x y(\tilde{\psi}(v(y)) - g(y)) \, dy + \int_x^1 (y - 1)(\tilde{\psi}(v(y)) - g(y)) \, dy,$$

$$(Tv)''(x) = \tilde{\psi}(v(x)) - g(x).$$

From this, for every $v \in L^1(0, 1)$, Tv and $(Tv)'$ are absolutely continuous on $[0, 1]$, and also

$$\begin{cases} |(Tv)(x)| \leq x(1 - x) \int_0^1 |\tilde{\psi}(v(y)) - g(y)| \, dy \leq \frac{3}{4}\|g\|_1, \\ |(Tv)'(x)| \leq \int_0^1 |\tilde{\psi}(v(y)) - g(y)| \, dy \leq 3\|g\|_1 \qquad (x \in [0, 1]), \end{cases} \tag{4.65}$$

where $|\tilde{\psi}(s)| \leq 2\|g\|_1$ $(s \in R^1)$ is used. Therefore,

$$T(L^1(0, 1)) \subset \{w \in L^1(0, 1) : w \in C^1[0, 1], \ \|w\|_\infty \leq \tfrac{3}{4}\|g\|_1, $$
$$\|w'\|_\infty \leq 3\|g\|_1\} .(^{13})$$

Now, the set of the above right-hand side is sequentially compact as a subset of $C[0, 1]$ by the Ascoli-Arzela theorem. So it is also sequentially compact as a subset of $L(0, 1)$, and is a convex set. Also $T : L^1(0, 1) \to L^1(0, 1)$ is a continuous operator. (In fact, let $\|v_n - v\|_1 \to 0$ as $n \to \infty$, and

---

$(^{12})$ See, for example, Theorem 5 on page 456 of [45].

$(^{13})$ $C^1[0, 1]$ is the space of all functions that are continuously differentiable on $[0, 1]$, and $\|w\|_\infty$ denotes the norm on $L^\infty(0, 1)$.

let $\{n'\}$ be an arbitrary subsequence of $\{n\}$. Since $\|v_{n'} - v\|_1 \to 0$, we can select a subsequence $\{n'_k\}$ of $\{n'\}$ such that $\lim_{k\to\infty} v_{n'_k}(x) = v(x)$ (a.e. x). Since $\tilde{\psi}$ is continuous, $\lim_{k\to\infty} \psi(v_{n'_k}(y)) = \tilde{\psi}(v(y))$ (a.e. y), and $|\tilde{\psi}(v_{n'_k}(y))| \leq 2\|g\|_1$. Therefore, by the convergence theorem we have

$$\|Tv_{n'_k} - Tv\|_1 \leq \int_0^1 |\tilde{\psi}(v_{n'_k}(y)) - \tilde{\psi}(v(y))|\, dy \to 0 \qquad \text{as } k \to 0.$$

After all, we have shown that an arbitrary subsequence $\{Tv_{n'}\}$ of $\{Tv_n\}$ ($\subset L^1(0, 1)$) has a subsequence that converges to (the same limit) Tv. Hence $\|Tv_n - Tv\|_1 \to 0$ as $n \to \infty$.) Therefore, from the fixed point theorem of Schauder, there exists $\tilde{v}$ such that $T\tilde{v} = \tilde{v}$. Then from (4.65) and those we investigated before (4.65) $\tilde{v}$ $(= T\tilde{v})$ and $\tilde{v}'$ $(= (T\tilde{v})')$ are absolutely continuous on $[0, 1]$, and

$$\tilde{\psi}(\tilde{v}) - g \; (= (T\tilde{v})'') = \tilde{v}'',$$
$$\tilde{v}(0) \; (= (T\tilde{v})(0)) = 0 \; (= (T\tilde{v})(1)) = \tilde{v}(1). \qquad (4.66)$$

Also, $|\tilde{v}(x)| \; (= |(T\tilde{v})(x)|) \leq (3/4)\|g\|_1$ $(x \in [0, 1])$ holds. Then from the definition of $\tilde{\psi}$, $\tilde{\psi}(\tilde{v}(x)) = \psi(\tilde{v}(x))$ $(x \in [0, 1])$, i.e., $\tilde{\psi}(\tilde{v}) = \psi(\tilde{v})$. From this and (4.66), $\tilde{v}$ satisfies (4.63) and is a function such that $\tilde{v}$ and $\tilde{v}'$ are absolutely continuous on $[0, 1]$.

EXAMPLE 4.10 (Carleman's equations of hyperbolic type). We consider the following problem given in Example 2.3:

$$\begin{cases} \begin{cases} u_t + u_x + u^2 - v^2 = 0, \\ v_t + v_y + v^2 - u^2 = 0 \qquad (t, x, y > 0), \end{cases} \\ u(0, x, y) = u_0(x, y), \quad v(0, x, y) = v_0(x, y) \qquad (x, y > 0), \\ u(t, 0, y) = v(t, x, 0) = 0 \qquad (x, y > 0). \end{cases} \qquad (4.67)$$

Let $(R^+)^2 = \{(x, y) : x, y \geq 0\}$ and set $X = L^1((R^+)^2) \times L^1((R^+)^2)$. Now, for $M > 0$ we set $D_M(A) = \{[u, v] : u, v, u_x, v_y \text{ are continuous on } (R^+)^2 \text{ and } u(0, y) = v(x, 0) = 0 \; (x, y \geq 0), \; 0 \leq u(x, y), v(x, y) \leq M \; (x, y \geq 0)\}$, and define an operator A by

$$D(A) = \left(\bigcup_{M>0} D_M(A) \right) \cap \{[u, v] \in X : [u_x + u^2 - v^2, v_y + v^2 - u^2] \in X\},$$

$$A[u, v] = [-u_x - u^2 + v^2, -v_y - v^2 + u^2] \qquad ([u, v] \in D(A)).$$

Then, as we saw in Example 2.3, A is a dissipative operator on X. Hence, the closure $\overline{A}$ of A is also a dissipative operator (see the remark after Lemma 2.16). Moreover, if we set $X_0 = \{[u, v] \in X : u, v \geq 0\}$, then $D(\overline{A})$ $(\subset X_0)$ is dense in X_0 and $\overline{A}$ satisfies

$$R(I - \lambda\overline{A}) \supset X_0 \qquad (\lambda > 0). \qquad (4.68)$$

From this, we can apply Theorem 4.2 and generate a semigroup of contractions on X_0 from A. We will use I and II below to prove (4.68).

I. Let f and g be continuous on $(R^+)^2$ with $0 \le f, g \le M$ and $0 < \lambda < 1/(4M)$. Then there exists a unique $[u, v] \in D_M(A)$ satisfying

$$\begin{cases} u + \lambda(u_x + u^2 - v^2) = f, \\ v + \lambda(v_y + v^2 - u^2) = g, \\ u(0, y) = v(x, 0) = 0 \qquad (x, y \ge 0). \end{cases} \qquad (4.69)$$

The proof of this is given in IV below.

II. Let Y be an arbitrary Banach space, let C be a closed convex subset of Y, and let B be a dissipative operator on Y. If $R(I - \mu B) \supset C$ and $(I - \mu B)^{-1}C \subset C$, where $\mu > 0$, then for arbitrary λ with $\lambda > \mu$, we have $R(I - \lambda B) \supset C$ and $(I - \lambda B)^{-1}C \subset C$.

PROOF OF II. Let $\lambda > \mu$ and set $J_\lambda = (I - \lambda B)^{-1}$ and $J_\mu = (I - \mu B)^{-1}$. Now let $x \in C$ and define

$$Tz = J_\mu \left(\frac{\mu}{\lambda}x + \frac{\lambda - \mu}{\lambda}z \right) \qquad (z \in C).$$

Since $T : C \to C$ is a strict contraction operator, it has a unique fixed point $z_0 \in C$. Hence $z_0 = J_\mu((\mu/\lambda)x + ((\lambda - \mu)/\lambda)z_0)$; that is, $(I - \mu B)z_0 \ni (\mu/\lambda)x + (1 - (\mu/\lambda))z_0$. From this, we obtain $x \in (I - \lambda B)z_o \subset R(I - \lambda B)$ and thus $J_\lambda x = z_0 \in C$. Since $x \in C$ is arbitrary, we have proved that $R(I - \lambda B) \supset C$ and $(I - \lambda B)^{-1}C \subset C$.

III (PROOF OF (4.68)). Choose an arbitrary $M > 0$, set $C_M = \{[f, g] \in X : 0 \le f, g \le M\}$, and let $0 < \lambda < 1/(4M)$. Then from I, for an arbitrary $[f, g] \in C_M \cap [C((R^+)^2) \times C((R^+)^2)]([14])$, there exists a unique $[u, v] \in D_M(A)$ that satisfies (4.69). Then note that $[u, v] \in X$, that is, $u, v \in L^1((R^+)^2)$. (This follows by the same reasoning as in the proof of (2.30)—set $[u_1, v_1] = [u, v]$ and $[u_2, v_2] = [0, 0]$, and apply the proof—to obtain

$$\int_0^T \int_0^T [|u| + |v|] \, dx \, dy \le \int_0^T \int_0^T |u + \lambda(u_x + u^2 - v^2)| \, dx \, dy$$
$$+ \int_0^T \int_0^T |v + \lambda(v_y + v^2 - u^2)| \, dx \, ds$$
$$= \int_0^T \int_0^T [|f| + |g|] \, dx \, dy$$
$$\le \int_0^\infty \int_0^\infty [|f| + |g|] \, dx \, dy < \infty \qquad (T > 0).)$$

Then we have $[u, v] \in D(A)$, $(I - \lambda A)[u, v] = [f, g]$, and $(I - \lambda A)^{-1}[f, g] = [u, v] \in D_M(A) \cap X \subset C_M$. Hence $R(I - \lambda A) \supset C_M \cap [C((R^+)^2) \times C((R^+)^2)]$

([14]) We denote the set of all continuous functions on $(R^+)^2$ by $C((R^+)^2)$.

and $(I - \lambda A)^{-1}(C_M \cap [C((R^+)^2) \times C((R^+)^2)] \subset C_M$. Note that $R(I - \lambda\overline{A})$ is a closed set (see the first part of the proof of Theorem 4.5) and $(I - \lambda\overline{A})^{-1}$ is a continuous operator. Thus, it follows that

$$R(I - \lambda\overline{A}) \supset C_M \quad \text{and} \quad (I - \lambda\overline{A})^{-1} C_M \subset C \quad (0 < \lambda < 1/(4M)).$$

Since C_M is a closed convex subset of X, from II we have $R(I - \lambda\overline{A}) \supset C_M$ for every $\lambda > 0$. Therefore, we obtain

$$R(I - \lambda\overline{A}) \supset \bigcup_{M>0} C_M \quad (\lambda > 0).$$

From this and since $\bigcup_{M>0} C_M$ is dense in X_0, we obtain (4.68). Finally, if we consider the set of all pairs of continuously differentiable, nonnegative functions that have compact support in the interior of $(R^+)^2$. Since this set is contained in $D(A)$ and is dense in X_0, it follows that $D(A)$ $(\subset X_0)$ is dense in X_0 and hence $D(\overline{A})$ is also dense in X_0.

Lastly,

IV. (PROOF OF I). $C[(R^+)^2]$ denotes the Banach space of all bounded and continuous real functions defined on $(R^+)^2$ (the norm of $u \in C[(R^+)^2]$ is $\|u\| = \sup\{|u(x, y)| : (x, y) \in (R^+)^2\}$). Now let $\varphi_M(s) = s^2$ $(|s| \leq M)$ and $\varphi_M(s) = M^2$ $(|s| > M)$, and set $\kappa = 1/\lambda$. Then, there is a unique pair $u, v \in C[(R^+)^2]$ satisfying

$$\begin{cases} u(x, y) = \displaystyle\int_0^x e^{\kappa(s-x)}[\kappa f(s, y) + \varphi_M(v(s, y)) - \varphi_M(u(s, y))] \, ds, \\ v(x, y) = \displaystyle\int_0^y e^{\kappa(s-y)}[\kappa g(x, s) + \varphi_M(u(x, s)) - \varphi_M(v(x, s))] \, ds. \end{cases} \quad (4.70)$$

In fact, for $w = [u, v] \in C[(R^+)^2] \times C[(R^+)^2]$, define $Tw = [\tilde{u}, \tilde{v}]$ by

$$\tilde{u}(x, y) = \int_0^x e^{\kappa(s-x)}[\kappa f(s, y) + \varphi_M(v(s, y)) - \varphi_M(u(s, y))] \, ds,$$

$$\tilde{v}(x, y) = \int_0^y e^{\kappa(s-y)}[\kappa g(x, s) + \varphi_M(u(x, s)) - \varphi_M(v(x, s))] \, ds.$$

Obviously, T is an operator from $C[(R^+)^2] \times C[(R^+)^2]$ into itself. From $|\varphi_M(s) - \varphi_M(t)|] \leq 2M|s - t|$ $(-\infty < s, t < \infty)$, we obtain

$$\|Tw_1 - Tw_2\| \leq 4M\lambda\|w_1 - w_2\| \quad (w_1, w_2 \in C[(R^+)^2] \times C[(R^+)^2]).$$

Since $0 < \lambda < 1/(4M)$, T is a strict contraction operator from $C[(R^+)^2] \times C[(R^+)^2]$ into itself. Hence by the fixed point theorem, there exists a unique $w = [u, v] \in C[(R^+)^2] \times C[(R^+)^2]$ such that $Tw = w$. Therefore, we have shown that there exists a unique pair $u, v \in C[(R^+)^2]$ satisfying (4.70).

Next, we show that the solution $u, v \in C[(R^+)^2]$ of (4.70) satisfies

$$0 \leq u(x, y) \leq M, \quad 0 \leq v(x, y) \leq M \quad ((x, y) \in (R^+)^2). \quad (4.71)$$

From (4.70) we have

$$\begin{cases} u_x(x, y) = -\kappa u(x, y) + \kappa f(x, y) + \varphi_M(v(x, y)) - \varphi_M(u(x, y)), \\ v_y(x, y) = -\kappa v(x, y) + \kappa g(x, y) + \varphi_M(u(x, y)) - \varphi_M(v(x, y)). \end{cases} \tag{4.72}$$

Now we choose $y \geq 0$ arbitrarily, and set $I = \{x \geq 0 : u(x, y) < 0\}$ and $J = \{x \geq 0 : u(x, y) > M\}$. We will show that $I = \varnothing$ and $J = \varnothing$. First, suppose $I \neq \varnothing$; then since $0 \notin I$ from $u(0, y) = 0$, I becomes an (nonempty) open set of R^1. Hence I may be expressed as the set sum of at most a countable number of open intervals $\{(a_k, b_k)\}$ that are pairwise disjoint. If we take $x_0 \in (a_1, b_1)$ $(\subset I)$ and integrate the first equation of (4.72) over $[a_1, x_0]$, then from $u(a_1, y) = 0$, $f \geq 0$, and $\varphi_M \geq 0$, we have

$$u(x_0, y) \geq \kappa \int_{a_1}^{x_0} [-u(x, y)] dx - \int_{a_1}^{x_0} \varphi_M(u(x, y)) dx.$$

Since $\kappa > 4M$ (> 0), $-u(x, y) \geq 0$, $\varphi_M(u(x, y)) \leq M|u(x, y)| = -Mu(x, y)$ $(x \in [a_1, x_0])$, from the inequality above we have

$$u(x_0, y) \geq 4M \int_{a_1}^{x_0} [-u(x, y)] dx + M \int_{a_1}^{x_0} u(x, y) dx$$
$$= 3M \int_{a_1}^{x_0} [-u(x, y)] dx \geq 0.$$

On the other hand, $u(x_0, y) < 0$ because $x_0 \in I$. This is a contradiction. Therefore $I = \varnothing$. Next, if we suppose $J \neq \varnothing$, then J becomes an (nonempty) open subset of R^1 and there exists a sequence $\{(\alpha_k, \beta_k)\}$ of at most a countable number of open intervals which are pairwise disjoint, such that $J = \bigcup_k (\alpha_k, \beta_k)$. We take $x_1 \in (\alpha_1, \beta_1)$ $(\subset J)$ and integrate the first equation of (4.72) on $[\alpha_1, x_1]$. Noting that $u(\alpha_1, y) = M$, $-u(x, y) \leq -M$, $\varphi_M(u(x, y)) = M^2$ $(x \in [\alpha_1, x_1])$, $f \leq M$, and $\varphi_M \leq M^2$, we obtain

$$u(x_1, y) - M = -\kappa \int_{\alpha_1}^{x_1} u(x, y) dx + \kappa \int_{\alpha_1}^{x_1} f(x, y) dx$$
$$+ \int_{\alpha_1}^{x_1} \varphi_M(v(x, y)) dx - \int_{\alpha_1}^{x_1} \varphi_M(u(x, y)) dx$$
$$\leq -\kappa M(x_1 - \alpha_1) + \kappa M(x_1 - \alpha_1) + M^2(x_1 - \alpha_1) - M^2(x_1 - \alpha_1)$$
$$= 0$$

Hence, $u(x_1, y) \leq M$. This contradicts $x_1 \in J$. Therefore $J = \varnothing$. Now, since $I = \varnothing$ and $J = \varnothing$, $0 \leq u(x, y) \leq M$ $(x \geq 0)$ for every $y \geq 0$. Similarly, from the second equation in (4.72), we obtain $0 \leq v(x, y) \leq M$ $(x, y \geq 0)$. Therefore (4.71) is proved.

As we have shown, there is a unique pair $u, v \in C[(R^+)^2]$ that satisfies (4.70) and this pair also satisfies (4.71). Moreover, the fact that the pair

u, $v \in C[(R^+)^2]$ is the solution of (4.70) is equivalent to saying the pair u, $v \in C[(R^+)^2]$ is the solution of (4.72) satisfying $u(0, y) = v(x, 0) = 0$ $(x, y \geq 0)$. Therefore, we have shown that there exists a unique solution u, $v \in C[(R^+)^2]$ of (4.72) satisfying $u(0, y) = v(x, 0) = 0$ $(x, y \geq 0)$, and this solution also satisfies (4.71). Now since u and v satisfy (4.71), from the definition of φ_M, we have $\varphi_M(v(x, y)) = v(x, y)^2$ and $\varphi_M(u(x, y)) = u(x, y)^2$, and hence $[u, v] \in D_M(A)$ is a unique solution of (4.69).

CHAPTER 5

Cauchy Problems for Evolution Equations

In this chapter we let X be a real or complex Banach space and consider Cauchy's problems mainly for the evolution equation $d(/dt)u(t) \in Au(t)$.

§1. Integral solutions of Cauchy problems

Let A be an (not necessarily single-valued) operator with $D(A) \subset X$, and $R(A) \subset X$, and let T be a positive number. We consider the Cauchy problems for A:

$$(\mathrm{CP};x) \qquad \begin{cases} (d/dt)u(t) \in Au(t) & (0 \le t \le T), \\ u(0) = x, \end{cases}$$

and

$$(\mathrm{CP};x)_\infty \qquad \begin{cases} (d/dt)u(t) \in Au(t) & (0 \le t < \infty), \\ u(0) = x, \end{cases}$$

where $x \ (\in X)$ is a given element.

Let A be a closed dissipative operator satisfying

$$R(I - \lambda A) \supset D(A) \quad \text{for all } \lambda > 0. \tag{R_1}$$

Then, by $T(t)x = \lim_{\lambda \to 0+}(I - \lambda A)^{-[t/\lambda]}x \ (x \in \overline{D(A)}, \ t \ge 0)$, we can construct a semigroup of contractions $\{T(t) : t \ge 0\}$ on $\overline{D(A)}$ (see Theorem 4.2). Now, if $T(t)x \ (x \in D(A))$ is strongly differentiable for a.e. $t \ge 0$, then $T(t)x \colon [0, \infty) \to X$ is a unique solution of the Cauchy problem $(\mathrm{CP};x)_\infty$ (see Theorem 4.10). Conversely, the following theorem holds.

THEOREM 5.1. *Let A be a closed dissipative operator satisfying (R_1) and $x \in \overline{D(A)}$. If $u(t) \colon [0, \infty) \to X$ is a solution of the Cauchy problem $(\mathrm{CP};x)_\infty$, then*

$$u(t) = \lim_{\lambda \to 0+}(I - \lambda A)^{[-t/\lambda]}x \qquad (t \ge 0).$$

PROOF. Choose an arbitrary $s > 0$, let $0 < 2\lambda < s$, and set $u_\lambda(t) = (I - \lambda A)^{[-t/\lambda]}x \ (t \ge 0)$ and $\lambda^{-1}[u(t) - u(t - \lambda)] - u'(t) = g_\lambda(t)$ (a.e. $t \ge \lambda$). Since $\lim_{\lambda \to 0+} g_\lambda(t) = 0$ (a.e. $t > 0$) and $\|g_\lambda(t)\| \le 2M$ $(\lambda \le t \le s)$ (where M is a Lipschitz constant of $u(t)$ on $[0, s]$), $\lim_{\lambda \to 0+} \int_\lambda^s \|g_\lambda(t)\| \, dt = 0$.

Next, from $u(t - \lambda) + \lambda g_\lambda(t) = u(t) - \lambda u'(t) \in (I - \lambda A)u(t)$, we have $u(t) = (I - \lambda A)^{-1}[u(t - \lambda) + \lambda g_\lambda(t)]$. Hence,

$$\|u_\lambda(t) - u(t)\| \leq \|(I - \lambda A)^{-[(t-\lambda)/\lambda]}x - u(t - \lambda) - \lambda g_\lambda(t)\|$$
$$\leq \|u_\lambda(t - \lambda) - u(t - \lambda)\| + \lambda \|g_\lambda(t)\| \quad (\text{a.e. } t \geq \lambda).$$

If we integrate this over $[\lambda, s]$, then

$$\lambda^{-1} \int_{s-\lambda}^{s} \|u_\lambda(t) - u(t)\| \, dt \leq \lambda^{-1} \int_0^\lambda \|u_\lambda(t) - u(t)\| \, dt + \int_\lambda^s \|g_\lambda(t)\| \, dt.$$

As $\lambda \to 0+$, $u_\lambda(t)$ converges uniformly to $T(t)x$ $(= \lim_{\lambda \to 0+}(I - \lambda A)^{-[t/\lambda]}x)$ on $[0, s]$. Since $\|T(t)x - u(t)\|$ is continuous on $[0, \infty)$, if we let $\lambda \to 0+$ in the above inequality, then we obtain

$$\|T(s)x - u(s)\| \leq 0.$$

Therefore, $u(s) = T(s)x = \lim_{\lambda \to 0+}(I - \lambda A)^{-[s/\lambda]}x$ $(s > 0)$. $\square$

However, the semigroup of contractions, formulated on $\overline{D(A)}$ through the method of Theorem 4.2 with closed dissipative operator A satisfying (R_1), does not necessarily always give a solution of the Cauchy problem for A. In fact, the semigroup of contractions $\{T(t) : t \geq 0\}$ on $C[-1, 1]$, described in Example 4.7, is constructed using $T(t)x = \lim_{\lambda \to 0+}(I - \lambda A)^{-[t/\lambda]}x$ $(x \in C[-1, 1], t \geq 0)$ from an m-dissipative operator A with $\overline{D(A)} = C[-1, 1]$, and yet, for every $x \in C[-1, 1]$, $T(t)x$ is not weakly differentiable for all $t \geq 0$. Hence, for any $x \in \overline{D(A)}$, $T(t)x$ is not a solution of $(\text{CP}; x)_\infty$. Also, from this and Theorem 5.1, for this m-dissipative operator A, we see that, for any $x \in \overline{D(A)}$, the Cauchy problem $(\text{CP}; x)_\infty$ (for A) does not have a solution.

Therefore, it is interesting to generalize the concept of solutions of $(\text{CP}; x)$ and $(\text{CP}; x)_\infty$ in such a way that a semigroup of contractions on $\overline{D(A)}$, which can be formulated through the method of Theorem 4.2 with a dissipative operator A satisfying (R_1), results in giving "a generalized solution in a sense" of the Cauchy problem for A.

DEFINITION 5.1. Let $x \in X$ and ω be a given real number. $u(t): [0, T] \to X$ is called an integral solution of type ω of the Cauchy problem $(\text{CP}; x)$ (for A) if $u(t)$ satisfies (i) through (iii):

(i) $u(0) = x$.

(ii) $u(t)$ is continuous on $[0, T]$.

(iii) For every $r, t \in [0, T]$ with $r < t$, and every $x_0 \in D(A)$, $y_0 \in Ax_0$

$$e^{-2\omega t}\|u(t) - x_0\|^2 - e^{-2\omega r}\|u(r) - x_0\|^2 \leq 2\int_r^t e^{-2\omega\xi}\langle y_0, u(\xi) - x_0\rangle_s \, d\xi. \quad (5.1)$$

$u(t): [0, \infty) \to X$ is called an integral solution of type ω of the Cauchy problem $(\text{CP}; x)_\infty$ if for every $T > 0$, the restriction $u_T(t)$ of $u(t)$ to $[0, T]$ is an integral solution of type ω of the Cauchy problem $(\text{CP}; x)$ on $[0, T]$.

REMARK. We sometimes call the above definition with (5.1) replaced by

$$e^{-\omega t}\|u(t) - x_0\| - e^{-\omega r}\|u(r) - x_0\| \leq \int_r^t e^{-\omega\xi}\tau_+(y_0, u(\xi) - x_0)\, d\xi \quad (5.1)'$$

an integral solution of type ω of the Cauchy problem $(\mathrm{CP}; x)$, but, in fact these two definitions are equivalent. We show this next.

$(5.1)' \Rightarrow (5.1)$: As we will prove the following in the remark below, "let $f(t)$ be a continuous and nonnegative function on $[0, T]$, and $g(t)$ be a (Lebesgue) integrable real-valued function on $[0, T]$. If $f(t) - f(s) \leq \int_s^t g(\xi)\, d\xi$ for every $s, t \in [0, T]$ with $s < t$, then $f(t)^2 - f(s)^2 \leq 2\int_s^t f(\xi)g(\xi)\, d\xi$ holds for every $s, t \in [0, T]$ with $s < t$."

Now, since $\langle y_0, u(\xi) - x_0\rangle_s = \|u(\xi) - x_0\|\tau_+(y_0, u(\xi) - x_0)$ (see (2.14)), we put $f(\xi) = e^{-\omega\xi}\|u(\xi) - x_0\|$ and $g(\xi) = e^{-\omega\xi}\tau_+(y_0, u(\xi) - x_0)$ $(0 \leq \xi \leq T)$, and apply the above result.

$(5.1) \Rightarrow (5.1)'$: From $|\langle y_0, u(\xi) - x_0\rangle_s| \leq \|y_0\|\|u(\xi) - x_0\|$ and (5.1),

$$(e^{-\omega t}\|u(t) - x_0\|)^2 - (e^{-\omega r}\|u(r) - x_0\|)^2 \leq 2\int_r^t \|y_0\|K e^{-\omega\xi}\|u(\xi) - x_0\|\, d\xi$$

$(0 \leq r \leq t \leq T)$, where $K = \max\{e^{-\omega\xi} : 0 \leq \xi \leq T\}$. Then, using the remark after Lemma 3.4, we obtain

$$e^{-\omega t}\|u(t) - x_0\| - e^{-\omega r}\|u(r) - x_0\| \leq \int_r^t \|y_0\|K\, d\xi$$
$$= \|y_0\|K(t - r) \quad (0 \leq r \leq t \leq T).$$

Now if we set $f(\xi) = e^{-\omega\xi}\|u(\xi) - x_0\| - \|y_0\|K\xi$ $(0 \leq \xi \leq T)$, then from the inequality above, $f(\xi)$ is a monotone decreasing function. Hence, $f(\xi)$ is differentiable a.e. $\xi \in [0, T]$ and its derivative $f'(\xi)$ is integrable on $[0, T]$ and

$$f(t) - f(r) \leq \int_r^t f'(\xi)\, d\xi \quad (0 \leq r \leq t \leq T).$$

From this, $e^{-\omega\xi}\|u(\xi) - x_0\|$ is also differentiable a.e. $\xi \in [0, T]$ and

$$(d/d\xi)e^{-\omega\xi}\|u(\xi) - x_0\| = f'(\xi) + \|y_0\|K \quad \text{a.e. } \xi \in [0, T].$$

From this and the inequality above, we obtain

$$e^{-\omega t}\|u(t) - x_0\| - e^{-\omega r}\|u(r) - x_0\| \leq \int_r^t (d/d\xi)e^{-\omega\xi}\|u(\xi) - x_0\|\, d\xi$$

$(0 \leq r \leq t \leq T)$.

Next, at a point r where $e^{-\omega r}\|u(r) - x_0\|$ is differentiable, we have that $e^{-2\omega r}\|u(r) - x_0\|^2$ is also differentiable, and since its derivative is

$$2e^{-\omega r}\|u(r) - x_0\|(d/dr)e^{-\omega r}\|u(r) - x_0\|,$$

from (5.1) we have

$$2e^{-\omega r}\|u(r) - x_0\|(d/dr)e^{-\omega r}\|u(r) - x_0\| \le 2e^{-2\omega r}\langle y_0, u(r) - x_0\rangle_s$$
$$= 2e^{-2\omega r}\|u(r) - x_0\|\tau_+(y_0, u(r) - x_0) \quad \text{(a.e. } r \in [0, T]).$$

Next we consider the set $E = \{r \in [0, T] : \|u(r) - x_0\| = 0\}$. Since the set of isolated points of E is at most a countable set, its measure is 0. Also, if $e^{-\omega r}\|u(r) - x_0\|$ is differentiable at a limiting point r_0 of E, then the derivative of $e^{-\omega r}\|u(r) - x_0\|$ at r_0 is 0 and we have

$$e^{-\omega r_0}\tau_+(y_0, u(r_0) - x_0) = e^{-\omega r_0}\tau_+(y_0, 0) = e^{-\omega r_0}\|y_0\| \ge 0$$
$$= [(d/dr)e^{-\omega r}\|u(r) - x_0\|]_{r=r_0}.$$

Thus,

$$(d/dr)e^{-\omega r}\|u(r) - x_0\| \le e^{-\omega r}\tau_+(y_0, u(r) - x_0) \quad \text{(a.e. } r \in [0, T]).$$

Integrating this over $[r, t]$ $(0 \le r \le t \le T)$ we obtain

$$\int_r^t d/d\xi)e^{-\omega\xi}\|u(\xi) - x_0\|\,d\xi$$
$$\le \int_r^t e^{-\omega\xi}\tau_+(y_0, u(\xi) - x_0)\,d\xi \quad (0 \le r \le t \le T).$$

From this and $e^{-\omega r}\|u(t) - x_0\| - e^{-\omega r}\|u(r) - x_0\| \le \int_r^t (d/d\xi)e^{-\omega\xi}\|u(\xi) - x_0\|\,d\xi$ $(0 \le r \le t \le T)$, it follows that $(5.1)'$ holds.

REMARK. Let $f(t)$ be a continuous nonnegative function on $[0, T]$ and let $g(t)$ be an integrable real-valued function on $[0, T]$. If for every s, $t \in [0, T]$ with $s < t$, $f(t) - f(s) \le \int_s^t g(\xi)\,d\xi$ holds, then we have

$$f(t)^2 - f(s)^2 \le 2\int_s^t f(\xi)g(\xi)\,d\xi$$

for every s, $t \in [0, T]$ with $s < t$. To prove this, we define $\varphi(\tau): R^1 \to [0, \infty)$ and $\psi(\tau): R^1 \to R^1$ as follows:

$$\varphi(\tau) = \begin{cases} f(\tau) & (\tau \in [0, T]), \\ 0 & (\tau \notin [0, T]), \end{cases} \qquad \psi(t) = \begin{cases} g(\tau) & (\tau \in [0, T]), \\ 0 & (\tau \notin [0, T]). \end{cases}$$

Choose an infinitely differentiable nonnegative function $\rho(\tau)$ on $R^1 = (-\infty, \infty)$ such that $\rho(\tau) = 0$ $(|\tau| > 1)$ and $\int_{-\infty}^\infty \rho(\tau)\,d\tau = 1$, and for every $\varepsilon > 0$, define

$$\rho_\varepsilon(\tau) = \varepsilon^{-1}\rho(\tau/\varepsilon)$$

and

$$\varphi_\varepsilon(\tau) = \int_{-\infty}^\infty \rho_\varepsilon(\sigma)\varphi(\tau - \sigma)\,d\sigma, \qquad \psi_\varepsilon(\tau) = \int_{-\infty}^\infty \rho_\varepsilon(\sigma)\psi(\tau - \sigma)\,d\sigma \qquad (\tau \in R^1).$$

Note that $\varphi_\varepsilon(\tau)$, $\psi_\varepsilon(\tau) \in C^\infty(R^1)$ and

$$\lim_{\varepsilon \to 0+} \varphi_\varepsilon(\tau) = \varphi(\tau) = f(\tau) \qquad (\tau \in (0, T)).$$

Let $0 < 2\varepsilon < T$ and $\varepsilon \le s < t \le T - \varepsilon$. Then from the assumption, for ξ such that $|\xi| \le \varepsilon$, we have

$$f(t - \xi) - f(s - \xi) \le \int_s^t g(\tau - \xi)\, d\tau.$$

By multiplying both sides of this inequality by $\rho_\varepsilon(\xi)$ and integrating with respect to ξ on $[-\varepsilon, \varepsilon]$, we have

$$\varphi_\varepsilon(t) - \varphi_\varepsilon(s) \le \int_s^t \psi_\varepsilon(\tau)\, d\tau.$$

Dividing both sides by $t - s$ and letting $s \to t - 0$, we have $(d/dt)\varphi_\varepsilon(t) \le \psi_\varepsilon(t)$. Hence we obtain

$$(d/dt)\varphi_\varepsilon(t)^2 \le 2\varphi_\varepsilon(t)\psi_\varepsilon(t) \qquad (t \in [\varepsilon, T - \varepsilon]).$$

Next we integrate this over $[s, t]$ and have

$$\varphi_\varepsilon(t)^2 - \varphi_\varepsilon(s)^2 \le 2 \int_s^t \varphi_\varepsilon(\tau)\psi_\varepsilon(\tau)\, d\tau \qquad (\varepsilon \le s \le t \le T - \varepsilon).$$

If we let $\varepsilon \to 0+$, then

$$|\varphi_\varepsilon(\tau)| \le M \ (\tau \in R^1), \quad \text{where } M = \sup\{|f(\tau)| : 0 \le \tau \le T\},$$

and from

$$\left| \int_s^t \varphi_\varepsilon(\tau)\psi_\varepsilon(\tau)\, d\tau - \int_s^t \varphi_\varepsilon(\tau)\psi(\tau)\, d\tau \right|$$

$$\le \int_{-1}^1 \rho(\sigma) \left[\int_s^t |\varphi_\varepsilon(\tau)(\psi(\tau - \varepsilon\sigma) - \psi(\tau))|\, d\tau \right] d\sigma$$

$$\le M \int_{-1}^1 \rho(\sigma) \left[\int_s^t |\psi(\tau - \varepsilon\sigma) - \psi(\tau)|\, d\tau \right] d\sigma \to 0,$$

$$\int_s^t \varphi_\varepsilon(\tau)\psi(\tau)\, d\tau \to \int_s^t \varphi(\tau)\psi(\tau)\, d\tau$$

we have

$$\varphi(t)^2 - \varphi(s)^2 \le 2 \int_s^t \varphi(\tau)\psi(\tau)\, d\tau \quad (0 < s \le t < T);$$

that is, we obtain

$$f(t)^2 - f(s)^2 \le 2 \int_s^t f(\tau)g(\tau)\, d\tau \qquad (0 < s \le t < T).$$

Here if we let $s \to 0+$ and $t \to T - 0$, then the inequality above holds for every s, $t \in [0, T]$ with $s \le t$.

THEOREM 5.2. *Let* $u(t): [0, T] \to X$ *be continuous and let* $u(0) = x \in X$. *Then a necessary and sufficient condition for* $u(t)$ *to be an integral solution of type* ω *of* (CP; x) *is that for every* r, $t \in [0, T]$ *with* $r < t$, *and every* $x_0 \in D(A)$, $y_0 \in Ax_0$

$$\|u(t) - x_0\|^2 - \|u(r) - x_0\|^2$$
$$\leq 2 \int_r^t \langle y_0, u(\xi) - x_0 \rangle_s \, d\xi + 2\omega \int_r^t \|u(\xi) - x_0\|^2 \, d\xi. \tag{5.2}$$

PROOF. Since the result is self-evident when $\omega = 0$, we consider only the case $\omega \neq 0$.

(Necessity) Let $u(t)$ be an integral solution of type ω of (CP; x), and $x_0 \in D(A)$, $y_0 \in Ax_0$. First, consider the case $\omega > 0$. For $0 \leq r < t \leq T$, from (5.1) we have

$$e^{-2\omega(t-r)}\|u(t) - x_0\|^2 - \|u(r) - x_0\|^2 \leq 2 \int_r^t e^{-2\omega(\xi-r)} \langle y_0, u(\xi) - x_0 \rangle_s \, d\xi.$$

By integrating this with respect to r over $[r, t]$ we have

$$(2\omega)^{-1}(1 - e^{-2\omega(t-r)})\|u(t) - x_0\|^2 - \int_r^t \|u(\xi) - x_0\|^2 \, d\xi$$
$$\leq 2 \int_r^t \left[\int_\eta^t e^{-2\omega(\xi-\eta)} \langle y_0, u(\xi) - x_0 \rangle_s \, d\xi \right] d\eta$$
$$= 2 \int_r^t \langle y_0, u(\xi) - x_0 \rangle_s \left[\int_r^\xi e^{-2\omega(\xi-\eta)} \, d\eta \right] d\xi$$
$$= 2 \int_r^t (2\omega)^{-1}(1 - e^{-2\omega(\xi-r)}) \langle y_0, u(\xi) - x_0 \rangle_s \, d\xi.$$

That is, we obtain

$$2 \int_r^t e^{-2\omega(\xi-r)} \langle y_0, u(\xi) - x_0 \rangle_s \, d\xi$$
$$\leq 2 \int_r^t \langle y_0, u(\xi) - x_0 \rangle_s \, d\xi + 2\omega \int_r^t \|u(\xi) - x_0\|^2 \, d\xi$$
$$+ (e^{-2\omega(t-r)} - 1)\|u(t) - x_0\|^2.$$

From this and the first inequality, we obtain (5.2). Next let $\omega < 0$. For $0 \leq r < t \leq T$, from (5.1) we have

$$\|u(t) - x_0\|^2 - e^{2\omega(t-r)}\|u(r) - x_0\|^2 \leq 2 \int_r^t e^{2\omega(t-\xi)} \langle y_0, u(\xi) - x_0 \rangle_s \, d\xi. \tag{5.3}$$

By integrating this with respect to t over $[r, t]$, we have

$$\int_r^t \|u(\xi) - x_0\|^2 \, d\xi - (2\omega)^{-1}(e^{2\omega(t-r)} - 1)\|u(r) - x_0\|^2$$
$$\leq 2 \int_r^t (2\omega)^{-1}(e^{2\omega(t-\xi)} - 1) \langle y_0, u(\xi) - x_0 \rangle_s \, d\xi.$$

Hence we obtain

$$2 \int_r^t e^{2\omega(t-\xi)} \langle y_0, u(\xi) - x_0 \rangle_s \, d\xi$$

$$\leq 2 \int_r^t \langle y_0, u(\xi) - x_0 \rangle_s \, d\xi + 2\omega \int_r^t \|u(\xi) - x_0\|^2 \, d\xi$$
$$- (e^{2\omega(t-r)} - 1) \|u(r) - x_0\|^2,$$

and from this and (5.3), (5.2) holds.

(Sufficiency). Let $x_0 \in D(A)$, $y_0 \in Ax_0$, and assume $u(t)$ satisfies (5.2). Consider the case $\omega > 0$. For $0 \leq r < t \leq T$, from (5.2) we have

$$e^{-2\omega t} (\|u(t) - x_0\|^2 - \|u(r) - x_0\|^2) \tag{5.4}$$

$$\leq 2e^{-2\omega t} \int_r^t \langle y_0, u(\xi) - x_0 \rangle_s \, d\xi + 2\omega e^{-2\omega t} \int_r^t \|u(\xi) - x_0\|^2 \, d\xi;$$

that is,

$$(d/dt) \left[e^{-2\omega t} \int_r^t \|u(\xi) - x_0\|^2 \, d\xi \right] - e^{-2\omega t} \|u(r) - x_0\|^2$$

$$\leq 2e^{-2\omega t} \int_r^t \langle y_0, u(\xi) - x_0 \rangle_s \, d\xi.$$

By integrating this with respect to t over $[r, t]$, we obtain the next equation:

$$2\omega e^{-2\omega t} \int_r^t \|u(\xi) - x_0\|^2 \, d\xi + (e^{-2\omega t} - e^{-2\omega r}) \|u(r) - x_0\|^2$$

$$\leq 2 \int_r^t e^{-2\omega \xi} \langle y_0, u(\xi) - x_0 \rangle_s \, d\xi - 2e^{-2\omega t} \int_r^t \langle y_0, u(\xi) - x_0 \rangle_s \, d\xi.$$

By adding respective sides of this to (5.4), we obtain (5.1). Now assume $\omega < 0$. By multiplying both sides of (5.2) by $e^{-2\omega r}$, we have

$$e^{-2\omega r} [\|u(t) - x_0\|^2 - \|u(r) - x_0\|^2] \tag{5.5}$$

$$\leq 2e^{-2\omega r} \int_r^t \langle y_0, u(\xi) - x_0 \rangle_s \, d\xi + 2\omega e^{-2\omega r} \int_r^t \|u(\xi) - x_0\|^2 \, d\xi$$

for $0 \leq r < t \leq T$. That is,

$$(d/dr) \left[e^{-2\omega r} \int_r^t \|u(\xi) - x_0\|^2 \, d\xi \right] + e^{-2\omega r} \|u(t) - x_0\|^2$$

$$\leq 2e^{-2\omega r} \int_r^t \langle y_0, u(\xi) - x_0 \rangle_s \, d\xi.$$

By integrating this with respect to r over $[r, t]$ and rearranging, we have

$$(e^{-2\omega t} - e^{-2\omega r}) \|u(t) - x_0\|^2 + 2\omega e^{-2\omega r} \int_r^t \|u(\xi) - x_0\|^2 \, d\xi$$

$$\leq 2 \int_r^t e^{-2\omega \xi} \langle y_0, u(\xi) - x_0 \rangle_s \, d\xi - 2e^{-2\omega r} \int_r^t \langle y_0, u(\xi) - x_0 \rangle_s \, d\xi.$$

By adding the respective sides of this to (5.5), we obtain (5.1). □

From this, we can derive the following corollary.

COROLLARY 5.3. *Let* $\omega_1 < \omega_2$. *An integral solution of type* ω_1 *of* $(CP; x)$ *is always an integral solution of type* ω_2 *of the same Cauchy problem. Also, the same holds for* $(CP; x)_\infty$.

Now we examine the relation between the integral solution and the semigroup.

THEOREM 5.4. *Let* X_0 *be a (nonempty) subset of* X. *Suppose that for each* $x \in X_0$, *the Cauchy problem* $(CP; x)_\infty$ *has a unique integral solution* $u(t; x) \in X_0$ $(t \geq 0)$ *of type* ω. *If we set*

$$T(t)x = u(t; x) \qquad (x \in X_0, t \geq 0),$$

then $\{T(t) : t \geq 0\}$ *is a semigroup on* X_0, *that is,* $T(0)x = x$ $(x \in X_0)$, $\lim_{t \to s} T(t)x = T(s)x$ $(s \geq 0, x \in X_0)$, *and* $T(t+s) = T(t)T(s)$ $(t, s \geq 0)$ *hold.*

PROOF. We give and fix arbitrary $x \in X_0$ and $s \geq 0$. Since $u(s; x) \in X_0$, there exists an integral solution $u(t; u(s; x))$ of type ω of the Cauchy problem $(CP; u(s; x))_\infty$ (for A) with $u(s; x)$ as its initial value. Now, if we set

$$v(t) = \begin{cases} u(t; x) & (0 \leq t \leq s), \\ u(t - s; u(s; x)) & (s < t < \infty), \end{cases}$$

then we see easily that $v(t): [0, \infty) \to X_0$ is an integral solution of type ω of $(CP; x)_\infty$. Since $u(t; x): [0, \infty) \to X_0$ is a unique integral solution of type ω of $(CP; x)_\infty$, it follows that $v(t) = u(t; x)$ $(t \geq 0)$, and hence we obtain $u(t; x) = u(t - s; u(s; x))$ $(s < t)$. Thus $T(t + s)x = T(t)T(s)x$ $(t, s \geq 0, x \in X_0)$ holds. Also, since $u(t; x): [0, \infty) \to X_0$ is continuous, $\lim_{t \to s} T(t)x = T(s)x$ $(s \geq 0, x \in X_0)$. $\square$

Here we consider a semigroup of contractions $\{T(t) : t \geq 0\}$ on $\overline{D(A)}$ which is formulated through Theorem 4.2 with a dissipative operator A satisfying condition (R_1). Then from (4.15), we see that for every $x \in \overline{D(A)}$, $T(t)x: [0, \infty) \to \overline{D(A)}$ is an integral solution of type 0 of the Cauchy problem $(CP; x)_\infty$ for A. Moreover, as we show later, $T(t)x$ is the unique integral solution of type 0 of $(CP; x)_\infty$ (see Corollary 5.13 and the remark following this corollary).

Finally, we examine the relation between the solution of the Cauchy problem for A and the integral solution of type ω, when $A - \omega I$ is a dissipative operator.

THEOREM 5.5. *Let* $A - \omega I$ $(\omega$ *a real number*$)$ *be a dissipative operator. Then, solutions of the Cauchy problems* $(CP; x)$ *and* $(CP; x)_\infty$ *for* A *are always integral solutions of type* ω *of the same Cauchy problem.*

PROOF. Let $u(t): [0, T] \to X$ be a solution of $(CP; x)$, and let $x_0 \in D(A)$, $y_0 \in Ax_0$. Then, from Lemma 2.8 we have

$$(d/dr)\|u(r) - x_0\|^2 = 2\langle (d/dr)u(r), u(r) - x_0 \rangle_i \qquad (\text{a.e. } r \in [0, T]).$$

$$\langle (d/dr)u(r), u(r) - x_0 \rangle_i = \langle (d/dr)u(r) - y_0 + y_0, u(r) - x_0 \rangle$$
$$\leq \langle (d/dr)u(r) - y_0, u(r) - x_0 \rangle_i + \langle y_0, u(r) - x_0 \rangle_s.$$

Since $(d/dr)u(r) \in Au(r)$ (a.e. $r \in [0, T]$) and $A - \omega I$ is a dissipative operator,

$$\langle (d/dr)u(r) - y_0, u(r) - x_0 \rangle_i$$
$$= \langle (d/dr)u(r) - \omega u(r) - (y_0 - \omega x_0), u(r) - x_0 \rangle_i + \omega \| u(r) - x_0 \|^2$$
$$\leq \omega \| u(r) - x_0 \|^2.$$

Hence,

$$(d/dr)\| u(r) - x_0 \|^2 \leq 2\langle y_0, u(r) - x_0 \rangle_s + 2\omega \| u(r) - x_0 \|^2 \quad \text{(a.e. } r \in [0, T]).$$

We integrate this over $[r, t]$ $(0 \leq r < t \leq T)$ and obtain (5.2). Therefore, $u(t)$ is an integral solution of type ω of $(\text{CP}; x)$. The same is true for $(\text{CP}; x)_\infty$. $\square$

THEOREM 5.6. *Let $u(t): [0, T] \to X$ be an integral solution of type ω of $(\text{CP}; x)$ for A and weakly differentiable a.e. $t \in [0, T]$.*

(i) *If $A - \omega I$ is a maximal dissipative operator, then $(w - d/dt)u(t) \in Au(t)$ (a.e. $t \in [0, T]$).*

(ii) *If $A - \omega I$ is a maximal dissipative operator on $\overline{D(A)}$ and $u(t) \in \overline{D(A)}$ $(0 \leq t \leq T)$, then $(w - d/dt)u(t) \in Au(t)$ (a.e. $t \in [0, T]$).*

(iii) *Suppose the assumption of either (i) or (ii) is satisfied. If $u(t)$ is Lipschitz continuous on $[0, t]$, then $u(t)$ is a unique solution of the Cauchy problem $(\text{CP}; x)$ (for A).*

Moreover, the same holds for the Cauchy problem $(\text{CP}; x)_\infty$.

PROOF. Let $x_0 \in D(A)$, $y_0 \in Ax_0$. From (5.2) and

$$\| u(t) - x_0 \|^2 - \| u(r) - x_0 \|^2 \geq 2\operatorname{Re}\langle u(t) - u(r), f \rangle$$

$(f \in F(u(r) - x_0))$, we have

$$\operatorname{Re}\left(\frac{u(t) - u(r)}{t - r}, f \right) \leq \frac{1}{t - r} \int_r^t \langle y_0, u(\xi) - x_0 \rangle_s \, d\xi + \frac{\omega}{t - r} \int_r^t \| u(\xi) - x_0 \|^2 \, d\xi$$

for every $0 \leq r < t \leq T$ and every $f \in F(u(r) - x_0)$. Now let $t \to r + 0$. Then from the upper semicontinuity of $\langle \cdot, \cdot \rangle_s: X \times X \to (-\infty, \infty)$, for every $r \in [0, T]$ and $f \in F(u(r) - x_0)$ we have

$$\limsup_{t \to r+0} \operatorname{Re}\left(\frac{u(t) - u(r)}{t - r}, f \right) \leq \langle y_0, u(r) - x_0 \rangle_s + \omega \| u(r) - x_0 \|^2.$$

Now, if we let $u(t)$ be weakly differentiable at $t = r$ $(\in (0, T))$ and let its weak derivative be $u'(r)$, then from the inequality above

$$\operatorname{Re}(u'(r), f) \leq \langle y_0, u(r) - x_0 \rangle_s + \omega \| u(r) - x_0 \|^2 \quad (f \in F(u(r) - x_0)).$$

Thus, since there exists $g \in F(u(r) - x_0)$ such that $\langle y_0, u(r) - x_0 \rangle_s = \mathrm{Re}(y_0, g)$, we have that

$$\mathrm{Re}((u'(r) - \omega u(r)) - (y_0 - \omega x_0), g)$$
$$= \mathrm{Re}(u'(r) - y_0, g) - \omega \| u(r) - x_0 \|^2 \le 0.$$

Hence, for arbitrary $x_0 \in D(A)$, $x_0 \in D(A)$, $y_0 \in Ax_0$,

$$\langle (u'(r) - \omega u(r)) - (y_0 - \omega x_0), u(r) - x_0 \rangle_i \le 0.$$

Thus, if $A - \omega I$ is the maximal dissipative operator, or is a maximal dissipative operator on $\overline{D(A)}$ with $u(r) \in \overline{D(A)}$, then $u(r) \in D(A)$ and $u'(r) - \omega u(r) \in (A - \omega I)u(r)$. That is, we have $u'(r) \in Au(r)$. Therefore, we have shown (i) and (ii).

Next, if $u(t)$ is Lipschitz continuous on $[0, T]$ and weakly differentiable a.e. $t \in [0, T]$, then $u(t)$ is strongly differentiable a.e. $t \in [0, T]$ and its strong derivative $(d/dt)u(t) = (w - d/dt)u(t)$ (a.e. $t \in [0, T]$) (see Theorem 1.16). Hence we see from either (i) or (ii) that $u(t)$ is a solution of $(\mathrm{CP}; x)$. Finally, to show the uniqueness of the solution, we suppose $v(t): [0, T] \to X$ is a solution of $(\mathrm{CP}; x)$ Then, by the same reasoning used in the proof of Lemma 4.9, we can show

$$(d/dt)\| u(t) - v(t) \|^2 \le 2\omega \| u(t) - v(t) \|^2 \quad (\text{a.e. } t \in [0, T]).$$

From this we obtain

$$\| u(t) - v(t) \|^2 \le 2\omega \int_0^t \| u(s) - v(s) \|^2 \, ds \quad (0 \le t \le T).$$

Therefore if $\omega \le 0$, then $v(t) = u(t)$ $(0 \le t \le T)$; also if $\omega > 0$, if $\omega > 0$, we put $M = \max\{\| u(t) - v(t) \|^2; 0 \le t \le T\}$, and then from the inequality above, we obtain $\| u(t) - v(t) \|^2 \le M(2\omega t)^n/n!$ $(0 \le t \le T, n = 1, 2, \ldots)$ and hence $v(t) = u(t)$ $(0 \le t \le T)$. [1] □

§2. Difference approximations for Cauchy problems

If $A - \omega I$, with ω a given real number, is a dissipative operator on X, then we call A a dissipative operator of type ω. Thus, the dissipative operators we dealt with previously were dissipative operators of type 0. In this section we deal with the Cauchy problem for a dissipative operator A of type ω:

$$(\mathrm{CP}; x) \qquad \begin{cases} (d/dt)u(t) \in Au(t) & (0 \le t \le T), \\ u(0) = x, \end{cases}$$

where T is a given positive number and x $(\in X)$ is a given element.

---

[1] As we see from this proof, if $A - \omega I$ is a dissipative operator, then the Cauchy problems $(\mathrm{CP}; x)$ and $(\mathrm{CP}; x)_\infty$ have at most one solution (Lemma 4.9 is the case $\omega = 0$).

We consider the following difference approximation in order to construct an integral solution of $(CP; x)$:

$$\begin{cases} \dfrac{x_k^\lambda - x_{k-1}^\lambda}{t_k^\lambda - t_{k-1}^\lambda} - \varepsilon_k^\lambda \in A x_k^\lambda & (k = 1, 2, \ldots, N_\lambda; \lambda > 0), \\[2mm] x_0^\lambda = x, \end{cases} \tag{5.6}$$

where $\{t_k^\lambda : k = 1, 2, \ldots, N_\lambda\}$ is a division of the interval $[0, T]$ into subintervals such that

$$\Delta_\lambda : 0 = t_0^\lambda < t_1^\lambda < \cdots < t_{N_\lambda - 1}^\lambda < T \leq t_{N_\lambda}^\lambda \tag{5.7}$$

satisfying the following condition:

$$|\Delta_\lambda| = \max\{t_k^\lambda - t_{k-1}^\lambda : k = 1, 2, \ldots, N_\lambda\} \to 0, \qquad (\lambda \to 0+). \tag{5.8}$$

Now, for $\lambda > 0$ we define a step function $u_\lambda(t) : [0, T] \to X$ by

$$u_\lambda(t) = \begin{cases} x & (t = 0), \\ x_k^\lambda & (t \in (t_{k-1}^\lambda, t_k^\lambda] \cap (0, T]), \qquad k = 1, 2, \ldots, N_\lambda. \end{cases}$$

Then the following theorem holds.

THEOREM 5.7. *Let A be a dissipative operator of type ω and $x \in \overline{D(A)}$. If $\{\varepsilon_k^\lambda : k = 1, 2, \ldots, N_\lambda\}$ of (5.6) satisfies*

$$\varepsilon_\lambda = \sum_{k=1}^{N_\lambda} \|\varepsilon_k^\lambda\|(t_k^\lambda - t_{k-1}^\lambda) \to 0 \qquad (\lambda \to 0+), \tag{5.9}$$

then there exists a continuous function $u(t) : [0, T] \to \overline{D(A)}$ that satisfies (i) *through* (iii):

(i) $u(t) = \lim_{\lambda \to 0+} u_\lambda(t)$ *(convergence is uniform on $[0, T]$).*

(ii) *If $v(t) : [0, T] \to X$ is an integral solution of type ω of the Cauchy problem $(CP; x)$ for A, where $z \in X$, then*

$$\|u(t) - v(t)\| \leq e^{\omega(t-s)}\|u(s) - v(s)\| \qquad (0 \leq s < t \leq T). \tag{5.10}$$

(iii) $u(t)$ *is a unique integral solution of type ω of the Cauchy problem $(CP; x)$ for A.*

REMARK. We consider a dissipative operator A that satisfies condition (R_1') (i.e., for all $\lambda > 0$, $R(I - \lambda A) \supset \overline{D(A)}$), and set $J_\lambda = (I - \lambda A)^{-1}$ $(\lambda > 0)$ and let $x \in \overline{D(A)}$. Since $J_\lambda^k x - J_\lambda^{k-1} x \in \lambda A J_\lambda^k x$ $(k = 1, 2, \ldots)$, for $\lambda > 0$ we select the smallest natural number N_λ such that $\lambda N_\lambda \geq T$, and if we consider $x_k^\lambda = J_\lambda^k x$ $t_k^\lambda = k\lambda$, and $\varepsilon_k^\lambda = 0$ $(k = 1, 2, \ldots, N_\lambda)$, then (5.6) through (5.8) hold. And then $u_\lambda(t) = J_\lambda^{[t/\lambda]+1} x$ $(t \in (0, T], t \neq k\lambda)$, and (5.9) holds. Hence from the above theorem, we obtain the convergence of (4.3) (on $[0, T]$), and $T(t)x : [0, T] \to \overline{D(A)}$ is a unique integral solution of type 0 of $(CP; x)$. We will touch on this more in detail in the next section.

We will use the next lemma in the proof of Theorem 5.7.

LEMMA 5.8. *We suppose that for each* $\lambda > 0$ $\{t_k^\lambda : 0 \le k \le N_\lambda\}$, $\{x_k^\lambda : 0 \le k \le N_\lambda\}$, *and* $\{\varepsilon_k^\lambda : 1 \le k \le N_\lambda\}$ *satisfy* (5.6) *and* (5.7), *and set* $|\Delta_\lambda| = \max\{t_k^\lambda - t_{k-1}^\lambda : k = 1, 2, \ldots, N_\lambda\}$ *and* $\varepsilon_\lambda = \sum_{k=1}^{N_\lambda} \|\varepsilon_k^\lambda\|(t_k^\lambda - t_{k-1}^\lambda)$. *Now let* A *be a dissipative operator of type* ω, $\omega_0 = \max\{0, \omega\}$, $\lambda > 0$, $\mu > 0$, $|\Delta_\lambda|\omega_0 \le 1/2$, $|\Delta_\mu|\omega_0 \le 1/2$, *and* $u \in D(A)$. *Then for all integers* i *and* j *such that* $0 \le i \le N_\lambda$ *and* $0 \le j \le N_\mu$ *we have*

$$\|x_i^\lambda - x_j^\mu\| \le [\exp(2\omega_0(t_i^\lambda + t_j^\mu))] \tag{5.11}$$
$$\times [2\|x - u\| + \varepsilon_\lambda + \varepsilon_\mu + ((t_i^\lambda - t_j^\mu)^2 + |\Delta_\lambda|t_i^\lambda + |\Delta_\mu|t_j^\mu)^{1/2}\||Au\||].$$

PROOF. First note that from the dissipativity of $A - \omega I$, for every $\lambda' > 0$, $\alpha > 0$, $\beta > 0$, and every $u_i \in D(A)$, $v_i \in Au_i$ $(i = 1, 2)$ we have

$$(1 - \lambda'\omega)\|u_1 - u_2\| \le \|(u_1 - \lambda'v_1) - (u_2 - \lambda'v_2)\|, \tag{5.12}$$
$$(\alpha + \beta - \alpha\beta\omega)\|u_1 - u_2\| \le \alpha\|u_1 - u_2 - \beta v_1\| + \beta\|u_1 - u_2 + \alpha v_2\|, \tag{5.13}$$
$$(1 - \lambda'\omega)\|u_1 - u_2\| \le \|u_1 - u_2 - \lambda'v_1\| + \lambda'\||Au_2\||. \tag{5.14}$$

In fact, (5.12) is self-evident when $1 - \lambda'\omega \le 0$. If $1 - \lambda'\omega > 0$, then from the dissipativity of $A - \omega I$ we have

$$\|u_1 - u_2\| \le \|u_1 - u_2 - \lambda'(1 - \lambda'\omega)^{-1}[(v_1 - \omega u_1) - (v_2 - \omega u_2)]\|$$
$$= (1 - \lambda'\omega)^{-1}\|u_1 - u_2 - \lambda'(v_1 - v_2)\|.$$

Hence we obtain (5.12). Next from (5.12)

$$(1 - \lambda'\omega)\|u_1 - u_2\| \le \|u_1 - u_2 - \lambda'v_1\| + \lambda'\|v_2\|.$$

Since this inequality holds for every $v_2 \in Au_2$, we obtain (5.14). If we put $\lambda' = \alpha\beta/(\alpha + \beta)$ in (5.12), then

$$[1 - \alpha\beta\omega(\alpha + \beta)^{-1}]\|u_1 - u_2\| \le \|u_1 - u_2 - \alpha\beta(\alpha + \beta)^{-1}(v_1 - v_2)\|$$
$$\le \alpha(\alpha + \beta)^{-1}\|u_1 - u_2 - \beta v_1\| + \beta(\alpha + \beta)^{-1}\|u_1 - u_2 + \alpha v_2\|.$$

From this we obtain (5.13).

Now, set $h_k^\lambda = t_k^\lambda - t_{k-1}^\lambda$ $(1 \le k \le N_\lambda)$, $h_k^\mu = t_k^\mu - t_{k-1}^\mu$ $(1 \le k \le N_\mu)$, and define $a_{i,j}$, $\gamma_{i,j}$ $(0 \le i \le N_\lambda, 0 \le j \le N_\mu)$ by the following:

$$a_{i,j} = \|x_i^\lambda - x_j^\mu\|, \qquad \gamma_{i,j} = \prod_{k=1}^i (1 - \omega_0 h_k^\lambda) \prod_{k=1}^j (1 - \omega_0 h_k^\mu),$$

where $\prod_{k=1}^0 (1 - \omega_0 h_k^\lambda) = 1$ and $\prod_{k=1}^0 (1 - \omega_0 h_k^\mu) = 1$. Since $\varepsilon_\lambda \ge \sum_{k=1}^i \|\varepsilon_k^\lambda\|h_k^\lambda$, $\varepsilon_\mu \ge \sum_{k=1}^j \|\varepsilon_k^\mu\|h_k^\mu$, and $\gamma_{i,j}^{-1} \le \exp(2\omega_0(t_i^\lambda + t_j^\mu))$ (here we use $0 \le \omega_0 h_k^\lambda \le \omega_0|\Delta_\lambda| \le 1/2$, $0 \le \omega_0 h_k^\mu \le \omega_0|\Delta\mu| \le 1/2$, $(1 - t)^{-1} \le \exp(2t)$ $(0 \le t \le$

$1/2))$, in order to prove (5.11) we need to show that

$$\gamma_{i,j}a_{i,j} \le 2\|x - u\| + \sum_{k=1}^{i} \|\varepsilon_k^\lambda\|h_k^\lambda + \sum_{k=1}^{j} \|\varepsilon_k^\mu\|h_k^\mu \qquad (5.15)$$
$$+ ((t_i^\lambda - t_j^\mu)^2 + |\Delta_\lambda|t_i^\lambda + |\Delta_\mu|t_j^\mu)^{1/2}\||Au\||$$

$(0 \le i \le N_\lambda,\ 0 \le j \le N_\mu).\,(^2)$ Then, applying induction with respect to (i, j), we show (5.15) is true for all integers i, j such that $0 \le i \le N_\lambda$ and $0 \le j \le N_\mu$.

First from (5.14), for every integer k such that $1 \le k \le N_\lambda$,

$$(1 - \omega_0 h_k^\lambda)\|x_k^\lambda - u\| \le \|x_k^\lambda - u - (x_k^\lambda - x_{k-1}^\lambda - h_k^\lambda\varepsilon_k^\lambda)\| + h_k^\lambda\||Au\||$$
$$\le \|x_{k-1}^\lambda - u\| + h_k^\lambda\|\varepsilon_k^\lambda\| + h_k^\lambda\||Au\||.$$

Thus (note that $\gamma_{k,0} = (1 - \omega_0 h_k^\lambda)\gamma_{k-1,0}$ and $0 < \gamma_{k-1,0} \le 1$) we obtain

$$\gamma_{k,0}\|x_k^\lambda - u\| \le \gamma_{k-1,0}\|x_{k-1}^\lambda - u\| + h_k^\lambda\|\varepsilon_k^\lambda\| + h_k^\lambda\||Au\||.$$

If we add these together from $k = 1$ to $k = i$ $(1 \le i \le N_\lambda)$, then

$$\gamma_{i,0}\|x_i^\lambda - u\| \le \|x - u\| + \sum_{k=1}^{i} h_k^\lambda\|\varepsilon_k^\lambda\| + t_i^\lambda\||Au\||.$$

Hence

$$\gamma_{i,0}a_{i,0} = \gamma_{i,0}\|x_i^\lambda - x\| \le \gamma_{i,0}\|x_i^\lambda - u\| + \|x - u\|$$
$$\le 2\|x - u\| + \sum_{k=1}^{i} \|\varepsilon_k^\lambda\|h_k^\lambda + t_i^\lambda\||Au\||.$$

Also, since $a_{0,0} = 0$, the above inequality holds for $i = 0$. Hence we have shown that $a_{i,0}$ satisfies (5.15) for every integer i such that $0 \le i \le N_\lambda$. Similarly, we see that $a_{0,j}$ satisfies (5.15) for every integer j such that $0 \le j \le N_\mu$.

Next, let $1 \le i \le N_\lambda$ and $1 \le j \le N_\mu$, and assume that $a_{i-1,j}$ and $a_{i,j-1}$ satisfy (5.15). Then we show $a_{i,j}$ also satisfies (5.15). Let $\alpha = h_j^\mu$, $\beta = h_i^\lambda$, $u_1 = x_i^\lambda$, $v_1 = (x_i^\lambda - x_{i-1}^\lambda)/h_i^\lambda - \varepsilon_i^\lambda$ $(\in Ax_i^\lambda$, see $(5.6))$, $u_2 = x_j^\mu$ and $v_2 = (x_j^\mu - x_{j-1}^\mu)/h_j^\mu - \varepsilon_j^\mu$ $(\in Ax_j^\mu)$, and apply (5.13) to obtain

$$(h_i^\lambda + h_j^\mu - \omega_0 h_i^\lambda h_j^\mu)a_{i,j} \le h_j^\mu\|x_i^\lambda - x_j^\mu - (x_i^\lambda - x_{i-1}^\lambda - h_i^\lambda\varepsilon_i^\lambda)\|$$
$$+ h_i^\lambda\|x_i^\lambda - x_j^\mu + (x_j^\mu - x_{j-1}^\mu - h_j^\mu\varepsilon_j^\mu)\|$$
$$\le h_j^\mu a_{i-1,j} + h_i^\lambda a_{i,j-1} + (\|\varepsilon_i^\lambda\| + \|\varepsilon_j^\mu\|)h_i^\lambda h_j^\mu.$$

$(^2)$ We set $\sum_{k=1}^{0} \|\varepsilon_k^\lambda\|h_k^\lambda = 0$ and $\sum_{k=1}^{0} \|\varepsilon_k^\mu\|h_k^\mu = 0$.

From this and the assumption that $a_{i-1,j}$ and $a_{i,j-1}$ satisfy (5.15), we have

$$
\begin{aligned}
(h_i^\lambda &+ h_j^\mu - \omega_0 h_i^\lambda h_j^\mu)\gamma_{i,j}a_{i,j} \\
&\le h_j^\mu(1 - \omega_0 h_i^\lambda)\gamma_{i-1,j}a_{i-1,j} + h_i^\lambda(1 - \omega_0 h_j^\mu)\gamma_{i,j-1}a_{i,j-1} \\
&\quad + \gamma_{i,j}(\|\varepsilon_i^\lambda\| + \|\varepsilon_j^\mu\|)h_i^\lambda h_j^\mu \\
&\le 2\|x - u\|[h_j^\mu[h_j^\mu(1 - \omega_0 h_j^\lambda) + h_i^\lambda(1 - \omega_0 h_j^\mu)] \\
&\quad + [h_j^\mu(1 - \omega_0 h_i^\lambda)((t_{i-1}^\lambda - t_j^\mu)^2 + |\Delta_\lambda|t_{i-1}^\lambda + |\Delta_\mu|t_j^\mu)^{1/2} \\
&\quad + h_i^\lambda(1 - \omega_0 h_j^\mu)((t_i^\lambda - t_{j-1}^\mu)^2 + |\Delta_\lambda|t_i^\lambda + |\Delta_\mu|t_{j-1}^\mu)^{1/2}]\||Au\|| \\
&\quad + \left[h_j^\mu(1 - \omega_0 h_i^\lambda)\left(\sum_{k=1}^{i-1}\|\varepsilon_k^\lambda\|h_k^\lambda + \sum_{k=1}^{j}\|\varepsilon_k^\mu\|h_k^\mu \right) \right. \\
&\quad \left. + h_i^\lambda(1 - \omega_0 h_j^\mu)\left(\sum_{k=1}^{i}\|\varepsilon_k^\lambda\|h_k^\lambda + \sum_{k=1}^{j-1}\|\varepsilon_k^\mu\|h_k^\mu \right) + \gamma_{i,j}(\|\varepsilon_i^\lambda\| + \|\varepsilon_j^\mu\|)h_i^\lambda h_j^\mu \right]
\end{aligned}
$$

We denote the three major terms on the right-hand side of the above equation by I_i $(i = 1, 2, 3)$, respectively, i.e., the right-hand side $= I_1 + I_2 + I_3$. Then from $h_j^\mu(1 - \omega_0 h_i^\lambda) + h_i^\lambda(1 - \omega_0 h_j^\mu) \le h_i^\lambda + h_j^\mu - \omega_0 h_i^\lambda h_j^\mu$, we have

$$
I_1 \le 2\|x - u\|(h_i^\lambda + h_j^\mu - \omega_0 h_i^\lambda h_j^\mu). \tag{5.16}
$$

Next, we evaluate I_2 by using the Schwarz inequality

$$
\begin{aligned}
I_2 &\le [h_j^\mu(1 - \omega_0 h_i^\lambda)^2 + h_i^\lambda(1 - \omega_0 h_j^\mu)^2]^{1/2} \\
&\quad \times [h_j^\mu((t_{i-1}^\lambda - t_j^\mu)^2 + |\Delta_\lambda|t_{i-1}^\lambda + |\Delta_\mu|t_j^\mu) + h_i^\lambda((t_i^\lambda - t_{j-1}^\mu)^2 \\
&\quad + |\Delta_\lambda|t_i^\lambda + |\Delta_\mu|t_{j-1}^\mu)]^{1/2}\||Au\||.
\end{aligned}
$$

Since

$$
\begin{aligned}
h_j^\mu&((t_{i-1}^\lambda - t_j^\mu)^2 + |\Delta_\lambda|t_{i-1}^\lambda + |\Delta_\mu|t_j^\mu) + h_i^\lambda((t_i^\lambda - t_{j-1}^\mu)^2 + |\Delta_\lambda|t_i^\lambda + |\Delta_\mu|t_{j-1}^\mu) \\
&= h_j^\mu[((t_i^\lambda - t_j^\mu) - h_i^\lambda)^2 + |\Delta_\lambda|t_{i-1}^\lambda + |\Delta_\mu|t_j^\mu] \\
&\quad + h_i^\lambda[((t_i^\lambda - t_j^\mu) + h_j^\mu)^2 + |\Delta_\lambda|t_i^\lambda + |\Delta_\mu|t_{j-1}^\mu] \\
&\le h_j^\mu(t_i^\lambda - t_j^\mu)^2 + |\Delta_\lambda|h_i^\lambda h_j^\mu + |\Delta_\lambda|t_{i-1}^\lambda h_j^\mu + |\Delta_\mu|t_j^\mu h_j^\mu \\
&\quad + h_i^\lambda(t_i^\lambda - t_j^\mu)^2 + |\Delta_\mu|h_i^\lambda h_j^\mu + |\Delta_\lambda|t_i^\lambda h_i^\lambda + |\Delta_\mu|t_{j-1}^\mu h_i^\lambda
\end{aligned}
$$

(here we used $h_i^\lambda = t_i^\lambda - t_{i-1}^\lambda \le |\Delta_\lambda|$ and $h_j^\mu = t_j^\mu - t_{j-1}^\mu \le |\Delta_\mu|$)

$$
= (h_i^\lambda + h_j^\mu)[(t_i^\lambda - t_j^\mu)^2 + |\Delta_\lambda|t_i^\lambda + |\Delta_\mu|t_j^\mu],
$$

and

$$(h_j^\mu + h_i^\lambda)[h_j^\mu(1 - \omega_0 h_i^\lambda)^2 + h_i^\lambda(1 - \omega_0 h_j^\mu)^2]$$

$$\leq [(h_i^\lambda + h_j^\mu - \omega_0 h_i^\lambda h_j^\mu) + \omega_0 h_i^\lambda h_j^\mu][h_j^\mu(1 - \omega_0 h_i^\lambda) + h_i^\lambda(1 - \omega_0 h_j^\mu)]$$

$$= (h_i^\lambda + h_j^\mu - \omega_0 h_i^\lambda h_j^\mu)^2 - (\omega_0 h_i^\lambda h_j^\mu)^2 \leq (h_i^\lambda + h_j^\mu - \omega_0 h_i^\lambda h_j^\mu)^2$$

we have

$$I_2 \leq (h_i^\lambda + h_j^\mu - \omega_0 h_i^\lambda h_j^\mu)[(t_i^\lambda - t_j^\mu)^2 + |\Delta_\lambda| t_i^\lambda + |\Delta_\mu| t_j^\mu]^{1/2} |||Au|||. \tag{5.17}$$

Finally, from $\gamma_{i,j} \leq 1 - \omega_0 h_i^\lambda$ and $\gamma_{i,j} \leq 1 - \omega_0 h_j^\mu$ we have

$$I_3 \leq [h_j^\mu(1 - \omega_0 h_i^\lambda) + h_i^\lambda(1 - \omega_0 h_j^\mu)]\left(\sum_{k=1}^{i} \|\varepsilon_k^\lambda\| h_k^\lambda + \sum_{k=1}^{j} \|\varepsilon_k^\mu\| h_k^\mu\right)$$

$$\leq (h_i^\lambda + h_j^\mu - \omega_0 h_i^\lambda h_j^\mu)\left(\sum_{k=1}^{i} \|\varepsilon_k^\lambda\| h_k^\lambda + \sum_{k=1}^{j} \|\varepsilon_k^\mu\| h_k^\mu\right). \tag{5.18}$$

From (5.16) through (5.18), we see that $a_{i,j}$ satisfies (5.15). Therefore, by induction the inequality (5.15) holds for all integers i, j such that $0 \leq i \leq N_\lambda$ and $0 \leq j \leq N_\mu$. □

REMARK. Let A be a dissipative operator of type ω, $\omega_0 = \max\{0, \omega\}$ and let λ, $\mu > 0$ satisfy $\lambda\omega_0 \leq 1/2$ and $\mu\omega_0 \leq 1/2$. Then, since $1 - \lambda\omega > 0$ and $1 - \mu\omega > 0$, from (5.12) we can define $J_\lambda = (I - \lambda A)^{-1}$ and $J_\mu = (I - \mu A)^{-1}$ as single-valued operators. Let $x \in D(J_\lambda^k) \cap D(J_\mu^k) \cap D(A)$ $(k = 1, 2, \ldots)$. Then for every natural number k, since $J_\lambda^k x - J_\lambda^{k-1} x \in \lambda A J_\lambda^k x$, (5.6) holds with $x_k^\lambda = J_\lambda^k x$, $t_k^\lambda = k\lambda$, and $\varepsilon_k^\lambda = 0$. The same holds with λ and μ interchanged. Hence, from (5.11),

$$\|J_\lambda^m x - J_\mu^n x\| \leq [\exp(2\omega_0(m\lambda + n\mu))][(m\lambda - n\mu)^2 + m\lambda^2 + n\mu^2]^{1/2} |||Ax|||$$
$$\tag{5.19}$$

holds for every nonnegative integers m and n. (Given an arbitrary $T > 0$, if we select the smallest natural numbers N_λ and N_μ such that $\lambda N_\lambda \geq T$ and $\mu N_\mu \geq T$, then from (5.11), we have that (5.19) holds for m and n such that $0 \leq m \leq N_\lambda$ and $0 \leq n \leq N_\mu$. Since $T > 0$ is arbitrary, (5.19) holds for all nonnegative integers m and n.) In particular, if A is a dissipative operator, we have

$$\|J_\lambda^m x - J_\mu^n x\| \leq [(m\lambda - n\mu)^2 + m\lambda^2 + n\mu^2]^{1/2} |||Ax||| \qquad (m, n \geq 0).$$

This estimation is better than (4.1).

PROOF OF THEOREM 5.7. Since $x \in \overline{D(A)}$, there is a sequence $\{x_p\} \subset D(A)$ such that $\lim_{p\to\infty} x_p = x$. Let $\lambda > 0$, $\mu > 0$, $|\Delta_\lambda|\omega_0 \leq 1/2$, $|\Delta_\mu|\omega_0 \leq 1/2$ $(\omega_0 = \max\{0, \omega\})$, and t, $s \in [0, T]$. Then from the definition of u_λ, $u_\lambda(t) = x = x_0^\lambda$ when $t = 0$, and we can select an integer i $(1 \leq i \leq N_\lambda)$ such that $t \in (t_{i-1}^\lambda, t_i^\lambda]$, for which $u_\lambda(t) = x_i^\lambda$ when $t > 0$. Similarly,

$u_\mu(s) = x = x_0^\mu$ when $s = 0$, and there exists an integer j $(1 \le j \le N_\mu)$ such that $s \in (t_{j-1}^\mu, t_j^\mu]$, for which $u_\mu(s) = x_j^\mu$ when $s > 0$. Hence if we let $u = x_p$ and apply Lemma 5.8, we obtain

$$\|u_\lambda(t) - u_\mu(s)\| \le [\exp(2\omega_0(|\Delta_\lambda| + |\Delta_\mu| + t + s))]$$
$$\times [2\|x - x_p\| + \varepsilon_\lambda + \varepsilon_\mu$$
$$+ ((|\Delta_\lambda| + |\Delta_\mu| + |t - s|)^2 + |\Delta_\lambda|(t + |\Delta_\lambda|) \tag{5.20}$$
$$+ |\Delta_\mu|(s + |\Delta_\mu|))^{1/2}|||Ax_p|||].$$

(Here we used $t_i^\lambda \le t + |\Delta_\lambda|$, $t_j^\mu \le s + |\Delta_\mu|$, $|t_i^\lambda - t_j^\mu| \le |t_i^\lambda - t| + |t - s| + |s - t_j^\mu| \le |\Delta_\lambda| + |t - s| + |\Delta_\mu|$.) Now we consider the case $s = t$. From $|\Delta_\lambda|$, $|\Delta_\mu| \to 0$, ε_λ, $\varepsilon_\mu \to 0$ as λ, $\mu \to 0+$ (by (5.8) and (5.9)), we have for every x_p

$$\limsup_{\lambda, \mu \to 0+} \left[\sup_{0 \le t \le T} \|u_\lambda(t) - u_\mu(t)\| \right] \le 2[\exp(4\omega_0 T)]\|x - x_p\|.$$

Since $\|x - x_p\| \to 0$ as $p \to \infty$, from the above inequality we obtain

$$\sup_{0 \le t \le T} \|u_\lambda(t) - u_\mu(t)\| \to 0 \quad (\lambda, \mu \to 0+)$$

and have shown that $u_\lambda(t)$ converges uniformly on $[0, T]$ as $\lambda \to 0+$. Now we set

$$u(t) = \lim_{\lambda \to 0+} u_\lambda(t) \qquad (0 \le t \le T).$$

Since $u_\lambda(t) \in \overline{D(A)}$ $(\lambda > 0, 0 \le t \le T)$, we have that $u(t): [0, T] \to \overline{D(A)}$.

Next, we show that this $u(t): [0, T] \to \overline{D(A)}$ is an integral solution of type ω of the Cauchy problem (CP; x) for A. Since $u_\lambda(0) = x$, we have that $u(0) = \lim_{\lambda \to 0+} u_\lambda(0) = x$. If we let $\lambda \to 0+$ and $\mu \to 0+$ in (5.20), then for every s, $t \in [0, T]$ and x_p,

$$\|u(t) - u(s)\| \le [\exp(2\omega_0(t + s))][2\|x - x_p\| + |t - s||||Ax_p|||]. \tag{5.21}$$

From this we obtain $\limsup_{t \to s} \|u(t) - u(s)\| \le 2[\exp(4\omega_0 s)]\|x - x_p\| \to 0$ as $p \to \infty$, and $u(t)$ is continuous on $[0, T]$.[3] To show $u(t)$ satisfies the inequality (5.2), we let $0 < r < t \le T$, $x_0 \in D(A)$ and $y_0 \in Ax_0$. From (5.6)

$$(x_k^\lambda - x_{k-1}^\lambda)/h_k^\lambda - \varepsilon_k^\lambda \in Ax_k^\lambda \qquad (k = 1, 2, \dots, N_\lambda),$$

where $h_k^\lambda = t_k^\lambda - t_{k-1}^\lambda$. Then, because $A - \lambda I$ is a dissipative operator, there exists $g \in F(x_k^\lambda - x_0)$ such that

$$\text{Re}((x_k^\lambda - x_{k-1}^\lambda)/h_k^\lambda - \varepsilon_k^\lambda - \omega x_k^\lambda - (y_0 - \omega x_0), g) \le 0,$$

---

[3] In particular, if the initial value x is an element of $D(A)$, then taking $x_p = x$ $(p = 1, 2, \dots)$, from (5.21) we obtain $\|u(t) - u(s)\| \le [\exp(2\omega_0(t + s))]|||Ax||||t - s|$ $(s, t \in [0, T])$. Hence $u(t)$ is Lipschitz continuous on $[0, T]$.

that is,

$$\text{Re}(x_k^\lambda - x_{k-1}^\lambda, g) \le [\text{Re}(\varepsilon_k^\lambda + y_0, g) + \omega\|x_k^\lambda - x_0\|^2]h_k^\lambda.$$

Now, since

$$\begin{aligned}
\text{Re}(x_k^\lambda - x_{k-1}^\lambda, g) &= \text{Re}(x_k^\lambda - x_0 - (x_{k-1}^\lambda - x_0), g) \\
&\ge \|x_k^\lambda - x_0\|^2 - \|x_{k-1}^\lambda - x_0\| \, \|x_k^\lambda - x_0\| \\
&\ge (1/2)(\|x_k^\lambda - x_0\|^2 - \|x_{k-1}^\lambda - x_0\|^2),
\end{aligned}$$

and

$$\text{Re}(\varepsilon_k^\lambda + y_0, g) \le \langle y_0, x_k^\lambda - x_0 \rangle_s + \|\varepsilon_k^\lambda\| \, \|x_k^\lambda - x_0\|,$$

we obtain the following inequality:

$$\begin{aligned}
\|x_k^\lambda &- x_0\|^2 - \|x_{k-1}^\lambda - x_0\|^2 \\
&\le 2\langle y_0, x_k^\lambda - x_0 \rangle_s h_k^\lambda + 2\omega\|x_k^\lambda - x_0\|^2 h_k^\lambda + 2\|\varepsilon_k^\lambda\| \, \|x_k^\lambda - x_0\|h_k^\lambda \\
&= 2\int_{t_{k-1}^\lambda}^{t_k^\lambda} \langle y_0, u_\lambda(\tau) - x_0 \rangle_s \, d\tau + 2\omega\int_{t_{k-1}^\lambda}^{t_k^\lambda} \|u_\lambda(\tau) - x_0\|^2 \, d\tau \qquad (5.22) \\
&\quad + 2\int_{t_{k-1}^\lambda}^{t_k^\lambda} \theta_\lambda(\tau)\|u_\lambda(\tau) - x_0\| \, d\tau \qquad (k = 1, 2, \ldots, N_\lambda),
\end{aligned}$$

where $\theta_\lambda(\tau) = \|\varepsilon_k^\lambda\|$ $(\tau \in (t_{k-1}^\lambda, t_k^\lambda]; k = 1, 2, \ldots, N_\lambda)$. Now, if we take i and l $(1 \le i \le l \le N_\lambda)$ such that $r \in (t_{i-1}^\lambda, t_i^\lambda]$ and $t \in (t_{l-1}^\lambda, t_l^\lambda]$, and add (5.22) from $k = i + 1$ to $k = l$, then we obtain

$$\begin{aligned}
\|u_\lambda(t) - x_0\|^2 &- \|u_\lambda(r) - x_0\|^2 = \|x_l^\lambda - x_0\|^2 - \|x_i^\lambda - x_0\|^2 \\
&\le 2\int_{t_i^\lambda}^{t_l^\lambda} \langle y_0, u_\lambda(\tau) - x_0 \rangle_s \, d\tau + 2\omega\int_{t_i^\lambda}^{t_l^\lambda} \|u_\lambda(\tau) - x_0\|^2 \, d\tau \\
&\quad + 2\int_{t_i^\lambda}^{t_l^\lambda} \theta_\lambda(\tau)\|u_\lambda(\tau) - x_0\| \, d\tau.
\end{aligned}$$

Here we consider $\limsup_{\lambda \to 0+}$. Then from $t_i^\lambda \to r$, $t_l^\lambda \to t$, and $u_\lambda(\tau) \to u(\tau)$ uniformly on $[0, T]$, the second term on the right-hand side of the above inequality goes to $2\omega\int_\tau^t \|u(\tau) - x_0\|^2 \, d\tau$, and since there is a constant M such that if λ (> 0) is sufficiently small, $\|u_\lambda(\tau) - x_0\| \le M$ $(0 \le \tau \le T)$, the third term on the right-hand side of the above inequality is $\le 2M\int_0^T \theta_\lambda(\tau) \, d\tau \le 2M\sum_{k=1}^{N_\lambda} \|\varepsilon_k^\lambda\|(t_k^\lambda - t_{k-1}^\lambda) \to 0$ (by (5.9)). Finally, from the Lebesgue Convergence Theorem and the upper semicontinuity of $\langle \cdot, \cdot \rangle_s : X \times X \to (-\infty, \infty)$, we have

$$\limsup_{\lambda \to 0+} \int_{t_i^\lambda}^{t_l^\lambda} \langle y_0, u_\lambda(\tau) - x_0 \rangle_s \, d\tau \le \int_r^t \langle y_0, u(\tau) - x_0 \rangle_s \, d\tau.$$

Hence, for every $0 < r < t \leq T$, $x_0 \in D(A)$ and $y_0 \in Ax_0$, we have

$$\|u(t) - x_0\|^2 - \|u(r) - x_0\|^2 \leq 2\int_r^t \langle y_0, u(\tau) - x_0\rangle_s \, d\tau + 2\omega \int_r^t \|u(\tau) - x_0\|^2 \, d\tau,$$

that is, (5.2) holds. Also, if we let $r \to 0+$ in this inequality, then the above inequality holds also for $r = 0$. Hence from Theorem 5.2, $u(t) \colon [0, T] \to \overline{D(A)}$ is an integral solution of type ω of the Cauchy problem (CP; x) for A. We obtain immediately by (ii) that this $u(t)$ is a unique integral solution of type ω of the Cauchy problem (CP; x). In fact, if we let $v(t) \colon [0, T] \to X$ be an integral solution of type ω of the Cauchy problem (CP; x), then by (5.10) we have

$$\|u(t) - v(t)\| \leq e^{\omega t} \|u(0) - v(0)\| = e^{\omega t} \|x - x\| = 0 \qquad (0 \leq t \leq T).$$

Therefore $v(t) = u(t)$ $(0 \leq t \leq T)$. Hence if we show (ii), then the proof is finished.

To prove (ii), let $v(t) \colon [0, T] \to X$ be an integral solution of type ω of the Cauchy problem (CP; x) for A. Then from (5.2), for every $0 \leq x < t \leq T$ and $k = 1, 2, \ldots, N_\lambda$

$$\|v(t) - x_k^\lambda\|^2 - \|v(r) - x_k^\lambda\|^2$$
$$\leq 2\int_r^t \langle y_k^\lambda, v(\tau) - x_k^\lambda\rangle_s \, d\tau + 2\omega \int_r^t \|v(\tau) - x_k^\lambda\|^2 \, d\tau$$

holds, where $y_k^\lambda = (x_k^\lambda - x_{k-1}^\lambda)/h_k^\lambda - \varepsilon_k^\lambda$ $(\in Ax_k^\lambda)$. From $y_k^\lambda = (x_k^\lambda - v(\tau))/h_k^\lambda + (v(\tau) - x_{k-1}^\lambda)/h_k^\lambda - \varepsilon_k^\lambda$, we have

$$2\langle y_k^\lambda, v(\tau) - x_k^\lambda\rangle_s$$
$$= 2(h_k^\lambda)^{-1}[-\|v(\tau) - x_k^\lambda\|^2 + \langle v(\tau) - x_{k-1}^\lambda - h_k^\lambda \varepsilon_k^\lambda, v(\tau) - x_k^\lambda\rangle_s]$$
$$\leq (h_k^\lambda)^{-1}(\|v(\tau) - x_{k-1}^\lambda\|^2 - \|v(\tau) - x_k^\lambda\|^2) + 2\|\varepsilon_k^\lambda\| \, \|v(\tau) - x_k^\lambda\|.$$

Hence,

$$(\|v(t) - x_k^\lambda\|^2 - \|v(r) - x_k^\lambda\|^2)h_k^\lambda$$
$$\leq \int_r^t (\|v(\tau) - x_{k-1}^\lambda\|^2 - \|v(\tau) - x_k^\lambda\|^2) \, d\tau$$
$$+ \left(2\|\varepsilon_k^\lambda\| \int_r^t \|v(\tau) - x_k^\lambda\| \, d\tau + 2\omega \int_r^t \|v(\tau) - x_k^\lambda\|^2 \, d\tau\right) h_k^\lambda.$$

Since $u_\lambda(\xi) = x_k^\lambda$ $(\xi \in (t_{k-1}^\lambda, t_k^\lambda])$, the above inequality can be written in the following form:

$$\int_{t_{k-1}^{\lambda}}^{t_k^{\lambda}} (\|v(t) - u_\lambda(\xi)\|^2 - \|v(r) - u_\lambda(\xi)\|^2)\, d\xi$$

$$+ \int_r^t (\|v(\tau) - x_k^\lambda\|^2 - \|v(\tau) - x_{k-1}^\lambda\|^2)\, d\tau \tag{5.23}$$

$$\leq 2 \int_r^t \left[\int_{t_{k-1}^{\lambda}}^{t_k^{\lambda}} \theta_\lambda(\xi) \|v(\tau) - u_\lambda(\xi)\|\, d\xi \right] d\tau$$

$$+ 2\omega \int_r^t \left[\int_{t_{k-1}^{\lambda}}^{t_k^{\lambda}} \|v(\tau) - u_\lambda(\xi)\|^2\, d\xi \right] d\tau,$$

where θ_λ is the step function defined following (5.22). Let $0 < \alpha < \beta \leq T$. If we choose natural numbers j and l such that $\alpha \in (t_{j-1}^\lambda, t_j^\lambda]$ and $\beta \in (t_{l-1}^\lambda, t_l^\lambda]$, and add (5.23) from $k = j+1$ to $k = l$, then we have

$$\int_{t_j^\lambda}^{t_l^\lambda} (\|v(t) - u_\lambda(\xi)\|^2 - \|v(r) - u_\lambda(\xi)\|^2)\, d\xi$$

$$+ \int_r^t (\|v(\tau) - u_\lambda(\beta)\|^2 - \|v(\tau) - u_\lambda(\alpha)\|^2]\, d\tau$$

$$\leq 2 \int_r^t \left[\int_{t_j^\lambda}^{t_l^\lambda} \theta_\lambda(\xi) \|v(\tau) - u_\lambda(\xi)\|\, d\xi \right] d\tau$$

$$+ 2\omega \int_r^t \left[\int_{t_j^\lambda}^{t_l^\lambda} \|v(\tau) - u_\lambda(\xi)\|^2\, d\xi \right] d\tau.$$

Since there exists a constant M such that $\|v(\tau) - u_\lambda(\xi)\| \leq M$ ($0 \leq \tau$, $\xi \leq T$, and $\lambda > 0$ is sufficiently small), from (5.9) the first term on the right-hand side of the above inequality is $\leq 2MT \sum_{k=1}^{N_\lambda} \|\xi_k^\lambda\| h_k^\lambda \to 0$ as $\lambda \to 0$. Hence, if we let $\lambda \to 0+$ in the above inequality, we obtain

$$\int_\alpha^\beta (\|v(t) - u(\xi)\|^2 - \|v(r) - u(\xi)\|^2)\, d\xi$$

$$+ \int_r^t (\|v(\tau) - u(\beta)\|^2 - \|v(\tau) - u(\alpha)\|^2)\, d\tau \tag{5.24}$$

$$\leq 2\omega \int_r^t \left[\int_\alpha^\beta \|v(\tau) - u(\xi)\|^2\, d\xi \right] d\tau.$$

If we let $\alpha \to 0+$, then (5.24) holds also for $\alpha = 0$. From the above, we have shown that (5.24) holds for every $0 \leq r < t \leq T$ and $0 \leq \alpha < \beta \leq T$. Then from the following lemma, (^{4}) we have

$$\|v(t) - u(t)\| \leq e^{\omega(t-s)} \|u(s) - v(s)\| \qquad (0 \leq s < t \leq T),$$

that is, (5.10) holds. □

(^{4}) We apply the lemma with $c = 2\omega$, $[a, b] = [0, T]$, and $f(t, \xi) = \|v(t) - u(\xi)\|^2$.

LEMMA 5.9. *Let* c *be a real number and let* $f(t, \xi): [a, b] \times [a, b] \to R^1$ $(= (-\infty, \infty))$ *be a continuous function. If for every* $s, t, \alpha,$ *and* β *such that* $a \leq s < t \leq b$ *and* $a \leq \alpha < \beta \leq b$

$$\int_\alpha^\beta (f(t, \xi) - f(s, \xi)) d\xi + \int_s^t (f(\tau, \beta) - f(\tau, \alpha)) d\tau$$

$$\leq c \int_s^t \left[\int_\alpha^\beta f(\tau, \xi) d\xi \right] d\tau$$

holds, then $e^{-ct} f(t, t)$ *is monotone decreasing on* $[a, b]$.

PROOF. We define

$$F(t, s, \beta, \alpha) = \int_s^t \left[\int_\alpha^\beta f(\tau, \xi) d\xi \right] d\tau$$

for $t, s, \beta,$ and α in $[a, b]$. Since

$$\partial F / \partial t + \partial F / \partial s = \int_\alpha^\beta f(t, \xi) d\xi - \int_\alpha^\beta f(s, \xi) d\xi$$

and

$$\partial F / \partial \beta + \partial F / \partial \alpha = \int_s^t f(\tau, \beta) d\tau - \int_s^t f(\tau, \alpha) d\tau,$$

from the assumption, for $s, t, \alpha,$ and β such that $a < s < t < b$, and $a < \alpha < \beta < b$, we obtain

$$(\partial F / \partial t)(t, s, \beta, \alpha) + (\partial F / \partial s)(t, s, \beta, \alpha)$$
$$+ (\partial F / \partial \beta)(t, s, \beta, \alpha) + (\partial F / \partial \alpha)(t, s, \beta, \alpha)$$
$$\leq c F(t, s, \beta, \alpha).$$

Now choose arbitrarily s and t such that $a < s < t < b$, and set

$$G(h) = F(t + h, s + h, t + h, s + h) \qquad (0 \leq h \leq b - t).$$

Then $G(h)$ is continuous on $[0, b - t]$ and differentiable on $(0, b - t)$, and we have

$$G'(h) = (\partial F / \partial t)(t + h, s + h, t + h, s + h)$$
$$+ (\partial F / \partial s)(t + h, s + h, t + h, s + h)$$
$$+ (\partial F / \partial \beta)(t + h, s + h, t + h, s + h)$$
$$+ (\partial F / \partial \alpha)(t + h, s + h, t + h, s + h)$$
$$\leq c F(t + h, s + h, t + h, s + h) = c G(h) \qquad (0 < h < b - t).$$

Hence, we obtain $(d/dh)[e^{-ch} G(h)] \leq 0$ $(0 < h < b - t)$, and thus $e^{-ch} G(h)$ is monotone decreasing on $[0, b - t]$. Hence for every $h \in (0, b - t]$, we have $e^{-ch} G(h) \leq G(0)$, that is,

$$e^{-ch} \int_s^t \left[\int_s^t f(\tau + h, \xi + h) d\xi \right] d\tau \leq \int_s^t \left[\int_s^t f(\tau, \xi) d\xi \right] d\tau.$$

If we divide both sides by $(t - s)^2$ and let $s \to t - 0$, then

$$e^{-ch} f(t + h, t + h) \leq f(t, t) \qquad (0 < h \leq b - t).$$

Thus, if $a < t < t + h \leq b$, then $e^{-c(t+h)} f(t + h, t + h) \leq e^{-ct} f(t, t)$ and $e^{-ct} f(t, t)$ is monotone decreasing on $(a, b]$. Then since $f(t, t)$ is continuous on $[a, b]$, $e^{-ct} f(t, t)$ is also monotone decreasing on $[a, b]$. $\square$

REMARK. Let $A - \omega I$ be a maximal dissipative operator on $\overline{D(A)}$ and $x \in D(A)$, and consider the integral solution $u(t): [0, T] \to \overline{D(A)}$ of type ω of the Cauchy problem (CP; x) for A constructed in Theorem 5.7. Then, if $u(t)$ is weakly differentiable a.e. $t \in [0, T]$, then $u(t)$ is a solution of (CP; x). In fact, since the initial value x belongs to $D(A)$, we have that $u(t)$ is Lipschitz continuous on $[0, T]$ (see the footnote following (5.21)). Therefore, we may apply part (iii) of Theorem 5.6.

§3. Existence and uniqueness of integral solutions; generation of nonlinear semigroups

In this section we consider the existence and uniqueness of the solution of the following Cauchy problem for a dissipative operator A of type ω:

$$(\text{CP}; x)_\infty \quad \begin{cases} (d/dt)u(t) \in Au(t) & (0 \leq t < \infty), \\ u(0) = x \ (\in X). \end{cases}$$

THEOREM 5.10. *Let A be a dissipative operator of type ω satisfying the following condition* (R):

(R) *For all $x \in \overline{D(A)}$, $\liminf_{\lambda \to 0+} \lambda^{-1} d(R(I - \lambda A), x) = 0$, where $d(R(I - \lambda A), x)$ denotes the distance between the set $R(I - \lambda A)$ and a point x.*

Then, for each $x \in \overline{D(A)}$, the Cauchy problem $(\text{CP}; x)_\infty$ for A has a unique integral solution $u(t; x): [0, \infty) \to \overline{D(A)}$ of type ω and the following inequalities hold:

$$\|u(t; x) - u(t; y)\| \leq e^{\omega(t-s)} \|u(s; x) - u(s; y)\| \tag{5.25}$$
$$(0 \leq s < t < \infty; \ x, y \in \overline{D(A)}),$$

$$\|u(t; x) - u(s; x)\| \leq [\exp(2\omega_0(t + s))]\|\|Ax\|\||t - s| \tag{5.26}$$
$$(0 \leq s, t < \infty; \ x \in D(A)).$$

where $\omega_0 = \max\{0, \omega\}$.

We will use the following lemma in the proof of Theorem 5.10.

LEMMA 5.11. *Let A be a dissipative operator of type ω satisfying condition* (R), *and let $x_0 \in \overline{D(A)}$ and $\varepsilon > 0$. Then there exist $\{t_k\}$, $\{x_k\}$, and $\{y_k\}$ which satisfy* (i) *through* (iii), *where $x_k \in D(A)$ and $y_k \in Ax_k$ ($k = 1, 2, \ldots$):*

(i) $0 = t_0 < t_1 < \cdots < t_k < t_{k+1} < \cdots$, *and* $\lim_{k \to \infty} t_k = \infty$,

(ii) $t_k - t_{k-1} \leq \varepsilon$ ($k = 1, 2, \ldots$),

(iii) $\|x_k - x_{k-1} - (t_k - t_{k-1})y_k\| < \varepsilon(t_k - t_{k-1})$ ($k = 1, 2, \ldots$).

PROOF. We may assume $2\omega_0\varepsilon < 1$, where $\omega_0 = \max\{0, \omega\}$. For each $x \in \overline{D(A)}$, from condition (R) we can choose λ so that

$$0 < \lambda \leq \varepsilon, \quad \text{and}$$

there exist $x_\lambda \in D(A)$ and $y_\lambda \in Ax_\lambda$ such that   (5.27)

$$\|x_\lambda - x - \lambda y_\lambda\| < \lambda\varepsilon.$$

Then for each $x \in \overline{D(A)}$, we define $\lambda(x)$ as the least upper bound of λ that satisfies (5.27). Obviously we have $0 < \lambda(x) \leq \varepsilon$.

Now, considering $\lambda(x_0)$, from the definition of $\lambda(x_0)$, we can select $x_1 \in D(A)$, $y_1 \in Ax_1$, and h_1 that satisfy "$\lambda(x_0)/2 < h_1 \leq \varepsilon$, $\|x_1 - x_0 - h_1 y_1\| < h_1\varepsilon$." Next, considering $\lambda(x_1)$, from the definition there exist $x_2 \in D(A)$, $y_2 \in Ax_2$, and h_2 that satisfy "$\lambda(x_1)/2 < h_2 \leq \varepsilon$, $\|x_2 - x_1 - h_2 y_2\| < h_2\varepsilon$." Continuing this argument, we can choose $\{h_k\}$, $\{x_k\}$, and $\{y_k\}$ satisfying (5.28) and (5.29) below, where $x_k \in D(A)$ and $y_k \in Ax_k$ $(k = 1, 2, \dots)$:

$$\lambda(x_{k-1})/2 < h_k \leq \varepsilon \quad (k = 1, 2, \dots), \tag{5.28}$$

$$\|x_k - x_{k-1} - h_k y_k\| < h_k\varepsilon \quad (k = 1, 2, \dots). \tag{5.29}$$

Define $t_k = \sum_{i=1}^{k} h_i$ $(k = 1, 2, \dots)$. If we can show $\lim_{k\to\infty} t_k = \infty$, then the proof of the lemma is complete. Suppose $\lim_{k\to\infty} t_k \neq \infty$ and let a be the limit of $\{t_k\}$. As we will show later

$$\|x_i - x_j\| \leq \prod_{p=k+1}^{i} (1 - \omega_0 h_p)^{-1} \prod_{p=k+1}^{j} (1 - \omega_0 h_p)^{-1}[(t_i - t_j)\||Ax_k\||$$

$$+ \varepsilon(t_i - t_k) + \varepsilon(t_j - t_k)] \tag{5.30}$$

holds for every $i \geq j \geq k$ (≥ 1), where we set $\prod_{p=k+1}^{k}(1 - \omega_0 h_p)^{-1} = 1$. Noting that $\prod_{p=k+1}^{l}(1 - \omega_0 h_p)^{-1} \leq \exp[2\omega_0(t_l - t_k)]$ $(l \geq k \geq 1)$ (where we use $0 \leq \omega_0 h_p \leq \omega_0\varepsilon < 1/2$ and $(1 - t)^{-1} \leq \exp(2t)$ $(0 \leq t \leq 1/2)$), we obtain from the above inequality that

$$\limsup_{i,j\to\infty} \|x_i - x_j\| \leq 2[\exp(4\omega_0(a - t_k))]\varepsilon(a - t_k) \to 0 \quad (k \to \infty).$$

Hence $\{x_i\}$ is a Cauchy sequence and converges. Now set $x = \lim_{i\to\infty} x_i$. Since $x \in \overline{D(A)}$, from condition (R), there exist $\mu \in (0, \varepsilon/2]$, $x_\mu \in D(A)$, and $y_\mu \in Ax_\mu$ satisfying

$$\|x_\mu - x - \mu y_\mu\| < (\varepsilon/2)\mu. \tag{5.31}$$

Since $0 < \lambda(x_i) < 2h_{i+1} = 2(t_{i+1} - t_i) \to 0$, $\lim_{i\to\infty} \lambda(x_i) = 0$. Hence we can choose a natural number i_0 such that if $i \geq i_0$ then $\lambda(x_i) < \mu$. Then from the definition of $\lambda(x_i)$ we obtain

$$\|x_\mu - x_i - \mu y_\mu\| \geq \mu\varepsilon \quad (i \geq i_0).$$

Here if we let $i \to \infty$ then $\|x_\mu - x - \mu y_\mu\| \geq \mu \varepsilon$. This contradicts (5.31). Therefore $\lim_{k \to \infty} t_k = \infty$.

Finally, we prove that (5.30) holds for every $i \geq j \geq k$ (≥ 1) Choose an arbitrary natural number k and define $a_{i,j}$ and $\gamma_{i,j}$ $(i \geq j \geq k)$ by the following:

$$a_{i,j} = \|x_i - x_j\|, \qquad \gamma_{i,j} = \prod_{p=k+1}^{i} (1 - \omega_0 h_p) \prod_{p=k+1}^{j} (1 - \omega_0 h_p).$$

Now if we let $\lambda' = h_i$ and apply (5.14), then for every i $(> k)$ we have

$$\begin{aligned}
(1 - \omega_0 h_i)\|x_i - x_k\| &\leq (1 - \omega h_i)\|x_i - x_k\| \\
&\leq \|x_i - x_k - h_i y_i\| + h_i \|\|A x_k\|\| \\
&\leq \|x_i - x_{i-1} - h_i y_i\| + \|x_{i-1} - x_k\| + h_i \|\|A x_k\|\|.
\end{aligned}$$

Hence from (5.29) we obtain (note that $\gamma_{i,k} = (1 - \omega_0 h_i)\gamma_{i-1,k}$ and $0 < \gamma_{i-1,k} \leq 1$)

$$\gamma_{i,k} a_{i,k} \leq \gamma_{i-1,k} a_{i-1,k} + h_i \varepsilon + h_i \|\|A x_k\|\| \qquad (i > k).$$

Adding from $i = k+1$ to $i = i$ we obtain

$$\gamma_{i,k} a_{i,k} \leq (t_i - t_k)\|\|A x_k\|\| + \varepsilon(t_i - t_k) \qquad (i > k).$$

Hence (5.30) holds for $(i \geq)$ $j = k$. Also, it is self-evident that (5.30) holds for $i = j$. Now, let $i > j > k$ and assume that $a_{i-1,j}$ $(= \|x_{i-1} - x_j\|)$ and $a_{i,j-1}$ satisfy (5.30). Then, if we show $a_{i,j}$ satisfies (5.30) then, by induction, (5.30) holds for arbitrary $i \geq j \geq k$. Let $\alpha = h_j$ and $\beta = h_i$, and apply (5.13) to obtain

$$\begin{aligned}
(h_i + h_j - \omega_0 h_i h_j)a_{i,j} &\leq (h_i + h_j - \omega h_i h_j)\|x_i - x_j\| \\
&\leq h_j \|x_i - x_j - h_i y_i\| + h_i \|x_i - x_j + h_j y_j\| \\
&\leq h_j(a_{i-1,j} + \|x_i - x_{i-1} - h_i y_i\|) + h_i(a_{i,j-1} + \|x_j - x_{j-1} - h_j y_j\|).
\end{aligned}$$

From this and (5.29)

$$(h_i + h_j - \omega_0 h_i h_j)a_{i,j} \leq h_j a_{i-1,j} + h_i a_{i,j-1} + 2h_i h_j \varepsilon.$$

Then by the inductive assumption, we obtain

$$\begin{aligned}
(h_i + h_j &- \omega_0 h_i h_j)\gamma_{i,j} a_{i,j} \\
&\leq h_j(1 - \omega_0 h_i)\gamma_{i-1,j} a_{i-1,j} + h_i(1 - \omega_0 h_j)\gamma_{i,j-1} a_{i,j-1} + 2\varepsilon\gamma_{i,j} h_i h_j \\
&\leq [h_j(1 - \omega_0 h_i)(t_{i-1} - t_j) + h_i(1 - \omega_0 h_j)(t_i - t_{j-1})]\|\|A x_k\|\| \qquad (5.32) \\
&\quad + [h_j(1 - \omega_0 h_i)(\varepsilon(t_{i-1} - t_k) + \varepsilon(t_j - t_k)) \\
&\qquad + h_i(1 - \omega_0 h_j)(\varepsilon(t_i - t_k) + \varepsilon(t_{j-1} - t_k)) + 2\varepsilon\gamma_{i,j} h_i h_j]
\end{aligned}$$

(we put the right-hand side above equal to $I_1 + I_2$). Next we evaluate I_1 and I_2. Since $h_i - h_j \le h_i - h_j + (t_{i-1} - t_{j-1}) = t_i - t_j$,

$$\begin{aligned}
I_1 &= [(h_j(1 - \omega_0 h_i) + h_i(1 - \omega_0 h_j))(t_i - t_j) + \omega_0 h_i h_j (h_i - h_j)]|||Ax_k||| \\
&= [(h_i + h_j - \omega_0 h_i h_j)(t_i - t_j) - \omega_0 h_i h_j (t_i - t_j) + \omega_0 h_i h_j (h_i - h_j)]|||Ax_k||| \\
&\le (h_i + h_j - \omega_0 h_i h_j)(t_i - t_j)|||Ax_k|||.
\end{aligned} \tag{5.33}$$

Next, since $\gamma_{i,j} \le 1 - \omega_0 h_i$ and $\gamma_{i,j} \le 1 - \omega_0 h_j$, we have

$$\begin{aligned}
I_2 &\le h_j(1 - \omega_0 h_i)(\varepsilon(t_{i-1} - t_k) + \varepsilon(t_j - t_k)) \\
&\quad + h_i(1 - \omega_0 h_j)(\varepsilon(t_i - t_k) + \varepsilon(t_{j-1} - t_k)) \\
&\quad + h_i h_j (\varepsilon(1 - \omega_0 h_i) + \varepsilon(1 - \omega_0 h_j)) \\
&= (h_j(1 - \omega_0 h_i) + h_i(1 - \omega_0 h_j))(\varepsilon(t_i - t_k) + \varepsilon(t_j - t_k)) \\
&\le (h_i + h_j - \omega_0 h_i h_j)(\varepsilon(t_i - t_k) + \varepsilon(t_j - t_k)).
\end{aligned}$$

From this and (5.33) and (5.32), we obtain

$$\gamma_{i,j} a_{i,j} \le (t_i - t_j)|||Ax_k||| + \varepsilon(t_i - t_k) + \varepsilon(t_j - t_k)$$

and $a_{i,j} = \|x_i - x_j\|$ satisfies (5.30). $\square$

PROOF OF THEOREM 5.10. Let $x \in \overline{D(A)}$. Then by Lemma 5.11, for every $\lambda > 0$, there exist $\{t_k^\lambda\}$, $\{x_k^\lambda\}$, and $\{y_k^\lambda\}$ satisfying (5.34) through (5.36) below, where $x_k^\lambda \in D(A)$ and $y_k^\lambda \in Ax_k^\lambda$ $(k = 1, 2, \dots)$:

$$0 = t_0^\lambda < t_1^\lambda < \cdots < t_k^\lambda < t_{k+1}^\lambda < \cdots, \quad \lim_{k \to \infty} t_k^\lambda = \infty, \tag{5.34}$$

$$t_k^\lambda - t_{k-1}^\lambda \le \lambda \quad (k = 1, 2, \dots), \tag{5.35}$$

$$\|x_k^\lambda - x_{k-1}^\lambda - (t_k^\lambda - t_{k-1}^\lambda) y_k^\lambda\| < \lambda(t_k^\lambda - t_{k-1}^\lambda) \quad (k = 1, 2, \dots), \tag{5.36}$$

where $x_0^\lambda = x$. We set $(x_k^\lambda - x_{k-1}^\lambda)/(t_k^\lambda - t_{k-1}^\lambda) - y_k^\lambda = \varepsilon_k^\lambda$ $(k = 1, 2, \dots)$. Now choose $T > 0$ arbitrarily. Then from (5.34), for each $\lambda > 0$ we can choose a natural number N_λ such that $t_{N_\lambda - 1}^\lambda < T \le t_{N_\lambda}^\lambda$. Obviously, $\{t_k^\lambda : k = 0, 1, \dots, N_\lambda\}$, $\{x_k^\lambda : k = 0, 1, \dots, N_\lambda\}$, and $\{\varepsilon_k^\lambda : k = 1, 2, \dots, N_\lambda\}$ satisfy (5.6) and (5.7). Also by (5.35) $|\Delta_\lambda| = \max\{t_k^\lambda - t_{k-1}^\lambda : k = 1, 2, \dots, N_\lambda\} \le \lambda$, and hence we have $|\Delta_\lambda| \to 0$ as $\lambda \to 0+$, that is, (5.8) holds. Furthermore, from (5.36) and (5.35) we obtain

$$\sum_{k=1}^{N_\lambda} \|\varepsilon_k^\lambda\|(t_k^\lambda - t_{k-1}^\lambda) < \lambda t_{N_\lambda}^\lambda < \lambda(T + \lambda) \to 0 \quad (\lambda \to 0+)$$

and (5.9) holds. Therefore, using Theorem 5.7 the Cauchy problem for A

$$(\text{CP}; x) \quad \begin{cases} (d/dt)u(t) \in Au(t) & (0 \le t \le T), \\ u(0) = x \end{cases}$$

has unique integral solution $u(t): [0, T] \to \overline{D(A)}$ of type ω.

For arbitrary $x \in \overline{D(A)}$ and $T > 0$, we have shown that the Cauchy problem (CP; x) for A (on $[0, T]$) has unique integral solution ($[0, T] \to \overline{D(A)}$) of type ω. We denote this integral solution by $u_T(t; x)$ (so that, $u_T(t; x)$: $[0, T] \to \overline{D(A)}$). Then by part (ii) of Theorem 5.7

$$\|u_T(t; x) - u_T(t; y)\| \le e^{\omega(t-s)} \|u_T(s; x) - u_T(s; y)\|$$
$$(0 \le s < t \le T; x, y \in \overline{D(A)}), \quad (5.37)$$

and also, from the remark described in the footnote following (5.21)

$$\|u_T(t; x) - u_T(s; x)\| \le e^{2\omega_0(t+s)} \||Ax|\| |t - s| \quad (0 \le s, t \le T, \ x \in D(A)).$$
$$(5.38)$$

Then

If $0 < T_1 < T_2$ then $u_{T_1}(t; x) = u_{T_2}(t; x)$ $(0 \le t \le T_1, \ x \in \overline{D(A)})$.
$$(5.39)$$

In fact, when we consider the restriction of $u_{T_2}(t; x)$ to $[0, T_1]$, since this is an integral solution of type ω of the Cauchy problem (CP; x) for A on $[0, T_1]$, from the fact that $u_{T_1}(t; x)$ is the unique integral solution of type ω of the Cauchy problem for A on $[0, T_1]$, we have that $u_{T_1}(t; x) = u_{T_2}(t; x)$ $(0 \le t \le T)$.

Now for arbitrary $x \in \overline{D(A)}$ and $t \ge 0$, we define $u(t; x)$ by

$$u(t; x) = u_T(t; x) \quad (\in \overline{D(A)}).$$

Here, T is an arbitrary positive number such that $T \ge t$. From (5.39) this definition makes sense, and we see that $u(t; x): [0, \infty) \to \overline{D(A)}$ is a unique integral solution of type ω of the Cauchy problem (CP; x)$_\infty$ for A. Moreover, from (5.37) and (5.38) we obtain (5.25) and (5.26). □

REMARK. From Corollary 5.3 and the fact that a dissipative operator of type ω is also a dissipative operator of type ω' $(> \omega)$, we see the following: $u(t; x)$ of Theorem 5.10 is not only determined uniquely as the integral solution of type ω of the Cauchy problem (CP; x)$_\infty$ for A, but also, if $v(t; x): [0, \infty) \to X$ is an integral solution of type ω' $(\ne \omega)$ of the same Cauchy problem, then $v(t; x) = u(t; x)$ $(t \ge 0)$.

As a corollary of Theorem 5.10, we obtain the following generation theorem of semigroups.

THEOREM 5.12. *Let A be a dissipative operator of type ω satisfying condition* (R). *Then there exists a unique semigroup $\{T(t): t \ge 0\}$ of type ω on $\overline{D(A)}$ such that for each $x \in \overline{D(A)}$, $T(t)x: [0, \infty) \to \overline{D(A)}$ is an integral solution of type ω of the Cauchy problem (CP; x)$_\infty$ for A. Moreover, for each $x \in \overline{D(A)}$, $T(t)x: [0, \infty) \to \overline{D(A)}$ is a unique integral solution of type ω of the Cauchy problem (CP; x)$_\infty$ for A. Moreover, if we set $\omega_0 = \max\{0, \omega\}$*

then

$$\|T(t)x - T(s)x\| \le e^{2\omega_0(t+s)}\|\|Ax\|\|\|t - s|\;(^5) \qquad (0 \le s, t < \infty; x \in D(A)).$$
(5.40)

REMARK. Let $X_0 \subset X$ and let ω be a real number. $\{T(t) : t \ge 0\}$ is called a semigroup of type ω on X_0, if (i) through (iv) hold:

(i) For each $t > 0$, $T(t): X_0 \to X_0$ and $T(0)x = x$ $(x \in X_0)$,
(ii) $\|T(t)x - T(t)y\| \le e^{\omega t}\|x - y\|$ $(x, y \in X_0,\; t \ge 0)$.
(iii) $T(t + s) = T(t)T(s)$ $(t, s \ge 0)$,
(iv) $\lim_{t \to 0+} T(t)x = x$ $(x \in X_0)$.

Hence, the semigroups of contractions treated previously are precisely the semigroups of type 0.

PROOF OF THEOREM 5.12. By Theorem 5.10, for each $x \in \overline{D(A)}$, the Cauchy problem $(CP; x)_\infty$ for A has a unique integral solution $u(t; x)$: $[0, \infty) \to \overline{D(A)}$ of type ω, and (5.25) and (5.26) hold. Now for each $t \ge 0$ we define an operator $T(t): \overline{D(A)} \to \overline{D(A)}$ by

$$T(t)x = u(t; x) \qquad (x \in \overline{D(A)}).$$

Then by Theorem 5.4, $\{T(t) : t \ge 0\}$ is a semigroup on $\overline{D(A)}$. That is, (i), (iii), and (iv) in the above remark hold. Also, (ii) holds by (5.25). Hence, $\{T(t) : t \ge 0\}$ is the required semigroup of type ω on $\overline{D(A)}$. (We can obtain (5.40) from (5.26).) $\square$

COROLLARY 5.13. *Let A be a dissipative operator of type ω satisfying the following condition* (R_0):

(R_0) *For a sufficiently small $\lambda > 0$, $R(I - \lambda A) \supset D(A)$.*

Then there exists a unique semigroup $\{T(t) : t \ge 0\}$ of type ω on $\overline{D(A)}$ such that for each $x \in \overline{D(A)}$, $T(t)x: [0, \infty) \to \overline{D(A)}$ is an integral solution of type ω of the Cauchy problem $(CP; x)_\infty$ for A. Moreover, for each $x \in \overline{D(A)}$, $T(t)x: [0, \infty) \to \overline{D(A)}$ is a unique integral solution of type ω of the Cauchy problem $(CP; x)_\infty$ for A, and

$$T(t)x = \lim_{\lambda \to 0+} (I - \lambda \overline{A})^{-[t/\lambda]} x \;(\text{uniformly on bounded subintervals of } [0, \infty).$$
(5.41)

Here $\overline{A}$ is the closure of A. Moreover, (5.40) holds.

PROOF. Considering the closure $\overline{A}$ of A, since $\overline{A}$ is a dissipative operator of type ω, and from (R_0) we have

$$\lim_{\lambda \to 0+} \lambda^{-1} d(R(I - \lambda\overline{A}), x) = 0 \quad \text{for all } x \in \overline{D(\overline{A})}.$$
(5.42)

$(^5)$ By the same reasoning as in the proof of Lemma 3.1, we can show $\|T(t+h)x - T(t)x\| \le e^{\omega t}\omega^{-1}(e^{\omega h} - 1)\|\|Ax\|\|$ $(t, h \ge 0)$.

(In fact, let $\lambda > 0$ be sufficiently small. Since the closed dissipativity of $\overline{A} - \omega I$, $R(I - \lambda\overline{A})$ is a closed set, and hence, from (R_0) we can obtain $R(I - \lambda\overline{A}) \supset \overline{D(A)} = \overline{D(\overline{A})}$.) Considering $\overline{A}$ instead of A and applying Theorem 5.12, there exists a semigroup $\{T(t) : t \geq 0\}$ of type ω on $\overline{D(A)}$ such that for each $x \in \overline{D(A)}$ $(= \overline{D(\overline{A})})$, $T(t)x : [0, \infty) \to \overline{D(A)}$ is a unique integral solution of type ω of the Cauchy problem $(CP; x)_\infty$ for $\overline{A}$, and

$$\|T(t)x - T(s)x\| \leq e^{2\omega_0(t+s)}\||\overline{A}x\||\,|t - s| \qquad (0 \leq s, t < \infty; \ x \in D(\overline{A})).$$

Now, the integral solution of type ω of the Cauchy problem $(CP; x)_\infty$ for A is always the integral solution of type ω of the Cauchy problem $(CP; x)_\infty$ for $\overline{A}$, and the converse is also true. Therefore, $\{T(t) : t \geq 0\}$ is the required semigroup of type ω satisfying (5.40).

Finally, we prove (5.41). As we described above, $\overline{A}$ is a dissipative operator of type ω, and there is a λ_0 (> 0) such that

$$\text{if } 0 < \lambda \leq \lambda_0, \text{ then } R(I - \lambda\overline{A}) \supset \overline{D(\overline{A})} = \overline{D(A)}. \qquad (5.43)$$

We may assume that λ_0 satisfies $\lambda_0\omega_0 < 1$. Choose $x \in \overline{D(A)}$ and $T > 0$ arbitrarily, and let $0 < \lambda \leq \lambda_0$. Then since $1 - \lambda\omega > 0$, from the dissipativity of $\overline{A} - \omega I$, we can define $\overline{J}_\lambda \equiv (I - \lambda\overline{A})^{-1}$ as a single-valued operator, and

$$\|\overline{J}_\lambda y - \overline{J}_\lambda z\| \leq (1 - \lambda\omega)^{-1}\|y - z\| \qquad (y, z \in R(I - \lambda\overline{A})), \qquad (5.44)$$

$$\|\overline{J}_\lambda y - y\| \leq \lambda(1 - \lambda\omega)^{-1}\||\overline{A}y\|| \qquad (y \in R(I - \lambda\overline{A}) \cap D(\overline{A})). \qquad (5.45)$$

From (5.43) $\overline{J}_\lambda^k x$ can be defined for every natural number k, and $\overline{J}_\lambda^k x - \overline{J}_\lambda^{k-1} x \in \lambda\overline{A}J_\lambda^k x$ $(k = 1, 2, \ldots)$. Hence, if we choose the smallest natural number N_λ such that $\lambda N_\lambda \geq T$ and consider $x_k^\lambda = \overline{J}_\lambda^k x$, $t_k^\lambda = k\lambda$, and $\varepsilon_k^\lambda = 0$ $(k = 1, 2, \ldots, N_\lambda)$, then (5.7) through (5.9) holds. Therefore, by Theorem 5.7, $u(t) = \lim_{\lambda \to 0+} u_\lambda(t)$ (the convergence is uniform on $[0, T]$) is a unique integral solution of type ω of the Cauchy problem $(CP; x)$ for $\overline{A}$. Now, from the definition of $u_\lambda(t)$

$$u_\lambda(t) = \begin{cases} \overline{J}_\lambda^{[t/\lambda]+1} x & (t \in (0, T], \ t \neq k\lambda \ (k = 1, 2, \ldots, N_\lambda)), \\ \overline{J}_\lambda^{[t/\lambda]} x & (t \in [0, T], \ t = k\lambda \ (k = 0, 1, \ldots, N_\lambda)). \end{cases}$$

From this

$$\|u_\lambda(t) - \overline{J}_\lambda^{[t/\lambda]} x\| \leq \|\overline{J}_\lambda^{[t/\lambda]+1} x - \overline{J}_\lambda^{[t/\lambda]} x\| \leq (1 - \lambda\omega)^{-[t/\lambda]}\|\overline{J}_\lambda x - x\|$$

$(0 \leq t \leq T)$. (Here we used (5.44).) Furthermore, from (5.45) we obtain that $\|\overline{J}_\lambda x - x\| \to 0$ as $\lambda \to 0+$. Hence

$$u(t) = \lim_{\lambda \to 0+} \overline{J}_\lambda^{[t/\lambda]} x = \lim_{\lambda \to 0+} (I - \lambda\overline{A})^{-[t/\lambda]} x$$

(the convergence is uniform on $[0, T]$). From this and $T(t)x = u(t)$ $(0 \leq t \leq T)$ (since the restriction of $T(t)x$ to $[0, T]$ is also an integral solution of type ω of the Cauchy problem $(CP; x)$ for $\overline{A}$, we have $T(t)x = u(t))$, we obtain (5.41). $\square$

REMARK $1°$. The semigroup of contractions $\{T(t) : t \geq 0\}$ on $\overline{D(A)}$ formulated with Theorem 4.2 is precisely the case of $\omega = 0$ in the above corollary. Hence from this corollary, for each $x \in \overline{D(A)}$, $T(t)x : [0, \infty) \to \overline{D(A)}$ is the unique integral solution of type 0 of the Cauchy problem $(CP; x)_\infty$ for A. Further, the convergence (4.3) of Theorem 4.2 can be obtained immediately from (5.41), because $(I - \lambda\overline{A})^{-1}x = (I - \lambda A)^{-1}x$ for $x \in R \cap \overline{D(A)}$. Part (ii) of Theorem 4.2 can also be obtained from (5.40).

$2°$. Let A be a closed operator satisfying the assumptions of Corollary 5.13, and let $\{T(t) : t \geq 0\}$ be the semigroup of type ω on $\overline{D(A)}$ in Corollary 5.13. By the same reasoning as used in the proof of Theorem 4.5, we can prove that "if $T(t)x$ $(x \in \overline{D(A)})$ is strongly differentiable at $t_0 > 0$, then $[(d/dt)T(t)x]_{t=t_0} \in A^0 T(t_0)x \subset A T(t_0)x$." Also from this, we see that if $T(t)x$ $(x \in D(A))$ is strongly differentiable a.e. $t \geq 0$, then $T(t)x : [0, \infty) \to \overline{D(A)}$ is a unique solution of the Cauchy problem $(CP; x)_\infty$ for A (see Theorem 4.10).

In Theorems 4.8 and 4.11, we can obtain the same conclusions with the condition (R_1) replaced by the weaker condition (R). In fact, the next theorem holds.

THEOREM 5.14. *Let A be a single-valued dissipative operator of type ω with closed domain $D(A)$ satisfying the condition (R) and let $\{T(t) : t \geq 0\}$ be the semigroup of type ω on $D(A)$ of Theorem 5.12.*

(i) If A is a demicontinuous operator, then A is the weak infinitesimal generator of $\{T(t) ; t \geq 0\}$. Moreover for each $x \in D(A)$, $T(t)x$ is a unique solution of the Cauchy problem $(CP; x)_\infty$ for A.

(ii) If A is a continuous operator, then A is the infinitesimal generator of $\{T(t) : t \geq 0\}$, and for each $x \in D(A)$, $T(t)x$ is strongly differentiable at all t (≥ 0) and $(d/dt)T(t)x = AT(t)x$ $(t \geq 0)$.

PROOF. Let $x \in D(A)$. Then from Lemma 5.11, for every $\lambda > 0$, there exist $\{t_k^\lambda\}$, $\{x_k^\lambda\} \subset D(A)$, and $\{y_k^\lambda (= Ax_k^\lambda)\}$ satisfying (5.34) through (5.36). We set $\varepsilon_k^\lambda = (x_k^\lambda - x_{k-1}^\lambda)/(t_k^\lambda - t_{k-1}^\lambda) - Ax_k^\lambda$ $(k = 1, 2, \ldots)$ and define $u_\lambda(t) : [0, \infty) \to D(A)$ as follows:

$$u_\lambda(t) = \begin{cases} x & (t = 0), \\ x_k^\lambda & (t \in (t_{k-1}^\lambda, t_k^\lambda]), \quad k = 1, 2, \ldots. \end{cases}$$

Then, as we see from the method of construction of $\{T(t) : t \geq 0\}$ (see Theorems 5.12, 5.10, 5.7 and their proofs), we have

$$T(t)x = \lim_{\lambda \to 0+} u_\lambda(t) \qquad (0 \leq t < \infty). \tag{5.46}$$

Here, the convergence is uniform on bounded subintervals of $[0, \infty)$. Now for $t > 0$, we choose i such that $t \in (t_{i-1}^\lambda, t_i^\lambda]$. Then from $x_k^\lambda - x_{k-1}^\lambda =$

$(t_k^\lambda - t_{k-1}^\lambda)Ax_k^\lambda + (t_k^\lambda - t_{k-1}^\lambda)\varepsilon_k^\lambda$ $(k = 1, 2, \ldots)$, we have

$$u_\lambda(t) - x = x_i^\lambda - x = \sum_{k=1}^{i}(x_k^\lambda - x_{k-1}^\lambda)$$

$$= \sum_{k=1}^{i}(t_k^\lambda - t_{k-1}^\lambda)Ax_k^\lambda + \sum_{k=1}^{i}(t_k^\lambda - t_{k-1}^\lambda)\varepsilon_k^\lambda$$

$$= \int_0^{t_i^\lambda} Au_\lambda(\tau)\,d\tau + \sum_{k=1}^{i}(t_k^\lambda - t_{k-1}^\lambda)\varepsilon_k^\lambda.$$

$$\int_0^{t_i^\lambda} Au_\lambda(\tau)\,d\tau = \int_0^t Au_\lambda(\tau)\,d\tau + \int_t^{t_i^\lambda} Au_\lambda(\tau)\,d\tau$$

$$= \int_0^t Au_\lambda(\tau)\,d\tau + (t_i^\lambda - t)Au_\lambda(t).$$

Thus,

$$u_\lambda(t) - x = \int_0^t Au_\lambda(\tau)\,d\tau + (t_i^\lambda - t)Au_\lambda(t) + \sum_{k=1}^{i}(t_k^\lambda - t_{k-1}^\lambda)\varepsilon_k^\lambda. \tag{5.47}$$

Also from (5.35) and (5.36)

$$|t_i^\lambda - t| \le \lambda, \qquad \left\|\sum_{k=1}^{i}(t_k^\lambda - t_{k-1}^\lambda)\varepsilon_k^\lambda\right\| \le \lambda t_i^\lambda \le \lambda(t + \lambda). \tag{5.48}$$

(i) Let A be a demicontinuous operator. Then from (5.46)

$$\text{w-}\lim_{\lambda\to 0+} Au_\lambda(t) = AT(t)x,$$

and hence from the uniform boundedness theorem we have $\|Au_\lambda(t)\| = O(1)$ $(\lambda \to 0+)$. From this and (5.48), if we let $\lambda \to 0+$ in (5.47), we obtain

$$T(t)x - x = \lim_{\lambda\to 0+}\int_0^t Au_\lambda(\tau)\,d\tau. \tag{5.49}$$

Let $x^* \in X^*$. From $\lim_{\lambda\to 0+} u_\lambda(\tau) = T(\tau)x$ (where the convergence is uniform on $[0, t]$) and from the demicontinuity of A, we have $\lim_{\lambda\to 0+} x^*(Au_\lambda(\tau)) = x^*(AT(\tau)x)$, where the convergence uniform on $[0, t]$. Hence from (5.49) we obtain

$$x^*[T(t)x - x] = \int_0^t x^*(AT(\tau)x)\,d\tau \qquad (t \ge 0). \tag{5.50}$$

Then since $x^*(AT(\tau)x)$ is continuous on $[0, \infty)$

$$\lim_{t\to 0+} x^*(t^{-1}[T(t)x - x]) = x^*(Ax).$$

This equation holds for arbitrary $x^* \in X^*$. Therefore we obtain

$$\text{w-}\lim_{t\to 0+} t^{-1}[T(t)x - x] = Ax \qquad (x \in D(A)),$$

and A is the weak infinitesimal generator of $\{T(t) : t \geq 0\}$. (Furthermore, as we see from this argument, the weak derivative $(w - d/dt)T(t)x = AT(t)x$ $(t \geq 0 : x \in D(A))$ holds.) Next, by noting that the function, which is weakly continuous on a bounded closed interval, is (Bochner) integrable on the interval (see Theorem 1.14), we obtain from (5.50) that

$$x^*[T(t)x - x] = x^* \left[\int_0^t AT(\tau)x \right] d\tau \qquad (t \geq 0).$$

Since $x^* \in X^*$ is arbitrary, from the above equation we obtain

$$T(t)x - x = \int_0^t AT(\tau)x \, d\tau \qquad (t \geq 0). \tag{5.51}$$

Hence $T(t)x$ is strongly differentiable a.e. $t \geq 0$ and $(d/dt)T(t)x = AT(t)x$ (a.e. $t \geq 0$). Also $T(t)x$ is Lipschitz continuous on every bounded subinterval of $[0, \infty)$ (see (5.40)). We have shown that $T(t)x : [0, \infty) \to D(A)$ is a unique solution of the Cauchy problem $(\mathrm{CP} ; x)_\infty$ for A.

(ii) From the continuity of $AT(t)x : [0, \infty) \to X$ and from (5.51), $T(t)x$ is strongly differentiable at every $t \geq 0$ and $(d/dt)T(t)x = AT(t)x$ $(t \geq 0)$. In particular, A is the infinitesimal generator of $\{T(t) : t \geq 0\}$. $\square$

Next, we examine the conditions for a continuous operator to be the infinitesimal generator of a semigroup of type ω.

THEOREM 5.15. *Let A be a continuous operator with closed domain $D(A)$ $(\subset X)$ and with values in X. Then* (i) *through* (iii) *are equivalent*:

(i) *A is the infinitesimal generator of a semigroup of type ω on $D(A)$.*

(ii) *A is a dissipative operator of type ω, and satisfies the following condition* (5.52):

For all $x \in D(A)$, $\displaystyle\liminf_{\lambda \to 0+} \lambda^{-1} d(D(A), x + \lambda Ax) = 0$. \tag{5.52}

(iii) *A is a dissipative operator of type ω satisfying condition* (R), *that is, for all $x \in D(A)$ $(= \overline{D(A)})$, $\liminf_{\lambda \to 0+} \lambda^{-1} d(R(I - \lambda A), x) = 0$.*

Moreover, if $D(A)$ is a closed convex set, then (iv) *below is equivalent to the above conditions*:

(iv) *A is a dissipative operator of type ω and satisfies the following condition* (R_0'):

(R_0') *$R(I - \lambda A) \supset D(A)$ for all $\lambda > 0$ such that $\lambda\omega < 1$.*

PROOF. (i) $\Rightarrow$ (ii). Let A be the infinitesimal generator of a semigroup $\{T(t) : t \geq 0\}$ of type ω on $D(A)$ and let $x \in D(A)$. From $\theta(\lambda) = \lambda^{-1}[T(\lambda)x - x] - Ax \to 0$ $(\lambda \to 0+)$, we obtain

$$\lambda^{-1} d(D(A), x + \lambda Ax) \leq \lambda^{-1}\|T(\lambda)x - (x + \lambda Ax)\| = \|\theta(\lambda)\| \to 0 \quad (\lambda \to 0+)$$

and we have $\lim_{\lambda \to 0+} \lambda^{-1} d(D(A), x + \lambda Ax) = 0$. Hence (5.52) holds. Next, if we set $A_h = h^{-1}(T(h) - I)$ $(h > 0)$, then for every $x, y \in D(A)$ we have

$$\text{Re}(A_h x - \omega x - (A_h y - \omega y), f) = \text{Re}(A_h x - A_h y, f) - \omega \|x - y\|^2$$
$$\leq h^{-1}(\|T(h)x - T(h)y\|\|x - y\| - \|x - y\|^2) - \omega \|x - y\|^2$$
$$\leq h^{-1}(e^{\omega h} - 1)\|x - y\|^2 - \omega \|x - y\|^2 \qquad (f \in F(x - y)).$$

If we let $h \to 0+$, then for every $x, y \in D(A)$

$$\text{Re}(Ax - \omega x - (Ay - \omega y), f) \leq 0 \qquad (f \in F(x - y)).$$

Therefore, $A - \omega I$ is a dissipative operator; that is, A is a dissipative operator of type ω.

(ii) $\Rightarrow$ (iii). Let $x \in D(A)$. For every $\lambda > 0$, there exists $x_\lambda \in D(A)$ such that $\|x_\lambda - (x + \lambda Ax)\| < \lambda^2 + d(D(A), x + \lambda Ax)$.

From $d(D(A), x + \lambda Ax) \leq \lambda \|Ax\|$, we have $\|x_\lambda - x\| < \lambda^2 + 2\lambda\|Ax\| \to 0$ $(\lambda \to 0+)$. Then from the continuity of A, $\|Ax_\lambda - Ax\| \to 0$ $(\lambda \to 0+)$. Now

$$d(R(I - \lambda A), x) \leq \|x_\lambda - \lambda Ax_\lambda - x\| \leq \|x_\lambda - (x + \lambda Ax)\| + \lambda\|Ax_\lambda - Ax\|$$
$$< \lambda^2 + d(D(A), x + \lambda Ax) + \lambda\|Ax_\lambda - Ax\|.$$

Hence, from (5.52) we obtain $\liminf_{\lambda \to 0+} \lambda^{-1} d(R(I - \lambda A), x) = 0$.

(iii) $\Rightarrow$ (i) follows from (ii) of Theorem 5.14.

Next, we consider the case where $D(A)$ is a closed convex set.

(iv) $\Rightarrow$ (iii) is self-evident. To prove (ii) $\Rightarrow$ (iv), we take an $\alpha > 0$ with $\alpha\omega < 1$ and $z \in D(A)$, and set

$$Bx = \alpha Ax - x + z \qquad (x \in D(A)).$$

Then B is a continuous operator whose domain $D(B)$ $(= D(A))$ is a closed set. Also, from the dissipativity of $A - \omega I$, we have

$$\langle [Bx - (\alpha\omega - 1)x] - [By - (\alpha\omega - 1)y], x - y \rangle_i$$
$$= \alpha\langle Ax - \omega x - (Ay - \omega y), x - y \rangle_i \leq 0 \qquad (x, y \in D(B) = D(A)).$$

Hence, B is a dissipative operator of type $(\alpha\omega - 1)$. Next, we show that $\liminf_{\lambda \to 0+} \lambda^{-1} d(D(B), x + \lambda Bx) = 0$ $(x \in D(B))$ holds. Let $x \in D(B)$. Then, for each $\lambda \in (0, 1)$, we can select $x_\lambda \in D(A) = D(B)$ such that

$$\|x_\lambda - (x + \lambda Ax)\| < \lambda^2 + d(D(A), x + \lambda Ax).$$

Now we set

$$y_\lambda = (1 + \alpha)^{-1}[\alpha x_\lambda + \lambda z + (1 - \lambda)x] \qquad (0 < \lambda < 1).$$

Since $D(B)$ $(= D(A))$ is a convex set, $y_\lambda \in D(B)$. Then

$$d\left(D(B), x + \frac{\lambda}{1+\alpha}Bx\right) \leq \left\| y_\lambda - \left(x + \frac{\lambda}{1+\alpha}Bx\right)\right\|$$
$$\leq \alpha(1+\alpha)^{-1}\|x_\lambda - (x + \lambda Ax)\|$$
$$\leq \alpha(1+\alpha)^{-1}(\lambda^2 + d(D(A), x + \lambda Ax)).$$

Hence from (5.52) we obtain $\liminf_{\lambda \to 0+} \lambda^{-1}d(D(B), x + \lambda Bx) = 0$. Then from the equivalence of (i) and (ii) (shown in the first half), B is the infinitesimal generator of a semigroup of type $(\alpha\omega - 1)$ on $D(B)$ $(= D(A))$. Now let this semigroup of type $(\alpha\omega - 1)$ be $\{S(t) : t \geq 0\}$. Since $S(t): D(B) \to D(B)$ (closed set) satisfies

$$\|S(t)x - S(t)y\| \leq e^{(\alpha\omega - 1)t}\|x - y\| \qquad (t \geq 0; \ x, y \in D(B))$$

and $e^{(\alpha\omega - 1)t} < 1$ $(t > 0)$, for each $t > 0$ there exists a unique $x_t \in D(B)$ such that $S(t)x_t = x_t$. Then $x_t = x_s$ $(t, s > 0)$ holds. In fact, from $S(s)x_t = S(s)S(t)x_t = S(t)S(s)x_t$, we have that $S(s)x_t = x_t$ and obtain $x_t = x_s$. Hence, there exists $y \in D(B) = D(A)$, and $S(t)y = y$ for all $t \geq 0$. Therefore, $By = \lim_{h \to 0+} h^{-1}(S(h)y - y) = 0$; that is, $z = (I - \alpha A)y \in R(I - \alpha A)$. Since α (where $\alpha > 0$ and $\alpha\omega < 1$) and $z \in D(A)$ are arbitrary, we have that $R(I - \alpha A) \supset D(A)$ for all $\alpha > 0$ such that $\alpha\omega < 1$. That is, (R_0') holds. $\square$

COROLLARY 5.16. *Let A be a continuous operator defined on all of X with values in X. If A is a dissipative operator of type ω, then A is the infinitesimal generator of a semigroup of type ω on X, and $A - \omega I$ is an m-dissipative operator.*

PROOF. Let A be a dissipative operator of type ω and set $B = A - \omega I$. Then B is a continuous dissipative operator with $D(B) = X$. Obviously $d(X, x + \lambda Bx) = 0$ $(\lambda > 0, x \in X)$, and thus $\lim_{\lambda \to 0+} \lambda^{-1}d(X, x + \lambda Bx) = 0$ $(x \in X)$. Hence, from Theorem 5.15 we obtain $R(I - \lambda B) = X$ $(\lambda > 0)$, and $B = A - \omega I$ is an m-dissipative operator. Also, from Theorem 5.15, A is the infinitesimal generator of a semigroup of type ω on X. $\square$

REMARK. In (iv) of the above theorem, if A is a dissipative operator, then condition (R_0') is precisely the condition (R_1) given in Theorem 4.2 (that is, $R(I - \lambda A) \supset D(A)$ for all $\lambda > 0$).

If $D(A)$ is not a convex set, then the latter half of the above theorem does not hold in general. In fact, as we see in the next example, there is a continuous dissipative operator with closed domain that satisfies condition (R) but not condition (R_1).

EXAMPLE 5.1. Let $X = R^2$ (the two-dimensional Euclidean space) and consider the following operator A:

$$D(A) = \{(\xi, \eta) \in R^2 : \xi^2 + \eta^2 = 1\},$$
$$A(\xi, \eta) = (\eta, -\xi) \qquad ((\xi, \eta) \in D(A)).$$

Then A is a continuous dissipative operator defined on a closed set $D(A)$. To show that A satisfies (5.52), let $x = (\xi, \eta) \in D(A)$. Since for every real number λ, $x + \lambda Ax = (\xi + \lambda\eta, \eta - \lambda\xi)$ and $\|x + \lambda Ax\|^2 = 1 + \lambda^2$, if we set $z = (1+\lambda^2)^{-1/2}(x+\lambda Ax)$, then $z \in D(A)$ and $\|z - (x+\lambda Ax)\| = \sqrt{1+\lambda^2} - 1$. From this we obtain

$$\lambda^{-1} d(D(A), x + \lambda Ax) \le \lambda^{-1}\left(\sqrt{1+\lambda^2} - 1\right) \to 0 \qquad (\lambda \to 0+)$$

and (5.52) holds. Hence by the theorem above, A satisfies condition (R). On the other hand, from $\|x - \lambda Ax\|^2 = 1 + \lambda^2$ $(x \in D(A))$, we have that $R(I - \lambda A) \cap D(A) = \varnothing$ $(\lambda > 0)$. Therefore, A does not satisfy condition (R_1). (By the theorem above, A is the infinitesimal generator of a certain semigroup of contractions $\{T(t) : t \ge 0\}$ on $D(A)$. In fact, $T(t)$ is the restriction of a linear contraction operator $U(t)$ on R^2 to $D(A)$, which can be defined by $U(t)(\xi, \eta) = (\xi \cos t + \eta \sin t, -\xi \sin t + \eta \cos t)$ $((\xi, \eta) \in R^2)$. Then we see easily that $\{U(t) : t \ge 0\}$ is a semigroup of linear contractions on R^2.)

§4. Cauchy's problem for $(d/dt)u(t) \in Au(t) + f(t)$

We apply Theorem 5.10 to examine the following Cauchy problem with an inhomogeneous term $f(t)$:

$$\begin{cases} (d/dt)u(t) \in Au(t) + f(t) & (0 \le t \le T), \\ u(0) = x, \end{cases} \qquad (5.53)$$

where A is not necessarily a single-valued operator with $D(A) \subset X$, $R(A) \subset X$, and $T > 0$, $f(t) : [0, T] \to X$, and $x \in X$.

DEFINITION 5.2. Let $x \in X$, let $f(t)$ be a (Bochner) integrable function on $[0, T]$, and let ω be a real number. $u(t) : [0, T] \to X$ is called an integral solution of type ω of the Cauchy problem (5.53), if (i) through (iii) hold:

(i) $u(0) = x$.

(ii) $u(t)$ is continuous on $[0, T]$.

(iii) For every r, $t \in [0, T]$ $(r < t)$, and every $x_0 \in D(A)$ and $y_0 \in Ax_0$, we have

$$e^{-2\omega t}\|u(t) - x_0\|^2 - e^{-2\omega r}\|u(r) - x_0\|^2 \le 2\int_r^t e^{-2\omega r}\langle f(\tau) + y_0, u(\tau) - x_0\rangle_s \, d\tau. (^6)$$
$$(5.54)$$

(Hence, if $f(t) \equiv 0$, then this coincides with Definition 5.1.)

For each $z \in X$, we define a (multivalued) operator A_z by

$$A_z x = Ax + z \ (= \{y + z : y \in Ax\}) \qquad (x \in D(A)).$$

(^{6}) From $\langle f(\tau) + y_0, u(\tau) - x_0\rangle_s = \|u(\tau) - x_0\| \lim_{\alpha \to 0+} \alpha^{-1}[\|u(\tau) - x_0 + \alpha(f(\tau) + y_0)\| - \|u(\tau) - x_0\|]$ (see (2.14) and (2.2)), $\langle f(\tau) + y_0, u(\tau) - x_0\rangle_s$ is a measurable function. From this and $|\langle f(\tau) + y_0, u(\tau) - x_0\rangle_s| \le \|f(\tau) + y_0\|\|u(\tau) - x_0\|$, $e^{-2\omega\tau}\langle f(\tau) + y_0, u(\tau) - x_0\rangle_s$ is (Lebesgue) integrable on $[0, T]$.

If A is a dissipative operator of type ω, then A_z is also a dissipative operator of type ω for every $z \in X$. To apply Theorem 5.10, we introduce the following set:

$$E(A) = \left\{ z \in X : \liminf_{\lambda \to 0+} \lambda^{-1} d(R(I - \lambda A_z), x) = 0 \text{ for all } x \in \overline{D(A)} \right\}.$$

For simplicity, we deal with the case where A is a dissipative operator.

LEMMA 5.17. *Let A be a dissipative operator and $z \in E(A)$. Then, since A_z is a dissipative operator satisfying condition* (R), *from Theorem* 5.12 *there exists a unique semigroup of contractions* $\{T(t; A_z) : t \geq 0\}$ *on* $\overline{D(A)}$ *such that for each* $x \in \overline{D(A)}$ $(= \overline{D(A_z)})$ $T(t; A_z)x : [0, \infty) \to \overline{D(A)}$ *is an integral solution of type* 0 *of the Cauchy problem* $(d/dt)u(t) \in A_z u(t)$ $(0 \leq t < \infty)$ *with* $u(0) = x$ *for* A_z.

Let $T > 0$ *and* $g(t) : [0, T] \to X$ *be a* (Bochner) *integrable function on* $[0, T]$. *If* $v(t) : [0, T] \to X$ *is an integral solution of type* 0 *of the Cauchy problem*

$$\begin{cases} (d/dt)v(t) \in Av(t) + g(t) & (0 \leq t \leq T), \\ v(0) = v \ (\in X), \end{cases} \tag{5.55}$$

then for every $x \in \overline{D(A)}$ *and* $0 \leq r < t \leq T$ *we have*

$$\|v(t) - T(t; A_z)x\|^2 \leq \|v(r) - T(r; A_z)x\|^2 \\ + 2 \int_r^t \langle g(\tau) - z, v(\tau) - T(\tau; A_z)x \rangle_s \, d\tau. \tag{5.56}$$

PROOF. Choose $x \in \overline{D(A)}$ arbitrarily and set $u(t) = T(t; A_z)x$ $(t \geq 0)$. From Lemma 5.11, for every $\lambda > 0$ there exist $\{t_k^\lambda\}$, $\{x_k^\lambda\}$, and $\{y_k^\lambda\}$ which satisfy (5.34) through (5.36), where $x_k^\lambda \in D(A)$ $(= D(A_z)$ and $y_k^\lambda \in A_z x_k^\lambda$ $(k = 1, 2, \ldots)$. If we define $u_\lambda(0) = x$ and $u_\lambda(t) = x_k^\lambda$ $(t \in (t_{k-1}^\lambda, t_k^\lambda]; k = 1, 2, \ldots)$, then as we see from the method of construction of $T(t; A_z)x$, we have

$$u(t) \ (= T(t; A_z)x) = \lim_{\lambda \to 0+} u_\lambda(t) \qquad (0 \leq t < \infty),$$

where the convergence is uniform on bounded subintervals of $[0, \infty)$ (see Theorems 5.12, 5.10, 5.7 and their proofs).

$v(t)$ is an integral solution of type 0 of the Cauchy problem (5.55); thus for every $0 \leq r < t \leq T$, $k = 1, 2, \ldots$, we have

$$\|v(t) - x_k^\lambda\|^2 - \|v(r) - x_k^\lambda\|^2 \leq 2 \int_r^t \langle g(\tau) + y_k^\lambda - z, v(\tau) - x_k^\lambda \rangle_s \, d\tau.$$

(note that $y_k^\lambda - z \in Ax_k^\lambda$ since $y_k^\lambda \in A_z x_k^\lambda = Ax_k^\lambda + z$) and we have

$$\|v(t) - x_k^\lambda\| - \|v(r) - x_k^\lambda\|^2 \\ \leq 2 \int_r^t \langle y_k^\lambda, v(\tau) - x_k^\lambda \rangle_s \, d\tau + 2 \int_r^t \langle g(\tau) - z, v(\tau) - x_k^\lambda \rangle_s \, d\tau.$$

Then by the same reasoning as was used to obtain (5.24), we have that for every $0 \le r < t \le T$ and $0 \le \alpha < \beta \le T$

$$\int_{\alpha}^{\beta} (\|v(t) - u(\xi)\|^2 - \|v(r) - u(\xi)\|^2) \, d\xi$$

$$+ \int_{r}^{t} (\|u(\tau) - u(\beta)\|^2 - \|v(\tau) - u(\alpha)\|^2) \, d\tau \qquad (5.57)$$

$$\le 2 \int_{\alpha}^{\beta} \int_{r}^{t} \langle g(\tau) - z, v(\tau) - u(\xi) \rangle_s \, d\tau \, d\xi$$

holds. Then from this we can derive

$$\|v(t) - u(t)\|^2 \le \|v(r) - u(r)\|^2 + 2 \int_{r}^{t} \langle g(\tau) - z, v(\tau) - u(\tau) \rangle_s \, d\tau \quad (5.58)$$

$(0 \le r < t \le T)$ and obtain (5.56). Now to show (5.58), we define $\varphi(\tau, \xi)$ and $\psi(\tau, \xi) \colon R^2 \to R^1$ by

$$\varphi(\tau, \xi) = \begin{cases} \|v(\tau) - u(\xi)\|^2, & (\tau, \xi) \in [0, T] \times [0, T], \\ 0, & (\tau, \xi) \notin [0, T] \times [0, T], \end{cases}$$

$$\psi(\tau, \xi) = \begin{cases} 2\langle g(\tau) - z, v(\tau) - u(\xi) \rangle_s, & (\tau, \xi) \in [0, T] \times [0, T], \\ 0, & (\tau, \xi) \notin [0, T] \times [0, T]. \end{cases}$$

For every $\varepsilon > 0$, we consider the function $\rho_\varepsilon(\tau)$ described in the remark before Theorem 5.2, and set $\rho_\varepsilon(\tau, \xi) = \rho_\varepsilon(\tau)\rho_\varepsilon(\xi)$ and define

$$\varphi_\varepsilon(\tau, \xi) = \iint_{R^2} \rho_\varepsilon(\sigma, \eta) \varphi(\tau - \sigma, \xi - \eta) \, d\sigma \, d\eta \qquad ((\tau, \xi) \in R^2),$$

$$\psi_\varepsilon(\tau, \xi) = \iint_{R^2} \rho_\varepsilon(\sigma, \eta) \psi(\tau - \sigma, \xi - \eta) \, d\sigma \, d\eta \qquad ((\tau, \xi) \in R^2).$$

Then we have $\varphi_\varepsilon(\tau, \xi), \psi_\varepsilon(\tau, \xi) \in C^\infty(R^2)$ and $\lim_{\varepsilon \to 0+} \varphi_\varepsilon(\tau, \xi) = \varphi(\tau, \xi)$ $((\tau, \xi) \in (0, T) \times (0, T))$. Now let $0 < 2\varepsilon < T$, $\varepsilon \le \alpha < \beta \le T - \varepsilon$, and $\varepsilon \le r < t \le T - \varepsilon$. Then for σ and η such that $|\sigma| \le \varepsilon$ and $|\eta| \le \varepsilon$, we obtain from (5.57)

$$\int_{\alpha}^{\beta} (\|v(t - \sigma) - u(\xi - \eta)\|^2 - \|v(r - \sigma) - u(\xi - \eta)\|^2) \, d\xi$$

$$+ \int_{r}^{t} (\|v(\tau - \sigma) - u(\beta - \eta)\|^2 - \|v(\tau - \sigma) - u(\alpha - \eta)\|^2) \, d\tau$$

$$\le 2 \int_{\alpha}^{\beta} \int_{r}^{t} \langle g(\tau - \sigma) - z, v(\tau - \sigma) - u(\xi - \eta) \rangle_s \, d\tau \, d\xi.$$

Multiplying both sides by $\rho_\varepsilon(\sigma, \eta)$ and integrating with respect to σ and η over $[-\varepsilon, \varepsilon] \times [-\varepsilon, \varepsilon]$, we have

$$\int_{\alpha}^{\beta} [\varphi_\varepsilon(t, \xi) - \varphi_\varepsilon(r, \xi)] \, d\xi + \int_{r}^{t} [\varphi_\varepsilon(\tau, \beta) - \varphi_\varepsilon(\tau, \alpha)] \, d\tau$$

$$\le \int_{\alpha}^{\beta} \int_{r}^{t} \psi_\varepsilon(\tau, \xi) \, d\tau \, d\xi.$$

Dividing both sides by $(\beta - \alpha)(t - r)$, and letting $\beta \to \alpha + 0$ and $r \to t - 0$, we have

$$(\partial/\partial t)\varphi_\varepsilon(t, \alpha) + (\partial/\partial \alpha)\varphi_\varepsilon(t, \alpha) \leq \psi_\varepsilon(t, \alpha).$$

From this we obtain $(d/dt)\varphi_\varepsilon(t, t) \leq \psi_\varepsilon(t, t)$ and thus

$$\varphi_\varepsilon(t, t) \leq \varphi_\varepsilon(r, r) + \int_r^t \psi_\varepsilon(\tau, \tau)\, d\tau.$$

Next, from

$$\psi_\varepsilon(\tau, \tau) = \int_{-1}^1 \int_{-1}^1 \rho(\sigma)\rho(\eta)\psi(\tau - \varepsilon\sigma, \tau - \varepsilon\eta)\, d\sigma\, d\eta,$$

$$\psi(\tau - \varepsilon\sigma, \tau - \varepsilon\eta)$$
$$\leq 2\|g(\tau - \varepsilon\sigma) - g(\tau)\|\, \|v(\tau - \varepsilon\sigma) - u(\tau - \varepsilon\eta)\|$$
$$+ 2\langle g(\tau) - z, v(\tau - \varepsilon\sigma) - u(\tau - \varepsilon\eta)\rangle_s$$
$$\leq 4M\|g(\tau - \varepsilon\sigma) - g(\tau)\| + 2\langle g(\tau) - z, v(\tau - \varepsilon\sigma) - u(\tau - \varepsilon\eta)\rangle_s$$

we have

$$\int_r^t \psi_\varepsilon(\tau, \tau)\, d\tau \leq 2M \int_{-1}^1 \int_{-1}^1 \rho(\sigma)\rho(\eta) \left[\int_r^t \|g(\tau - \varepsilon\sigma) - g(\tau)\|\, d\tau \right] d\sigma\, d\eta$$
$$+ 2 \int_{-1}^1 \int_{-1}^1 \rho(\sigma)\rho(\eta) \left[\int_r^t \langle g(\tau) - z, v(\tau - \varepsilon\sigma) \right.$$
$$\left. - u(\tau - \varepsilon\eta)\rangle_s\, d\tau \right] d\sigma\, d\eta,$$

where $M = \max_{0 \leq \tau \leq T}[\|v(\tau)\| + \|u(\tau)\|]$. Now, since

$$\lim_{\varepsilon \to 0+} \int_r^t \|g(\tau - \varepsilon\sigma) - g(\tau)\|\, d\tau = 0$$

and

$$\limsup_{\varepsilon \to 0+} \langle g(\tau) - z, v(\tau - \varepsilon\sigma) - u(\tau - \varepsilon\eta)\rangle_s \leq \langle g(\tau) - z, v(\tau) - u(\tau)\rangle_s$$

(this is due to the continuity of $u(\tau)$ and $v(\tau)$, and the upper semicontinuity of $\langle \cdot, \cdot \rangle_s : X \times X \to R^1$), we have by the Lebesgue convergence theorem

$$\limsup_{\varepsilon \to 0+} \int_r^t \psi_\varepsilon(\tau, \tau)\, d\tau$$
$$\leq 2 \int_{-1}^1 \int_{-1}^1 \rho(\sigma)\rho(\eta) \left[\int_r^t \langle g(\tau) - z, v(\tau) - u(\tau)\rangle_s\, d\tau \right] d\sigma\, d\eta$$
$$= 2 \int_r^t \langle g(\tau - z, v(\tau) - u(\tau)\rangle_s\, d\tau.$$

From this and $\lim_{\varepsilon \to 0+} \varphi_\varepsilon(\tau, \tau) = \varphi(\tau, \tau)$ $(0 < \tau < T)$, we obtain

$$\varphi(t, t) \leq \varphi(r, r) + 2 \int_r^t \langle g(\tau) - z, v(\tau) - u(\tau)\rangle_s\, d\tau \qquad (0 < r < t < T),$$

and thus (5.58) holds for $0 < r < t < T$. Here if we let $r \to 0+$ and $t \to T - 0$, then (5.58) holds also for $r = 0$ and $t = T$. $\square$

REMARK. For $v(t)$ and $T(t; A_z)$ in the above lemma, the following inequality holds:

$$\|v(t) - T(t - \alpha; A_z)x\|^2$$
$$\leq \|v(r) - T(r - \alpha; A_z)x\|^2 + 2\int_r^t \langle g(\tau) - z, v(\tau) - T(\tau - \alpha; A_z)x \rangle_s \, d\tau$$

$$(5.59)$$

for every $0 \leq \alpha \leq r < t \leq T$ and $x \in \overline{D(A)}$. In fact, if we set $v_1(t) = v(t+\alpha)$ and $g_1(t) = g(t+\alpha)$ $(0 \leq t \leq T-\alpha)$, then $v_1(t)$ becomes an integral solution of type 0 of the Cauchy problem $(d/dt)v_1(t) \in Av_1(t) + g_1(t)$ $(0 \leq t \leq T-\alpha)$ with $v_1(0) = v(\alpha)$. Thus, from the above lemma (considering $v_1(t)$ and $g_1(t)$ instead of $v(t)$ and $g(t)$) we have

$$\|v_1(t) - T(t; A_z)x\|^2$$
$$\leq \|v_1(r) - T(r; A_z)x\|^2 + 2\int_r^t \langle g_1(\tau) - z, v_1(\tau) - T(\tau; A_z)x \rangle_s \, d\tau,$$

and hence we obtain

$$\|v(t + \alpha) - T(t; A_z)x\|^2 \leq \|v(r + \alpha) - T(r; A_z)x\|^2$$
$$+ 2\int_{r+\alpha}^{t+\alpha} \langle g(\tau) - z, v(\tau) - T(\tau - \alpha; A_z)x \rangle_s \, d\tau$$

$(x \in \overline{D(A)}, \ 0 \leq r < t \leq T - \alpha)$. Here we may replace $t + \alpha$ and $r + \alpha$ by t and r, respectively.

THEOREM 5.18. *Let A be a dissipative operator and $f(t): [0, T] \to \overline{E(A)}$ be a (Bochner) integrable function on $[0, T]$. Then for every $x \in \overline{D(A)}$, the Cauchy problem (5.53) has a unique integral solution of type 0.*

PROOF. The first step (the existence of integral solutions). Given any $\varepsilon > 0$, there exists a step function $g_\varepsilon(t): [0, T] \to \overline{E(A)}$ satisfying $\int_0^T \|g_\varepsilon(t) - f(t)\| \, dt \leq \varepsilon/2$. [7] Now, let $0 = a_0^\varepsilon < a_1^\varepsilon < \cdots < a_{n-1}^\varepsilon < a_n^\varepsilon = T$ and $g_\varepsilon(t) = y_k^\varepsilon$ $(a_{k-1}^\varepsilon < t < a_k^\varepsilon)$. Then we select $z_k^\varepsilon \in E(A)$ $(k = 1, 2, \ldots, n)$ which satisfy $\|y_k^\varepsilon - z_k^\varepsilon\| \leq \varepsilon/(2T)$, and set

$$f_\varepsilon(t) = \begin{cases} z_k^\varepsilon & (a_{k-1}^\varepsilon \leq t < a_k^\varepsilon), \quad k = 1, 2, \ldots, n-1, \\ z_n^\varepsilon & (a_{n-1}^\varepsilon \leq t \leq a_n^\varepsilon = T). \end{cases}$$

Then $\int_0^T \|f_\varepsilon(t) - f(t)\| \, dt \leq \varepsilon$. For simplicity, we denote $A_{z_k^\varepsilon}$ $(A_{z_k^\varepsilon}x = Ax + z_k^\varepsilon \ (x \in D(A)))$ by A_k. Then from Theorem 5.12, each A_k generates

---

[7] For example, we apply Corollary 1.13

a semigroup of contractions $\{T_k(t) : 0 \le t < \infty\}$ on $\overline{D(A)}$, and $T_k(t)x$ $(x \in \overline{D(A)})$ is an integral solution of type 0 of $(d/dt)u(t) \in A_k u(t)$ $(0 \le t < \infty)$ with $u(0) = x$. Hence for $0 \le r < t < \infty$, $x_0 \in D(A)$ and $y_0 \in Ax_0$

$$\|T_k(t)x - x_0\|^2 - \|T_k(r)x - x_0\|^2 \le 2 \int_r^t \langle z_k^\varepsilon + y_0, T_k(\tau)x - x_0 \rangle_s d\tau \quad (5.60)$$

$(x \in \overline{D(A)};\ k = 1, 2, \ldots, n)$ holds. (Note that $z_k^\varepsilon + y_0 \in A_{z_k^\varepsilon} x_0 = A_k x_0$.) Also from the remark following Lemma 5.17 (see (5.59)), if we let $v(t)$ be an integral solution of type 0 of the Cauchy problem (5.55), then for every $0 \le r < t \le T$ and $z \in \overline{D(A)}$

$$\|v(t) - T_k(t-r)z\|^2 \le \|v(r) - z\|^2 + 2 \int_r^t \langle g(\tau) - z_k^\varepsilon, v(\tau) - T_k(\tau-r)z \rangle_s d\tau$$
$$(5.61)$$

$(k = 1, 2, \ldots, n)$ holds.

Now choose $x \in \overline{D(A)}$ arbitrarily, and for an every $\varepsilon > 0$ define $u_\varepsilon(t) : [0, T] \to \overline{D(A)}$ by

$$\begin{cases} u_\varepsilon(0) = x, \\ u_\varepsilon(t) = T_k(t - a_{k-1}^\varepsilon)u_\varepsilon(a_{k-1}^\varepsilon) \quad (t \in [a_{k-1}^\varepsilon, a_k^\varepsilon]), \quad k = 1, 2, \ldots, n. \end{cases}$$

Obviously $u_\varepsilon(t)$ is continuous on $[0, T]$. Let $x_0 \in D(A)$, $y_0 \in Ax_0$, and $0 \le r < t \le T$. Choose natural numbers j and i such that $r \in [a_{j-1}^\varepsilon, a_j^\varepsilon]$ and $t \in [a_{i-1}^\varepsilon, a_i^\varepsilon]$. From (5.60) and the definitions of $u_\varepsilon(s)$ and $f_\varepsilon(s)$, we have

$$\|u_\varepsilon(t) - x_0\|^2 - \|u_\varepsilon(a_{i-1}^\varepsilon) - x_0\|^2 \le 2 \int_{a_{i-1}^\varepsilon}^t \langle f_\varepsilon(\tau) + y_0, u_\varepsilon(\tau) - x_0 \rangle_s d\tau,$$

$$\|u_\varepsilon(a_k^\varepsilon) - x_0\|^2 - \|u_\varepsilon(a_{k-1}^\varepsilon) - x_0\|^2 \le 2 \int_{a_{i-1}^\varepsilon}^{a_k^\varepsilon} \langle f_\varepsilon(\tau) + y_0, u_\varepsilon(\tau) - x_0 \rangle_s d\tau$$

$$(k = j + 1, \ldots, i - 1),$$

$$\|u_\varepsilon(a_j^\varepsilon) - x_0\|^2 - \|u_\varepsilon(r) - x_0\|^2 \le 2 \int_r^{a_j^\varepsilon} \langle f_\varepsilon(\tau) + y_0, u_\varepsilon(\tau) - x_0 \rangle_s d\tau.$$

If we add the respective sides of these inequalities, we obtain

$$\|u_\varepsilon(t) - x_0\|^2 - \|u_\varepsilon(r) - x_0\|^2 \le 2 \int_r^t \langle f_\varepsilon(\tau) + y_0, u_\varepsilon(\tau) - x_0 \rangle_s d\tau. \quad (5.62)$$

Hence $u_\varepsilon(t) : [0, T] \to \overline{D(A)}$ is an integral solution of type 0 of the Cauchy problem $(d/dt)u_\varepsilon(t) \in Au_\varepsilon(t) + f_\varepsilon(t)$ $(0 \le t \le T)$ with $u_\varepsilon(0) = x$. Next, we show that $u_\varepsilon(t)$ converges uniformly on $[0, T]$ as $\varepsilon \to 0+$. Consider $u_{\varepsilon'}(t)$ $(\varepsilon \ne \varepsilon' > 0)$. Then, as we have shown above, $u_{\varepsilon'}(t) : [0, T] \to \overline{D(A)}$ is an integral solution of type 0 of the Cauchy problem

$$\begin{cases} (d/dt)u_{\varepsilon'}(t) \in Au_{\varepsilon'}(t) + f_{\varepsilon'}(t) \quad (0 \le t \le T), \\ u_{\varepsilon'}(0) = x (\in \overline{D(A)}). \end{cases}$$

Note that $\int_0^T \|f_{\varepsilon'}(t) - f(t)\| \, dt \le \varepsilon'$. Thus if we let $v(t) = u_{\varepsilon'}(t)$, $g(t) = f_{\varepsilon'}(t)$, $r = a_{k-1}^\varepsilon$, $z = u_\varepsilon(a_{k-1}^\varepsilon)$ and apply (5.61), then for every $a_{k-1}^\varepsilon \le t \le T$ we obtain

$$\|u_{\varepsilon'}(t) - T_k(t - a_{k-1}^\varepsilon)u_\varepsilon(a_{k-1}^\varepsilon)\|^2$$
$$\le \|u_{\varepsilon'}(a_{k-1}^\varepsilon) - u_\varepsilon(a_{k-1}^\varepsilon)\|^2$$
$$+ 2\int_{a_{k-1}^\varepsilon}^t \langle f_{\varepsilon'}(\tau) - z_k^\varepsilon, u_{\varepsilon'}(\tau) - T_k(\tau - a_{k-1}^\varepsilon)u_\varepsilon(a_{k-1}^\varepsilon)\rangle_s \, d\tau$$

$(k = 1, 2, \dots, n)$. Now if we let $0 < t \le T$ and choose a natural number i such that $t \in [a_{i-1}^\varepsilon, a_i^\varepsilon]$, then from (5.63) and the definitions of $u_\varepsilon(t)$ and $f_\varepsilon(t)$, we have

$$\|u_{\varepsilon'}(t) - u_\varepsilon(t)\|^2 = \|u_{\varepsilon'}(t) - T_i(t - a_{i-1}^\varepsilon)u_\varepsilon(a_{k-1}^\varepsilon)\|^2$$
$$\le \|u_{\varepsilon'}(a_{i-1}^\varepsilon) - u_\varepsilon(a_{i-1}^\varepsilon)\|^2 + 2\int_{a_i^\varepsilon - 1}^t \langle f_{\varepsilon'}(\tau) - f_\varepsilon(\tau), u_{\varepsilon'}(\tau) - u_\varepsilon(\tau)\rangle_s \, d\tau,$$
$$\|u_{\varepsilon'}(a_{k-1}^\varepsilon) - u_\varepsilon(a_{k-1}^\varepsilon)\|^2 = \|u_{\varepsilon'}(a_{k-1}^\varepsilon) - T_{k-1}(a_{k-1}^\varepsilon - a_{k-2}^\varepsilon)u_\varepsilon(a_{k-2}^\varepsilon)\|^2$$
$$\le \|u_{\varepsilon'}(a_{k-2}^\varepsilon) - u_\varepsilon(a_{i-2}^\varepsilon)\|^2 + 2\int_{a_{k-2}^\varepsilon}^{a_{k-1}^\varepsilon} \langle f_{\varepsilon'}(\tau) - f_\varepsilon(\tau), u_{\varepsilon'}(\tau) - u_\varepsilon(\tau)\rangle_s \, d\tau$$

$(k = 2, 3, \dots, i)$. We add the respective sides of these inequalities and obtain

$$\|u_{\varepsilon'}(t) - u_\varepsilon(t)\|^2 \le 2\int_0^t \langle f_{\varepsilon'}(\tau) - f_\varepsilon(\tau), u_{\varepsilon'}(\tau) - u_\varepsilon(\tau)\rangle_s \, d\tau$$
$$\le 2\int_0^t \|f_{\varepsilon'}(\tau) - f_\varepsilon(\tau)\| \|u_{\varepsilon'}(\tau) - u_\varepsilon(\tau)\| \, d\tau \qquad (0 \le t \le T). \tag{5.64}$$

Hence, for every $t \in [0, T]$

$$\|u_{\varepsilon'}(t) - u_\varepsilon(t)\| \le \int_0^t \|f_{\varepsilon'}(\tau) - f_\varepsilon(\tau)\| \, d\tau$$
$$\le \int_0^T \|f_{\varepsilon'}(\tau) - f(\tau)\| \, d\tau + \int_0^T \|f_\varepsilon(\tau) - f(\tau)\| \, d\tau \le \varepsilon' + \varepsilon$$

holds. (See the remark following Lemma 3.4). Therefore, $u_\varepsilon(t)$ converges uniformly on $[0, T]$ as $\varepsilon \to 0+$. Thus if we set

$$u(t) = \lim_{\varepsilon \to 0+} u_\varepsilon(t) \qquad (0 \le t \le T), \tag{5.65}$$

then $u(0) = x$, and $u(t): [0, T] \to \overline{D(A)}$ is a continuous function. Moreover for every $0 \le r < t \le T$, $x_0 \in D(A)$, and $y_0 \in Ax_0$, we see from (5.62) that

$$\|u(t) - x_0\|^2 - \|u(r) - x_0\|^2 \le 2\int_r^t \langle f(\tau) + y_0, u(\tau) - x_0\rangle_s \, d\tau.$$

(Note that

$$\langle f_\varepsilon(\tau) + y_0, u_\varepsilon(\tau) - x_0 \rangle_s \leq \|f_\varepsilon(\tau) - f(\tau)\| \, \|u_\varepsilon(\tau) - x_0\| + \langle f(\tau) + y_0, u_\varepsilon(\tau) - x_0 \rangle_s.)$$

Therefore, $u(t): [0, T] \to \overline{D(A)}$ is the solution of type 0 of the Cauchy problem (5.53).

The second step (the uniqueness of solution). Let $u(t)$ be the function defined by (5.65). Then, for an integral solution $v(t)$ of type 0 of the Cauchy problem (5.55), the following inequality holds:

$$\|v(t) - u(t)\| \leq \|v - x\| + \int_0^t \|g(\tau) - f(\tau)\| \, d\tau \qquad (0 \leq t \leq T). \qquad (5.66)$$

In fact, using the same reasoning as in obtaining (5.64), we can derive from (5.61) the inequality

$$\|v(t) - u_\varepsilon(t)\|^2 \leq \|v - x\|^2 + 2 \int_0^t \langle g(\tau) - f_\varepsilon(\tau), v(\tau) - u_\varepsilon(\tau) \rangle_s \, d\tau$$

$$\leq \|v - x\|^2 + 2 \int_0^t \|g(\tau) - f_\varepsilon(\tau)\| \, \|v(\tau) - u_\varepsilon(\tau)\| \, ds \qquad (0 \leq t \leq T).$$

From this we obtain

$$\|v(t) - u_\varepsilon(t)\| \leq \|v - x\| + \int_0^t \|g(\tau) - f_\varepsilon(\tau)\| \, d\tau \quad (0 \leq t \leq T)$$

and thus (5.66) holds.

In particular, if $v(t)$ is an integral solution of type 0 of the Cauchy problem (5.53), then from the above inequality (5.66) we obtain $v(t) = u(t)$ $(0 \leq t \leq T)$, and hence $u(t)$ defined by (5.65) is the unique of type 0 of the Cauchy problem (5.53). $\square$

COROLLARY 5.19. *Let A be an m-dissipative operator and let $f(t): [0, T] \to X$ be a (Bochner) integrable function on $[0, T]$. Then for every $x \in \overline{D(A)}$, the Cauchy problem (5.53) has a unique integral solution of type 0.*

PROOF. Let A be an m-dissipative operator. Since A_z becomes an m-dissipative operator for each $z \in X$, we obtain $E(A) = X$. Hence, we can apply Theorem 5.18 to prove this. $\square$

REMARK 1°. By the same reasoning as in the proof of Theorem 5.2, we can show the following: Let $u(t): [0, T] \to X$ be continuous, $u(0) = x \in X$ and let $f(t): [0, T] \to X$ be (Bochner) integrable. Then a necessary and sufficient condition for $u(t)$ to be an integral solution of type ω of the Cauchy problem (5.53) is that for every $r, t \in [0, T]$ with $r < t$, and every $x_0 \in D(A)$ and $y_0 \in Ax_0$

$$\|u(t) - x_0\|^2 - \|u(r) - x_0\|^2$$

$$\leq 2 \int_r^t \langle f(\tau) + y_0, u(\tau) - x_0 \rangle_s \, d\tau + 2\omega \int_r^t \|u(\tau) - x_0\|^2 \, d\tau$$

holds.

REMARK $2°$. If A is a dissipative operator of type ω in Theorem 5.18, then we see that for every $x \in \overline{D(A)}$, the Cauchy problem (5.53) has a unique integral solution of type ω. (We leave the proof to the reader.) Also from this, if $A - \omega I$ (ω real) is an m-dissipative operator and $f(t): [0, T] \to X$ is (Bochner) integrable, then for every $x \in \overline{D(A)}$ the Cauchy problem (5.53) has a unique integral solution of type ω. (In fact, this is because of the fact that if $A - \omega I$ is an m-dissipative operator, then for $\lambda > 0$ such that $\lambda \omega < 1$, $R(I - \lambda A_z) = X$ holds ($z \in X$ arbitrary) and thus $E(A) = X$.)

Finally, we describe the following theorem which is an extension of Theorem 5.15.

THEOREM 5.20. *Let* A *be a continuous dissipative operator with closed domain* $D(A)$ ($\subset X$) *and with values in* X, *and let* $f(t): [0, T] \to X$ *be a continuous function. Then* (i) *through* (iii) *are equivalent:*

(i) *For every* $s \in [0, T]$ *and* $x \in D(A)$, *there exists a continuous function* $u_s(t; x): [s, T] \to D(A)$ *satisfying*

$$\begin{cases} (d/dt)u_s(t; x) = Au_s(t; x) + f(t) & (s \le t \le T), \\ u_s(s; x) = x. \end{cases}$$

(ii) *If we set*

$$E_1(A) = \{z \in X : \liminf_{\lambda \to 0+} \lambda^{-1} d(D(A), x + \lambda A_z x) = 0 \ (x \in D(A))\},$$

then $f(s) \in \overline{E_1(A)}$ ($s \in [0, T]$).

(iii) $f(s) \in \overline{E(A)}$ ($s \in [0, T]$).

We leave the proof to the reader. In the above theorem, in particular for the case of $D(A) = X$, since A is an m-dissipative operator (see Corollary 5.16), $E(A) = X$. Hence from the above theorem, we obtain the following corollary:

COROLLARY 5.21. *Let* $A: X \to X$ *be a continuous dissipative operator and let* $f(t): [0, T] \to X$ *be a continuous function. Then for every* $x \in X$, *the Cauchy problem*

$$\begin{cases} (d/dt)u(t) = Au(t) + f(t) & (0 \le t \le T), \\ u(0) = x \end{cases}$$

has a unique strongly continuously differentiable solution on $[0, T]$.

An extension of this corollary will be given in the next chapter (see Corollary 6.22).

CHAPTER 6

Convergence and Perturbation
of Nonlinear Semigroups

Let A be a dissipative operator of type ω satisfying the following condition (R):

(R) For all $x \in \overline{D(A)}$, $\liminf_{\lambda \to 0+} \lambda^{-1} d(R(I - \lambda A), x) = 0$.

Then from Theorem 5.12, there exists a unique semigroup $\{T(t): t \geq 0\}$ of type ω on $\overline{D(A)}$ such that for each $x \in \overline{D(A)}$, $T(t)x: [0, \infty) \to \overline{D(A)}$ is the integral solution of type ω of the Cauchy problem $(\mathrm{CP}; x)_\infty$ for A. Hereafter, we call this semigroup $\{T(t): t \geq 0\}$ the semigroup generated by A. Also we say that A generates the semigroup $\{T(t): t \geq 0\}$.

In this chapter we consider "the relation between the convergence of A_n ($n = 1, 2, \ldots$) and the convergence of the sequence of semigroups $\{T_n(t): t \geq 0\}$, when the semigroups $\{T_n(t): t \geq 0\}$ are generated by dissipative operators A_n of type ω_n satisfying condition (R)," and also "the problem of whether $A + B$ generates a semigroup or not when B is an operator and A generates a semigroup." As before, we assume that X is a Banach space.

§1. Convergence of dissipative operators

DEFINITION 6.1. Let $\varnothing \neq D_n \subset X$ $(n = 1, 2, \ldots)$. We write $x \in \mathrm{Lim}_{n \to \infty} D_n$ if there exist $x_n \in D_n$ such that $\lim_{n \to \infty} x_n = x$.

LEMMA 6.1. *Let* $\varnothing \neq D_n \subset X$ $(n = 1, 2, \ldots)$. (i), (ii), *and* (iii) *are equivalent*:

(i) $x \in \mathrm{Lim}_{n \to \infty} D_n$.

(ii) $x \in \bigcap_{\varepsilon > 0} \bigcup_{m \geq 1} \bigcup_{n \geq m} (D_n)_\varepsilon$, *where* $(D_n)_\varepsilon = \{x \in X: there\ exists\ x' \in D_n\ with\ \|x - x'\| < \varepsilon\}$ $(\varepsilon > 0)$.

(iii) $\lim_{n \to \infty} d(x, D_n) = 0$.

PROOF. (i) $\Rightarrow$ (ii). Assume (i). Then there exist $x_n \in D_n$ $(n = 1, 2, \ldots)$ satisfying $\lim_{n \to \infty} \|x_n - x\| = 0$. Given $\varepsilon > 0$, choose a natural number m_ε such that $\|x - x_n\| < \varepsilon$ $(n \geq m_\varepsilon)$. Then $x \in (D_n)_\varepsilon$ for n with $n \geq m_\varepsilon$, and hence $x \in \bigcap_{n \geq m_\varepsilon} (D_n)_\varepsilon \subset \bigcup_{m \geq 1} \bigcap_{n \geq m} (D_n)_\varepsilon$.

(ii) $\Rightarrow$ (iii). Given $\varepsilon > 0$, we can choose a natural number m_ε such that $x \in (D_n)_\varepsilon$ for $n \geq m_\varepsilon$. Hence if $n \geq m_\varepsilon$, then $d(x, D_n) < \varepsilon$, and thus we obtain (iii).

(iii) $\Rightarrow$ (i). This is because for each natural number n there exists an $x_n \in D$ such that $\|x_n - x\| < d(x, D_n) + 1/n$. $\square$

LEMMA 6.2. *Let* $\varnothing \neq D_n \subset X$ $(n = 1, 2, \ldots)$.
(i) $\mathrm{Lim}_{n \to \infty} D_n = \mathrm{Lim}_{n \to \infty} \overline{D}_n$.
(ii) $D \subset \mathrm{Lim}_{n \to \infty} D_n$ *if and only if* $\overline{D} \subset \mathrm{Lim}_{n \to \infty} D_n$.

PROOF. (i) Since $(\overline{D}_n)_\varepsilon = (D_n)_\varepsilon$ for every $\varepsilon > 0$, we can apply Lemma 6.1.

(ii) Let $D \subset \mathrm{Lim}_{n \to \infty} D_n$ and $x \in \overline{D}$. Given $\varepsilon > 0$ choose $x_\varepsilon \in D$ such that $\|x_\varepsilon - x\| < \varepsilon$. Since $x_\varepsilon \in \mathrm{Lim}_{n \to \infty} D_n$, $\lim_{n \to \infty} d(x_\varepsilon, D_n) = 0$ by Lemma 6.1. Now $d(x, D_n) \leq \|x - x_\varepsilon\| + d(x_\varepsilon, D_n)$. Hence we have that $\limsup_{n \to \infty} d(x, D_n) \leq \varepsilon$, that is, $\lim_{n \to \infty} d(x, D_n) = 0$. We apply Lemma 6.1 again and obtain $x \in \mathrm{Lim}_{n \to \infty} D_n$. $\square$

As an extension of the concept of the convergence of usual (single-valued) operators, we define the convergence of a sequence of multivalued operators as follows:

DEFINITION 6.2. Let A_n $(n = 1, 2, \ldots)$ be not necessarily single-valued operators with $D(A_n) \subset X$ and $R(A_n) \subset X$. We denote by $\mathrm{Lim}_{n \to \infty} A_n$ the operator (not necessarily single-valued), that has as its graph the limit $\mathrm{Lim}_{n \to \infty} \mathfrak{G}(A_n)$ (in the sense of Definition 6.1) of the graphs $\mathfrak{G}(A_n) \subset X \times X$ of A_n. Therefore, by $x \in D(A)$ and $y \in Ax$ we mean that there exist $x_n \in D(A_n)$ and $y_n \in A_n x_n$ $(n = 1, 2, \ldots)$ such that $\lim_{n \to \infty} x_n = x$ and $\lim_{n \to \infty} y_n = y$.

For example, if A_n $(n = 0, 1, 2, \ldots)$ are single-valued operators with $D(A_0) \subset D(A_n)$ $(n = 1, 2, \ldots)$ and $A_0 x = \lim_{n \to \infty} A_n x$ $(x \in D(A_0))$, then $A_0 \subset \mathrm{Lim}_{n \to \infty} A_n$.

From this definition and Lemma 6.2 we obtain easily the following corollary:

COROLLARY 6.3. *Let* A_n $(n = 1, 2, \ldots)$ *and* A *be not necessarily single-valued operators, and let their domains and ranges be contained in* X. *Then the following* (i) *through* (iii) *hold*:
(i) $\mathrm{Lim}_{n \to \infty} A_n = \mathrm{Lim}_{n \to \infty} \overline{A}_n$.
(ii) $A \subset \mathrm{Lim}_{n \to \infty} A_n$ *if and only if* $\overline{A} \subset \mathrm{Lim}_{n \to \infty} A_n$.
(iii) *If* $A \subset \mathrm{Lim}_{n \to \infty} A_n$, *then*

$$\overline{D(A)} \subset \mathrm{Lim}_{n \to \infty} D(A_n) \quad and \quad \overline{R(A)} \subset \mathrm{Lim}_{n \to \infty} R(A_n).$$

Next we examine the relation between the convergence of dissipative operators A_n and the convergence of their resolvents $(I - \lambda A_n)^{-1}$.

LEMMA 6.4. *Let* A_n $(n = 0, 1, 2, \ldots)$ *be dissipative operators, and let* A_0 *satisfy*

$$R(I - \lambda A_0) \supset X_0 \quad \text{for all } \lambda > 0, \tag{6.1}$$

where X_0 is (independent of λ) a subset of X. We put $J_\lambda^{(n)} = (I - \lambda A_n)^{-1}$ $(\lambda > 0, \ n = 0, 1, 2, \ldots)$, and let B_0 be the operator with the following set as its graph.

$$\{[J_\lambda^{(0)}x, \lambda^{-1}(J_\lambda^{(0)}x - x)]: x \in X_0, \ \lambda > 0\}.$$

Then (i) and (ii) are equivalent:
  (i) $B_0 \subset \mathrm{Lim}_{n \to \infty} A_n$.
  (ii) For each $\lambda > 0$, $X_0 \subset \mathrm{Lim}_{n \to \infty} R(I - \lambda A_n)$ and (6.2) holds:

$$\begin{cases} \text{For all } z_n \in R(I - \lambda A_n) \text{ and } x \in X_0 \text{ such that } \lim_{n \to \infty} z_n = x, \\ \lim_{n \to \infty} J_\lambda^{(n)} z_n = J_\lambda^{(0)} x. \end{cases} \tag{6.2}$$

PROOF. (i) $\Rightarrow$ (ii). Let $\lambda > 0$. Then from (i), we have $I - \lambda B_0 \subset$ $\mathrm{Lim}_{n \to \infty}(I - \lambda A_n)$. Hence $R(I - \lambda B_0) \subset \mathrm{Lim}_{n \to \infty} R(I - \lambda A_n)$ by (iii) of Corollary 6.3. Now, since $x = J_\lambda^{(0)}x - \lambda[\lambda^{-1}(J_\lambda^{(0)}x - x)] \in (I - \lambda B_0)J_\lambda^{(0)}x$ $(x \in X_0)$, we have $X_0 \subset R(I - \lambda B_0)$. Thus we obtain $X_0 \subset \mathrm{Lim}_{n \to \infty} R(I - \lambda A_n)$. Next, let $z_n \in R(I - \lambda A_n)$, $x \in X_0$, and $\lim_{n \to \infty} z_n = x$. Since

$$[J_\lambda^{(0)}x, \lambda^{-1}(J_\lambda^{(0)}x - x)] \in \mathfrak{G}(B_0) \subset \mathrm{Lim}_{n \to \infty} \mathfrak{G}(A_n),$$

we can choose $u_n \in D(A_n)$ and $v_n \in A_n u_n$ such that $\|u_n - J_\lambda^{(0)}x\| \to 0$ and $\|v_n - \lambda^{-1}(J_\lambda^{(0)}x - x)\| \to 0$ as $n \to \infty$. Also, since $z_n \in R(I - \lambda A_n)$, there exist $x_n \in D(A_n)$ and $y_n \in A_n x_n$ such that $z_n = x_n - \lambda y_n$. (Hence, we have $x_n = J_\lambda^{(n)} z_n$.) Then from the dissipativity of A_n, we have

$$\|u_n - x_n\| \le \|u_n - \lambda v_n - (x_n - \lambda y_n)\| = \|u_n - \lambda v_n - z_n\|$$
$$\to \|J_\lambda^{(0)}x - (J_\lambda^{(0)}x - x) - x\| = 0 \qquad (n \to \infty).$$

Therefore,

$$\lim_{n \to \infty} J_\lambda^{(n)} z_n = \lim_{n \to \infty} x_n = \lim_{n \to \infty} u_n = J_\lambda^{(0)}x.$$

(ii) $\Rightarrow$ (i). Let $[u, v] \in \mathfrak{G}(B_0)$. Then from the definition of B_0, there exist $\lambda > 0$ and $x \in X_0$ such that $u = J_\lambda^{(0)}x$ and $v = \lambda^{-1}(J_\lambda^{(0)}x - x)$. Since $x \in X_0 \subset \mathrm{Lim}_{n \to \infty} R(I - \lambda A_n)$, we can select $x_n \in D(A_n)$ and $y_n \in A_n x_n$ such that $\lim_{n \to \infty}(x_n - \lambda y_n)$. Then from (6.2), we have $\lim_{n \to \infty} x_n = \lim_{n \to \infty} J_\lambda^{(n)}(x_n - \lambda y_n) = J_\lambda^{(0)}x = u$. Also from this, $\lim_{n \to \infty} y_n = \lambda^{-1}(u - x) = v$. Therefore, we obtain $[u, v] \in \mathrm{Lim}_{n \to \infty} \mathfrak{G}(A_n)$, and thus $\mathfrak{G}(B_0) \subset \mathrm{Lim}_{n \to \infty} \mathfrak{G}(A_n)$. $\square$

THEOREM 6.5. We let A_n $(n = 1, 2, \ldots)$ and A be dissipative operators satisfying the condition (R_1') (that is, for all $\lambda > 0$, $R(I - \lambda A_n) \supset \overline{D(A_n)}$ and $R(I - \lambda A) \supset \overline{D(A)}$), and set $J_\lambda^{(n)} = (I - \lambda A_n)^{-1}$ and $J_\lambda = (I - \lambda A)^{-1}$ $(\lambda > 0)$. Also, let B be the operator that has $\{[J_\lambda x, \lambda^{-1}(J_\lambda x - x)]: x \in \overline{D(A)}, \ \lambda > 0\}$ as its graph. Then (i) and (ii) are equivalent:
  (i) $B \subset \mathrm{Lim}_{n \to \infty} A_n$.

(ii) $\overline{D(A)} \subset \text{Lim}_{n\to\infty} \overline{D(A_n)}$; *and for all* $x_n \in \overline{D(A_n)}$ *and* $x \in \overline{D(A)}$ *with* $\lim_{n\to\infty} x_n = x$, *we have* $\lim_{n\to\infty} J_\lambda^{(n)} x_n = J_\lambda x$ $(\lambda > 0)$.

In particular, if $\overline{D(A_n)} \supset \overline{D(A)}$ $(n \geq 1)$, *then the above are equivalent to the following*

(iii) $\lim_{n\to\infty} J_\lambda^{(n)} x = J_\lambda x$ $(x \in \overline{D(A)}, \lambda > 0)$.

PROOF. (i) $\Rightarrow$ (ii). Assume (i). Then from (iii) of Corollary 6.3, $\overline{D(B)} \subset \text{Lim}_{n\to\infty} \overline{D(A_n)}$. Also from $\lim_{\lambda\to 0+} J_\lambda x = x$ $(x \in \overline{D(A)})$ we have $\overline{D(B)} = \overline{D(A)}$. Hence $\overline{D(A)} \subset \text{Lim}_{n\to\infty} \overline{D(A_n)}$. Next, we set $A_0 = A$ and $X_0 = \overline{D(A)}$ and apply Lemma 6.4. Then for each $\lambda > 0$ the following (6.3) holds:

$$\begin{cases} \text{For all } z_n \in R(I - \lambda A_n) \text{ and } x \in \overline{D(A)} \text{ such that } \lim_{n\to\infty} z_n = x, \\ \lim_{n\to\infty} J_\lambda^{(n)} z_n = J_\lambda x. \end{cases} \quad (6.3)$$

Now, let $x_n \in \overline{D(A_n)}$, $x \in \overline{D(A)}$, and $\lim_{n\to\infty} x_n = x$. From $R(I - \lambda A_n) \supset \overline{D(A_n)}$ we have $x_n \in R(I - \lambda A_n)$. Hence from (6.3), $\lim_{n\to\infty} J_\lambda^{(n)} x_n = J_\lambda x$ $(\lambda > 0)$.

(ii) $\Rightarrow$ (i). Let $[u, v] \in \mathfrak{G}(B)$. From the definition of B, there exist $x \in \overline{D(A)}$ and $\lambda > 0$ such that $u = J_\lambda x$ and $v = \lambda^{-1}(J_\lambda x - x)$. Since $x \in \overline{D(A)} \subset \text{Lim}_{n\to\infty} \overline{D(A_n)}$, we can choose $x_n \in \overline{D(A_n)}$ such that $\lim_{n\to\infty} x_n = x$. Then from the assumption, we have $\lim_{n\to\infty} J_\lambda^{(n)} x_n = J_\lambda x = u$ and $\lim_{n\to\infty} \lambda^{-1}(J_\lambda^{(n)} x_n - x_n) = \lambda^{-1}(J_\lambda x - x) = v$; also $\lambda^{-1}(J_\lambda^{(n)} x_n - x_n) \in A_n J_\lambda^{(n)} x_n$. Hence we obtain $[u, v] \in \text{Lim}_{n\to\infty} \mathfrak{G}(A_n)$ and thus $\mathfrak{G}(B) \subset \text{Lim}_{n\to\infty} \mathfrak{G}(A_n)$.

Finally, we assume $\overline{D(A_n)} \supset \overline{D(A)}$ $(n \geq 1)$. Then (ii) $\Rightarrow$ (iii) is self-evident.

(iii) $\Rightarrow$ (ii). It is evident that $\overline{D(A)} \subset \text{Lim}_{n\to\infty} \overline{D(A_n)}$. Let $x_n \in \overline{D(A_n)}$, $x \in \overline{D(A)}$, and $\lim_{n\to\infty} x_n = x$. Then for each $\lambda > 0$

$$\|J_\lambda^{(n)} x_n - J_\lambda x\| \leq \|x_n - x\| + \|J_\lambda^{(n)} x - J_\lambda x\|.$$

Therefore, we obtain (ii) from (iii). $\square$

THEOREM 6.6. *Let* A_n $(n = 1, 2, \dots)$ *and* A *be m-dissipative operators and set* $J_\lambda^{(n)} = (I - \lambda A_n)^{-1}$ *and* $J_\lambda = (I - \lambda A)^{-1}$ $(\lambda > 0)$. *Then* (i) *through* (v) *are equivalent*:

(i) $A = \text{Lim}_{n\to\infty} A_n$.

(ii) $A \subset \text{Lim}_{n\to\infty} A_n$.

(iii) *For all* $x_n, x \in X$ *such that* $\lim_{n\to\infty} x_n = x$, $\lim_{n\to\infty} J_\lambda^{(n)} x_n = J_\lambda x$ $(\lambda > 0)$.

(iv) $\lim_{n\to\infty} J_\lambda^{(n)} x = J_\lambda x$ $(\lambda > 0, x \in X)$.

(v) *For some* $\lambda_0 > 0$, $\lim_{n\to\infty} J_{\lambda_0}^{(n)} x = J_{\lambda_0} x$ $(x \in X)$.

PROOF. (i) $\Rightarrow$ (ii) and (iii) $\Rightarrow$ (iv) $\Rightarrow$ (v) are self-evident.

(ii) $\Rightarrow$ (iii). Since $R(I - \lambda A_n) = X = R(I - \lambda A)$ $(\lambda > 0, \ n = 1, 2, \ldots)$, we set $X_0 = X$ and $A_0 = A$, and apply Lemma 6.4.

(v) $\Rightarrow$ (ii). Let $[x, y] \in \mathfrak{G}(A)$. Then from (v), $\lim_{n \to \infty} J_{\lambda_0}^{(n)}(x - \lambda_0 y) = J_{\lambda_0}(x - \lambda_0 y) = x$ and $\lim_{n \to \infty} \lambda_0^{-1}[J_{\lambda_0}^{(n)}(x - \lambda_0 y) - (x - \lambda_0 y)] = y$. Since $\lambda_0^{-1}[J_{\lambda_0}^{(n)}(x - \lambda_0 y) - (x - \lambda_0 y)] \in A_n J_{\lambda_0}^{(n)}(x - \lambda_0 y)$, $[x, y] \in \text{Lim}_{n \to \infty} \mathfrak{G}(A_n)$. Therefore, $\mathfrak{G}(A) \subset \text{Lim}_{n \to \infty} \mathfrak{G}(A_n)$.

(ii) $\Rightarrow$ (i). It suffices to show $A \supset \text{Lim}_{n \to \infty} A_n$, that is,

$$\mathfrak{G}(A) \supset \text{Lim}_{n \to \infty} \mathfrak{G}(A_n).$$

Let $[x, y] \in \text{Lim}_{n \to \infty} \mathfrak{G}(A_n)$. Then there exist $[x_n, y_n] \in \mathfrak{G}(A_n)$ such that $\|x_n - x\| + \|y_n - y\| \to 0$ as $n \to \infty$. Let $\lambda > 0$. Since $x_n - \lambda y_n \to x - \lambda y$ as $n \to \infty$, from (iii) we have $x_n = J_\lambda^{(n)}(x_n - \lambda y_n) \to J_\lambda(x - \lambda y)$ as $n \to \infty$. Hence $x = J_\lambda(x - \lambda y) \in D(A)$. Also

$$y = \lambda^{-1}[x - (x - \lambda y)] = \lambda^{-1}[J_\lambda(x - \lambda y) - (x - \lambda y)] \in A J_\lambda(x - \lambda y) = Ax.$$

From this we have $[x, y] \in \mathfrak{G}(A)$ and thus $\text{Lim}_{n \to \infty} \mathfrak{G}(A_n) \subset \mathfrak{G}(A)$. $\square$

REMARK. We obtain results similar to the above theorem when A_n and A are dissipative operators of type ω.

§2. Convergence of semigroups

We will use the following lemma to study the convergence of semigroups.

LEMMA 6.7. *Let A be a dissipative operator of type ω satisfying condition* (R) *and let $\{T(t): t \geq 0\}$ be the semigroup generated by A. Let $T > 0$ and $x \in \overline{D(A)}$ be given arbitrarily. For each $\lambda > 0$, suppose $\{t_k^\lambda: 0 \leq k \leq N_\lambda\}$, $\{x_k^\lambda: 0 \leq k \leq N_\lambda\}$, and $\{\varepsilon_k^\lambda: 1 \leq k \leq N_\lambda\}$ satisfy*

$$0 = t_0^\lambda < t_1^\lambda < \cdots < t_{N_\lambda - 1}^\lambda < T \leq t_{N_\lambda}^\lambda, \tag{6.4}$$

and

$$\begin{cases} \dfrac{x_k^\lambda - x_{k-1}^\lambda}{t_k^\lambda - t_{k-1}^\lambda} - \varepsilon_k^\lambda \in A x_k^\lambda & (k = 1, 2, \ldots, N_\lambda), \\ x_0^\lambda = x, \end{cases} \tag{6.5}$$

and define the step function $u_\lambda(t; x): [0, T] \to \overline{D(A)}$ by

$$u_\lambda(t; x) = \begin{cases} x & (t = 0), \\ x_k^\lambda & (t \in (t_{k-1}^\lambda, t_k^\lambda] \cap (0, T]), \end{cases} \quad k = 1, 2, \ldots, N_\lambda.$$

We set $\omega_0 = \max\{0, \omega\}$, $|\Delta_\lambda| = \max\{t_k^\lambda - t_{k-1}^\lambda: 1 \leq k \leq N_\lambda\}$ and $\varepsilon_\lambda = \sum_{k=1}^{N_\lambda} \|\varepsilon_k^\lambda\|(t_k^\lambda - t_{k-1}^\lambda)$.

Then if $|\Delta_\lambda|\omega_0 \leq 1/2$, the inequality

$$\|u_\lambda(s; x) - T(t)x\| \leq \exp(2\omega_0(t + s + |\Delta_\lambda|))$$
$$\times [2\|x - u\| + \varepsilon_\lambda \tag{6.6}$$
$$+ \{(|t - s| + |\Delta_\lambda|)^2 + |\Delta_\lambda|(s + |\Delta_\lambda|)\}^{1/2}\|Au\|]$$

holds for every $u \in D(A)$ and $0 \leq s, t \leq T$.

PROOF (The first step). We suppose that for each $\mu > 0$, $\{\hat{t}_k^\mu : 0 \leq k \leq \widehat{N}_\mu\}$, $\{\hat{x}_k^\mu : 0 \leq k \leq \widehat{N}_\mu\}$, and $\{\hat{\varepsilon}_k^\mu : 1 \leq k \leq \widehat{N}_\mu\}$ satisfy

$$0 = \hat{t}_0^\mu < \hat{t}_1^\mu < \cdots < \hat{t}_{N_\mu-1}^\mu < T \leq \hat{t}_{N_\mu}^\mu, \tag{6.7}$$

$$\begin{cases} \dfrac{\hat{x}_k^\mu - \hat{x}_{k-1}^\mu}{\hat{t}_k^\mu - \hat{t}_{k-1}^\mu} - \hat{\varepsilon}_k^\mu \in A\hat{x}_k^\mu & (k = 1, 2, \ldots, \widehat{N}_\mu), \\ \hat{x}_0^\mu = x, \end{cases} \tag{6.8}$$

and set $|\hat{\Delta}_\mu| = \max\{\hat{t}_k^\mu - \hat{t}_{k-1}^\mu : 1 \leq k \leq \widehat{N}_\mu\}$ and $\hat{\varepsilon}_\mu = \sum_{k=1}^{\widehat{N}_\mu} \|\hat{\varepsilon}_k^\mu\|(\hat{t}_k^\mu - \hat{t}_{k-1}^\mu)$.

If $|\Delta_\lambda|\omega_0 \leq 1/2$ and $|\hat{\Delta}_\mu|\omega_0 \leq 1/2$, then for every $u \in D(A)$ and $0 \leq i \leq N_\lambda$ and $0 \leq j \leq \widehat{N}_\mu$ we have

$$\|x_i^\lambda - \hat{x}_j^\mu\| \leq \exp(2\omega_0(t_i^\lambda + \hat{t}_j^\mu))\,[2\|x - u\| + \varepsilon_\lambda + \hat{\varepsilon}_\mu$$
$$+ \{(t_i^\lambda - \hat{t}_j^\mu)^2 + |\Delta_\lambda|t_i^\lambda + |\hat{\Delta}_\mu|\hat{t}_j^\mu\}^{1/2}\|Au\|]. \tag{6.9}$$

The proof is exactly the same as the proof of Lemma 5.8. (We replace $\{t_k^\mu\}$, $\{x_k^\mu\}$, and $\{\varepsilon_k^\mu\}$ by $\{\hat{t}_k^\mu\}$, $\{\hat{x}_k^\mu\}$, and $\{\hat{\varepsilon}_k^\mu\}$, respectively, in the proof of Lemma 5.8.) Next, we set

$$\hat{u}_\mu(t; x) = \begin{cases} x & (t = 0), \\ \hat{x}_k^\mu & (t \in (\hat{t}_{k-1}^\mu, \hat{t}_k^\mu] \cap (0, T]), \end{cases} \quad k = 1, 2, \ldots, \widehat{N}_\mu.$$

Then, if $|\Delta_\lambda|\omega_0 \leq 1/2$ and $|\hat{\Delta}_\mu|\omega_0 \leq 1/2$, then for every $u \in D(A)$ and $0 \leq s, t \leq T$

$$\|u_\lambda(s; x) - \hat{u}_\mu(t; x)\|$$
$$\leq \exp(2\omega_0(t + s + |\Delta_\lambda| + |\hat{\Delta}_\mu|)$$
$$\times [2\|x - u\| + \varepsilon_\lambda + \hat{\varepsilon}_\mu + \{(|t - s| + |\Delta_\lambda| + |\hat{\Delta}_\mu|)^2 \tag{6.10}$$
$$+ |\Delta_\lambda|(s + |\Delta_\lambda|) + |\hat{\Delta}_\mu|(t + |\hat{\Delta}_\mu|)\}^{1/2}\|Au\|].$$

In fact, we let $u \in D(A)$ and $0 < s, t \leq T$, and choose i and j such that $t_{i-1}^\lambda < s \leq t_i^\lambda$ and $\hat{t}_{j-1}^\mu < t \leq \hat{t}_j^\mu$. Then $u_\lambda(s; x) = x_i^\lambda$ and $\hat{u}_\mu(t; x) = \hat{x}_j^\mu$. Also, $t_i^\lambda \leq s + |\Delta_\lambda|$, $\hat{t}_j^\mu \leq t + |\hat{\Delta}_\mu|$, and $|t_i^\lambda - \hat{t}_j^\mu| \leq |t - s| + |\Delta_\lambda| + |\hat{\Delta}_\mu|$. Hence, if $|\Delta_\lambda|\omega_0 \leq 1/2$ and $|\hat{\Delta}_\mu|\omega_0 \leq 1/2$, then we obtain (6.10) from (6.9). Similarly, (6.10) holds when either $s = 0$ or $t = 0$.

(The second step) Let $|\Delta_\lambda|\omega_0 \leq 1/2$, $u \in D(A)$, and $0 \leq s \leq t \leq T$. Then we show (6.6) holds. Since A is a dissipative operator of type ω satisfying condition (R), from Lemma 5.11, we have that for every $\mu > 0$ there exist $\{\hat{t}_k^\mu\}$, $\{\hat{x}_k^\mu\}$, and $\{\hat{y}_k^\mu\}$ satisfying (6.11) through (6.13), where $\hat{x}_k^\mu \in D(A)$

and $\hat{y}_k^\mu \in A\hat{x}_k^\mu$ $(k = 1, 2, \ldots)$:

$$0 = \hat{t}_0^\mu < \hat{t}_1^\mu < \cdots < \hat{t}_k^\mu < \hat{t}_{k+1}^\mu < \cdots, \qquad \lim_{k \to \infty} \hat{t}_k^\mu = \infty, \qquad (6.11)$$

$$\hat{t}_k^\mu - \hat{t}_{k-1}^\mu \le \mu \qquad (k = 1, 2, \ldots), \qquad (6.12)$$

$$\|\hat{x}_k^\mu - \hat{x}_{k-1}^\mu - (\hat{t}_k^\mu - \hat{t}_{k-1}^\mu)\hat{y}_k^\mu\| < \mu(\hat{t}_k^\mu - \hat{t}_{k-1}^u) \qquad (k = 1, 2, \ldots), \quad (6.13)$$

where $\hat{x}_0^\mu = x$. We set $\hat{\varepsilon}_k^\mu = (\hat{x}_k^\mu - \hat{x}_{k-1}^\mu)/(\hat{t}_k^\mu - \hat{t}_{k-1}^\mu) - \hat{y}_k^\mu$ $(k = 1, 2, \ldots)$. From (6.11) it follows that for each $\mu > 0$ we can choose a natural number $\widehat{N}_\mu$ such that $\hat{t}_{\widehat{N}_\mu - 1} < T \le \hat{t}_{\widehat{N}_\mu}$. Obviously $\{\hat{t}_k^\mu : 0 \le k \le \widehat{N}_\mu\}$, $\{\hat{x}_k^\mu : 0 \le k \le \widehat{N}_\mu\}$, and $\{\hat{\varepsilon}_k^\mu : 1 \le k \le \widehat{N}_\mu\}$ satisfy (6.7) and (6.8). Also from (6.12) $|\hat{\Delta}_\mu| = \max\{\hat{t}_k^\mu - \hat{t}_{k-1}^\mu : 1 \le k \le \widehat{N}_\mu\} \le \mu$. Hence, let $\hat{u}_\mu(t; x)$ be defined as described above, and if $\mu\omega_0 \le 1/2$, then (6.10) holds. As we see from the construction of the semigroup $\{T(t) : t \ge 0\}$ (see Theorem 5.10, the proof of Theorem 5.12, and Theorem 5.7), $T(t)x = \lim_{\mu \to 0+} \hat{u}_\mu(t; x)$ $(0 \le t \le T)$, and hence we obtain (6.6) by letting $\mu \to 0+$ in (6.10) (note that $\hat{\varepsilon}_\mu < \mu(T + \mu)$ from (6.13) and (6.12), and thus $\lim_{\mu \to 0+} \hat{\varepsilon}_\mu = 0$).

THEOREM 6.8. *Let A_n $(n = 1, 2, \ldots)$ be dissipative operators of type ω_n satisfying condition* (R) *and let $\{T_n(t) : t \ge 0\}$ be the semigroups generated by A_n. Also, let A be a (not necessarily single-valued) operator satisfying condition* (R). *We assume* (a_1) *and* (a_2):

(a_1) *There is a constant $\alpha \ge 0$ such that $\omega_n \le \alpha$ $(n = 1, 2, \ldots)$,*

(a_2) *$A \subset \mathrm{Lim}_{n \to \infty} A_n$.*

Then, (i) *and* (ii) *hold:*

(i) *A is a dissipative operator of type α and A generates a semigroup of type α. We let $\{T(t) : t \ge 0\}$ be this semigroup. Then*

(ii) *For every $x \in \overline{D(A)}$, there exist $x_n \in D(A_n)$ such that $\lim_{n \to \infty} x_n = x$, and*

$$\lim_{n \to \infty} T_n(t)x_n = T(t)x \qquad (t \ge 0). \qquad (6.14)$$

Moreover, for every $x_n \in \overline{D(A_n)}$ and $x \in \overline{D(A)}$ such that $\lim_{n \to \infty} x_n = x$, (6.14) holds uniformly on bounded subintervals of $[0, \infty)$.

PROOF. (i) Since A_n is a dissipative operator of type ω_n with $\omega_n \le \alpha$, all the A_n are dissipative operators of type α. Let $x, x' \in D(A)$, $y \in Ax$, and $y' \in Ax'$ be given arbitrarily. Then from (a_2), we can choose x_n, $x_n' \in D(A_n)$, $y_n \in A_n x_n$, and $y_n' \in A_n x_n'$ such that

$$\|x_n - x\| + \|y_n - y\| \to 0 \quad \text{and} \quad \|x_n' - x'\| + \|y_n' - y'\| \to 0 \quad \text{as } n \to \infty.$$

Now, from the dissipativity of $A_n - \alpha I$ we have

$$\|x_n - \lambda(y_n - \alpha x_n) - [x_n' - \lambda(y_n' - \alpha x_n')]\| \ge \|x_n - x_n'\| \qquad (\lambda > 0, \ n \ge 1).$$

If we let $n \to \infty$, then

$$\|x - \lambda(y - \alpha x) - [x' - \lambda(y' - \alpha x')]\| \ge \|x - x'\| \qquad (\lambda > 0).$$

Hence, $A - \alpha I$ is a dissipative operator; in other words, A is a dissipative operator of type α. Also, A satisfies condition (R) (the assumption). Therefore, A generates a semigroup $\{T(t): t \geq 0\}$ of type α.

(ii) Let $x_n \in \overline{D(A_n)}$, $x \in \overline{D(A)}$, and $\lim_{n \to \infty} x_n = x$. We will show that (6.14) holds uniformly on bounded subintervals of $[0, \infty)$. Let $T > 0$ be given arbitrarily. Since A is a dissipative operator of type α satisfying condition (R), from Lemma 5.11, it follows that for every $\lambda > 0$ there exist t_k^λ, $x_k^\lambda \in D(A)$ and $y_k^\lambda \in Ax_k^\lambda$ $(k = 1, 2 \ldots)$ satisfying (6.15) through (6.17):

$$0 = t_0^\lambda < t_1^\lambda < \cdots < t_k^\lambda < t_{k+1}^\lambda < \cdots, \qquad \lim_{k \to \infty} t_k^\lambda = \infty, \qquad (6.15)$$

$$t_k^\lambda - t_{k-1}^\lambda \leq \lambda \qquad (k = 1, 2, \ldots), \qquad (6.16)$$

$$\|x_k^\lambda - x_{k-1}^\lambda - (t_k^\lambda - t_{k-1}^\lambda)y_k^\lambda\| < \lambda(t_k^\lambda - t_{k-1}^\lambda) \qquad (k = 1, 2, \ldots), \quad (6.17)$$

where $x_0^\lambda = x$. Now for each $\lambda > 0$, choose a natural number N_λ such that $t_{N_\lambda - 1}^\lambda < T \leq t_{N_\lambda}^\lambda$ and set $\varepsilon_k^\lambda = (x_k^\lambda - x_{k-1}^\lambda)/(t_k^\lambda - t_{k-1}^\lambda) - y_k^\lambda$ $(k = 1, 2, \ldots, N_\lambda)$. Obviously $\{t_k^\lambda: 0 \leq k \leq N_\lambda\}$, $\{x_k^\lambda: 0 \leq k \leq N_\lambda\}$, and $\{\varepsilon_k^\lambda: 1 \leq k \leq N_\lambda\}$ satisfy (6.4) and (6.5). Also, from (6.16) and (6.17) we have $\varepsilon_\lambda = \sum_{k=1}^{N_\lambda} \|\varepsilon_k^\lambda\|(t_k^\lambda - t_{k-1}^\lambda) < \lambda(T + \lambda)$ and $|\Delta_\lambda| = \max\{t_k^\lambda - t_{k-1}^\lambda ; 1 \leq k \leq N_\lambda\} \leq \lambda$. Hence, if we define a step function $u_\lambda(t; x): [0, T] \to \overline{D(A)}$ in the same way as in Lemma 6.7, then from the lemma it follows that if $\lambda\alpha \leq 1/2$ for every $u \in D(A)$ and $0 \leq t \leq T$, then we have

$$\|u_\lambda(t; x) - T(t)x\| \leq \exp(2\alpha(2t+\lambda))[2\|x - u\| + \lambda(T+\lambda) + (2\lambda^2 + \lambda t)^{1/2}\|Au\|].$$
$$(6.18)$$

Now, we choose $\lambda > 0$ such that $\lambda\alpha \leq 1/2$. (Hence, for every $u \in D(A)$ and $0 \leq t \leq T$, (6.18) holds.) Since $x_k^\lambda \in D(A)$, $y_k^\lambda \in Ax_k^\lambda$, and $A \subset \text{Lim}_{n \to \infty} A_n$ (the assumption (a_2)), for each natural number k there exist $x_k^{\lambda, n} \in D(A_n)$ and $y_k^{\lambda, n} \in A_n x_k^{\lambda, n}$ $(n = 1, 2, \ldots)$ such that k, $\|x_k^{\lambda, n} - x_k^\lambda\| + \|y_k^{\lambda, n} - y_k^\lambda\| \to 0$ as $n \to \infty$. Hence, for a given $\varepsilon > 0$, we can select a natural number $n(\varepsilon)$ such that if $n \geq n(\varepsilon)$ then

$$\|x_k^{\lambda, n} - x_k^\lambda\| < \varepsilon, \quad \|y_k^{\lambda, n} - y_k^\lambda\| < \varepsilon \qquad (1 \leq k \leq N_\lambda). \qquad (6.19)$$

Then from (6.16), (6.17), and (6.19), we have

$$\sum_{k=1}^{N_\lambda} \|x_k^{\lambda, n} - x_{k-1}^{\lambda, n} - (t_k^\lambda - t_{k-1}^\lambda)y_k^{\lambda, n}\|$$

$$\leq \sum_{k=1}^{N_\lambda} [\|x_k^\lambda - x_{k-1}^\lambda - (t_k^\lambda - t_{k-1}^\lambda)y_k^\lambda\|$$

$$+ \|x_k^\lambda - x_k^{\lambda, n}\| + \|x_{k-1}^\lambda - x_{k-1}^{\lambda, n}\| + (t_k^\lambda - t_{k-1}^\lambda)\|y_k^\lambda - y_k^{\lambda, n}\|]$$

$$\leq \lambda(T + \lambda) + 2N_\lambda\varepsilon + \|x - x_n\| + \varepsilon(T + \lambda) \qquad (n \geq n(\varepsilon)),$$

where $x_0^{\lambda,n} = x_n$. And also $|\Delta_\lambda| \max\{0, \omega_n\} \le \lambda\alpha \le 1/2$. Therefore, by Lemma 6.7 (instead of A, x, x_k^λ, and ε_k^λ, we use A_n, x_n, $x_k^{\lambda,n}$, and $\varepsilon_k^{\lambda,n} = (x_k^{\lambda,n} - x_{k-1}^{\lambda,n})/(t_k^\lambda - t_{k-1}^\lambda) - y_k^{\lambda,n}$, respectively), if $n \ge n(\varepsilon)$, then the next inequality holds:

$$\|u_{\lambda,n}(t; x_n) - T_n(t)x_n\| \tag{6.20}$$
$$\le \exp(2\alpha(2t + \lambda))[2\|x_n - w\| + \delta_{n,\lambda,\varepsilon} + (2\lambda^2 + \lambda t)^{1/2}\|A_n w\|]$$
$$(0 \le t \le T, \ w \in D(A_n)),$$

where $\delta_{n,\lambda,\varepsilon} = \lambda(T + \lambda) + 2N_\lambda\varepsilon + \|x - x_n\| + \varepsilon(T + \lambda)$ and

$$u_{\lambda,n}(t; x_n) = \begin{cases} x_n & (t = 0), \\ x_k^{\lambda,n} & (t \in (t_{k-1}^\lambda, t_k^\lambda] \cap (0, T]), \end{cases} \quad k = 1, 2, \ldots, N_\lambda.$$

From the definitions of functions u_λ and $u_{\lambda,n}$, and (6.19)

$$\|u_\lambda(t; x) - u_{\lambda,n}(t; x_n)\| \le \varepsilon \quad (n \ge n(\varepsilon), \ 0 < t \le T)$$

and

$$\|u_\lambda(0; x) - u_{\lambda,n}(0; x_n)\| = \|x - x_n\|.$$

From this

$$\|u_\lambda(t; x) - u_{\lambda,n}(t; x_n)\| \le \varepsilon + \|x - x_n\| \quad (n \ge n(\varepsilon), \ 0 \le t \le T). \tag{6.21}$$

Then from (6.18), (6.20), and (6.21), if $n \ge n(\varepsilon)$ then we obtain

$$\|T_n(t)x_n - T(t)x\|$$
$$\le \exp(2\alpha(2t + \lambda))[2\|x_n - w\| + \delta_{n,\lambda,\varepsilon} + 2\|x - u\|$$
$$+ \lambda(T + \lambda) + (2\lambda^2 + \lambda t)^{1/2}(\|A_n w\| + \|Au\|)] \tag{6.22}$$
$$+ \varepsilon + \|x - x_n\|$$

$(0 \le t \le T, \ w \in D(A_n), \text{ and } u \in D(A))$.

Now, let $u \in D(A)$ and $v \in Au$ be given arbitrarily. Then from the assumption (a_2), there exist $w_n \in D(A_n)$ and $w_n' \in A_n w_n$ $(n = 1, 2, \ldots)$ such that $\|w_n - u\| + \|w_n' - v\| \to 0$ as $n \to \infty$. If we set $w = w_n$ in (6.22) and let $n \to \infty$, we obtain the following inequality:

$$\limsup_{n \to \infty} \left[\sup_{0 \le t \le T} \|T_n(t)x_n - T(t)x\|\right]$$
$$\le \exp(2\alpha(2T + \lambda))[4\|x - u\| + \delta_{\lambda,\varepsilon} + \lambda(T + \lambda) \tag{6.23}$$
$$+ (2\lambda^2 + \lambda T)^{1/2}(\|v\| + \|Au\|)] + \varepsilon,$$

where $\delta_{\lambda,\varepsilon} = \lambda(T+\lambda) + 2N_\lambda\varepsilon + \varepsilon(T+\lambda)$. (Here we used $\|A_n w_n\| \le \|w_n'\| \to \|v\|$ as $n \to \infty$.) The above inequality holds for every $\varepsilon, \lambda > 0$. Now in (6.23), if we let $\varepsilon \to 0+$ first and $\lambda \to 0+$ next, then we have

$$\limsup_{n \to \infty} \left[\sup_{0 \le t \le T} \|T_n(t)x_n - T(t)x\|\right] \le 4e^{4\alpha T}\|x - u\|.$$

Here since $u \in D(A)$ is arbitrary and $x \in \overline{D(A)}$, we have shown that for every $T > 0$, $\lim_{n \to \infty} T_n(t)x_n = T(t)x$ (uniformly on $[0, T]$).

Finally, let $x \in \overline{D(A)}$. Since we obtain $\overline{D(A)} \subset \operatorname{Lim}_{n \to \infty} D(A_n)$ from (a_2) ((iii) of Corollary 6.3), there exist $x_n \in D(A_n)$ such that $\lim_{n \to \infty} x_n = x$. Then, as shown above, (6.14) holds for this $\{x_n\}$ and x. $\square$

From Theorems 6.8 and 6.5 we obtain the following corollary:

COROLLARY 6.9. *Let* A_n $(n = 1, 2, \ldots)$ *and* A *be dissipative operators, all of which satisfy condition* (R_1'), *and let* $\{T_n(t): t \geq 0\}$ *and* $\{T(t): t \geq 0\}$ *be the semigroups of contractions generated by* A_n *and* A, *respectively. Set* $J_\lambda^{(n)} = (I - \lambda A_n)^{-1}$ *and* $J_\lambda = (I - \lambda A)^{-1}$ $(\lambda > 0)$. *Then* (i) *and* (ii) *hold*:

(i) *If we assume* (ii) *of Theorem 6.5, then* (ii) *of Theorem 6.8 holds.*

(ii) *If* $\overline{D(A_n)} \supset \overline{D(A)}$ $(n = 1, 2, \ldots)$ *and* $\lim_{n \to \infty} J_\lambda^{(n)} x = J_\lambda x$ $(x \in \overline{D(A)}, \lambda > 0)$, *then for every* $x \in \overline{D(A)}$ *we have*

$$\lim_{n \to \infty} T_n(t)x = T(t)x \qquad (t \geq 0). \tag{6.24}$$

Moreover (6.24) *holds uniformly on bounded subintervals of* $[0, \infty)$.

PROOF. (i) Let an operator B be defined in the same way as in Theorem 6.5. Then from Theorem 6.5

$$B \subset \operatorname*{Lim}_{n \to \infty} A_n. \tag{6.25}$$

From $\lambda^{-1}(J_\lambda x - x) \in AJ_\lambda x$ $(\lambda > 0)$, we have $B \subset A$. Hence B is also a dissipative operator. Furthermore, from

$$x = J_\lambda x - \lambda[\lambda^{-1}(J_\lambda x - x)] \in ((I - \lambda B)J_\lambda x) \qquad (\lambda > 0, \ x \in \overline{D(A)}),$$

we have

$$R(I - \lambda B) \supset \overline{D(A)} = \overline{D(B)} \qquad (\lambda > 0);$$

that is, B satisfies condition (R_1'). Hence B generates a semigroup of contractions on $\overline{D(A)}$. We denote this semigroup by $\{S(t): t \geq 0\}$. Then, noting (6.25) it follows from Theorem 6.8 that, for every $x \in \overline{D(A)}$ $(= \overline{D(B)})$ there exist $x_n \in D(A_n)$ such that $\lim_{n \to \infty} x_n = x$, and

$$\lim_{n \to \infty} T_n(t)x_n = S(t)x \qquad (t \geq 0).$$

Moreover, this holds for every $x_n \in \overline{D(A_n)}$ and $x \in \overline{D(A)}$ satisfying $\lim_{n \to \infty} x_n = x$, and the convergence is uniform on bounded subintervals of $[0, \infty)$. Now we have $S(t)x = T(t)x$ $(t \geq 0, \ x \in \overline{D(A)})$. In fact, let $x \in \overline{D(A)}$; then $S(t)x$ is a unique integral solution of type 0 of the Cauchy problem $(\mathrm{CP}; x)_\infty$ for B. On the other hand, since that $T(t)x$ is an integral solution of type 0 of $(\mathrm{CP}; x)_\infty$ for A, it follows from $B \subset A$, that $T(t)x$ is also the integral solution of type 0 of $(\mathrm{CP}; x)_\infty$ for B. Therefore, $T(t)x = S(t)x$. Thus we have shown (i) from the above. (ii) is obvious from Theorem 6.5 and (i). $\square$

The next corollary is an extension of the Trotter-Kato theorem on the convergence of linear (contraction) semigroups to nonlinear semigroups.

COROLLARY 6.10. *Let* A_n $(n = 1, 2, \ldots)$ *be m-dissipative operators, and let* $\{T_n(t) \colon t \geq 0\}$ *be the semigroups of contractions generated by* A_n. *We set* $J_\lambda^{(n)} = (I - \lambda A_n)^{-1}$ $(\lambda > 0)$. *For some* $\lambda_0 > 0$, *we assume that there exists* $\lim_{n\to\infty} J_{\lambda_0}^{(n)} x$ *for all* $x \in X$, *and denote the limit by* $J_{\lambda_0} x$. *Then the following hold*:

(i) *The operator* A *having* $\{[J_{\lambda_0} x, \lambda_0^{-1}(J_{\lambda_0} x - x)] \colon x \in X\}$ *as its graph is an m-dissipative operator. Hence* A *generates a semigroup of contractions. Now let* $\{T(t) \colon t \geq 0\}$ *be this semigroup. Then*

(ii) *For every* $x \in \overline{R(J_{\lambda_0})}$ $(= \overline{D(A)})$, *there exist* $x_n \in D(A_n)$ *such that* $\lim_{n\to\infty} x_n = x$, *and*

$$\lim_{n\to\infty} T_n(t) x_n = T(t) x \qquad (t \geq 0).$$

Furthermore, the above equation holds for every $x_n \in \overline{D(A_n)}$ *and* $x \in \overline{R(J_{\lambda_0})}$ *satisfying* $\lim_{n\to\infty} x_n = x$, *and the convergence is uniform on arbitrary bounded subintervals of* $[0, \infty)$.

In particular, if $\overline{D(A_n)} \supset \overline{R(J_{\lambda_0})}$ $(n = 1, 2, \ldots)$, *for every* $x \in \overline{R(J_{\lambda_0})}$ *we have*

$$\lim_{n\to\infty} T_n(t) x = T(t) x \qquad (t \geq 0).$$

Moreover, the convergence is uniform on bounded subintervals of $[0, \infty)$.

PROOF. (i) Let $u_i \in D(A)$ and $v_i \in Au_i$ $(i = 1, 2)$. Then from the definition of A we can write

$$u_1 = J_{\lambda_0} x, \quad v_1 = \lambda_0^{-1}(J_{\lambda_0} x - x), \quad u_2 = J_{\lambda_0} y, \quad \text{and} \quad v_2 = \lambda_0^{-1}(J_{\lambda_0} y - y),$$

where $x, y \in X$. From the assumption, $\lim_{n\to\infty} J_{\lambda_0}^{(n)} x = J_{\lambda_0} x = u_1$ and $\lim_{n\to\infty} J_{\lambda_0}^{(n)} y = J_{\lambda_0} y = u_2$. Also from this, $\lim_{n\to\infty} \lambda_0^{-1}(J_{\lambda_0}^{(n)} x - x) = v_1$ and $\lim_{n\to\infty} \lambda_0^{-1}(J_{\lambda_0}^{(n)} y - y) = v_2$. Since $\lambda_0^{-1}(J_{\lambda_0}^{(n)} x - x) \in A_n J_{\lambda_0}^{(n)} x$ and $\lambda_0^{-1}(J_{\lambda_0}^{(n)} y - y) \in A_n J_{\lambda_0}^{(n)} y$, from the dissipativity of A_n we have

$$\|J_{\lambda_0}^{(n)} x - J_{\lambda_0}^{(n)} y - \lambda[\lambda_0^{-1}(J_{\lambda_0}^{(n)} x - x) - \lambda_0^{-1}(J_{\lambda_0}^{(n)} y - y)]\| \geq \|J_{\lambda_0}^{(n)} x - J_{\lambda_0}^{(n)} y\|$$

$$(\lambda > 0).$$

Here if we let $n \to \infty$, then

$$\|u_1 - u_2 - \lambda(v_1 - v_2)\| \geq \|u_1 - u_2\| \qquad (\lambda > 0).$$

Therefore A is a dissipative operator. Next from the definition of A, every $x \in X$ we have $x = J_\lambda x - \lambda_0[\lambda_0^{-1}(J_{\lambda_0} x - x)] \in (I - \lambda_0 A) J_{\lambda_0} x$. (From this, $J_{\lambda_0} x = (I - \lambda_0 A)^{-1} x$ $(x \in X)$.) Hence, we obtain $R(I - \lambda_0 A) = X$ and thus A is an m-dissipative operator (see Lemma 2.13).

(ii) Since A is an m-dissipative operator and $J_{\lambda_0} x = (I - \lambda_0 A)^{-1} x$ $(x \in X)$, from Theorem 6.6, our assumption $\lim_{n\to\infty} J_{\lambda_0}^{(n)} x = J_{\lambda_0} x$ $(x \in X)$ is equivalent to $A = \mathrm{Lim}_{n \to \infty} A_n$. Also, from $D(A) = R(J_{\lambda_0})$ we have $\overline{D(A)} = \overline{R(J_{\lambda_0})}$. Therefore, from Theorem 6.8 we obtain the required conclusion. □

COROLLARY 6.11. *Let X be a Hilbert space, and let A_n $(n = 1, 2, \ldots)$ and A be maximal dissipative operators (hence from Corollary 2.27, A_n and A are m-dissipative operators). Let $\{T_n(t): t \geq 0\}$ and $\{T(t): t \geq 0\}$ be the semigroups of contractions generated by A_n and A respectively. If $\overline{D(A)} \subset \overline{D(A_n)}$ $(n = 1, 2, \ldots)$, then* (i) *and* (ii) *are equivalent*:

(i) *For every $x \in \overline{D(A)}$ and $\lambda > 0$, $\lim_{n \to \infty} J_\lambda^{(n)} x = J_\lambda x$, where $J_\lambda^{(n)}$ and J_λ are the same as those in Corollary 6.9.*

(ii) *For every $x \in \overline{D(A)}$, $\lim_{n \to \infty} T_n(t)x = T(t)x$ $(t \geq 0)$ uniformly on bounded subintervals of $[0, \infty)$.*

PROOF. From part (ii) of Corollary 6.9, we obtain (i) $\Rightarrow$ (ii). We show (ii) $\Rightarrow$ (i) next. From Theorem 2.20, $\overline{D(A_n)}$ and $\overline{D(A)}$ are all closed convex sets. From this and the fact that $T_n(t): \overline{D(A_n)} \to \overline{D(A_n)}$ and $T(t): \overline{D(A)} \to \overline{D(A)}$ are contraction operators, $A_{h,n} = h^{-1}(T_n(h) - I)$ and $A_{h,0} = h^{-1}(T(h) - I)$ $(h > 0)$ are all dissipative operators, and also for all $\lambda > 0$

$$R(I - \lambda A_{h,n}) \supset \overline{D(A_n)} \quad \text{and} \quad R(I - \lambda A_{h,0}) \supset \overline{D(A)}$$

hold (see Example 2.4).

Let $x \in \overline{D(A)}$ and $\lambda > 0$. Then from $\overline{D(A)} \subset \overline{D(A_n)}$ we have $x \in \overline{D(A_n)}$. Hence from above, we can define $(I - \lambda A_{h,n})^{-1} x$ for $n = 0, 1, 2, \ldots$ and $h > 0$. Now we set $y_{h,n} = (I - \lambda A_{h,n})^{-1} x$ $(h > 0; n = 0, 1, 2, \ldots)$. Then from Theorem 3.13, $\lim_{h \to 0+} y_{h,n}$ $(n = 0, 1, 2, \ldots)$ exist and we have

$$\lim_{h \to 0+} y_{h,n} = J_\lambda^{(n)} x \quad (n = 1, 2, \ldots), \qquad \lim_{h \to 0+} y_{h,0} = J_\lambda x. \qquad (6.26)$$

(By Theorem 3.13, for every $x \in \overline{D(A)}$ and $\lambda > 0$, $\lim_{h \to 0}(I - \lambda A_{h,0})^{-1} x$ exists. We denote this limit by $J_\lambda' x$. Then, as we mentioned in the remark after Corollary 3.12, the operator B, which can be defined by $D(B) = \bigcup_{\lambda > 0} R(J_\lambda')$ and $Bx = \bigcup_{\lambda \in \Lambda(x)} \lambda^{-1}(I - J_\lambda'^{-1})x$ $(x \in D(B))$, is a dissipative operator and satisfies $\overline{D(B)} = \overline{D(A)}$ and $R(I - \lambda B) \supset \overline{D(A)}$. Also from Corollary 3.19, $T(t)x = \lim_{\lambda \to 0+} J_\lambda'^{[t/\lambda]} x$ $(t \geq 0, x \in \overline{D(A)})$. (Note that from (3.48), $J_\lambda' x = (I - \lambda B)^{-1} x$ $(x \in \overline{D(A)})$.) Now for this B, from Theorem 2.26, it follows that an m-dissipative operator $\widetilde{B}$ satisfying $D(\widetilde{B}) \subset \overline{D(A)}$ and $B \subset \widetilde{B}$ exists. Then from $J_\lambda' x = (I - \lambda B)^{-1} x = (I - \lambda \widetilde{B})^{-1} x$ $(\lambda > 0, x \in \overline{D(A)})$, we have $T(t)x = \lim_{\lambda \to 0+}(I - \lambda \widetilde{B})^{-[t/\lambda]} x$ $(t \geq 0, x \in \overline{D(A)})$. This shows that $\{T(t): t \geq 0\}$ is the semigroup generated by $\widetilde{B}$. Hence, two m-dissipative operators A and $\widetilde{B}$ generate the same semigroup of contractions. Therefore, by Corollary 4.17, $A = \widetilde{B}$, and thus $J_\lambda' x = (I - \lambda A)^{-1} x = J_\lambda x$ $(\lambda > 0, x \in \overline{D(A)})$. Hence we obtain $\lim_{h \to 0+} y_{h,0} =$

$J_\lambda x$. We also obtain $\lim_{h\to 0+} y_{h,n} = J_\lambda^{(n)} x$ $(n = 1, 2, \ldots)$ in the same way.)

Now let $\varepsilon > 0$ be given arbitrarily. Then $\lim_{n\to\infty} T_n(t)x = T(t)x$ uniformly on bounded subintervals of $[0, \infty)$, and from $\lim_{h\to 0+} T_n(h)x = x$ $(n \geq 1)$ and $\lim_{h\to 0+} T(h)x = x$, we can choose a $\delta(\varepsilon) > 0$ such that if $0 \leq h \leq \delta(\varepsilon)$ then

$$\|T_n(h)x - x\| < \varepsilon \quad (n = 1, 2, \ldots) \quad \text{and} \quad \|T(h)x - x\| < \varepsilon.$$

Then from $(3.20)'$ and the remark after it, for h such that $0 < h < \delta(\varepsilon)/2$, we have $\|y_{n,h} - x\| \leq K \max\{2\lambda\varepsilon/\delta(\varepsilon), \varepsilon\}$ $(n = 0, 1, 2, \ldots)$, and also for h such that $h \geq \delta(\varepsilon)/2$, $\|y_{n,h} - x\| \leq (2\lambda/\delta(\varepsilon))\|T_n(h)x - x\|$ $(n = 0, 1, 2, \ldots)$ holds, where $K = (7 + \sqrt{53})/2$ and $T_0(h) = T(h)$. Hence $M = \sup\{\|y_{n,h} - x\| : 0 < h \leq 1, n = 0, 1, 2, \ldots\} < \infty$. Then, as we see from the proof of (i) of Theorem 3.13 (the last part), if $0 < s, t \leq \delta(\varepsilon)$, then $\|y_{t,n} - y_{s,n}\| \leq \sqrt{2M\varepsilon}$ $(n = 0, 1, 2, \ldots)$. Here if we let $s \to 0+$, then from (6.26) we obtain

$$\|y_{t,n} - J_\lambda^{(n)} x\| \leq 2\sqrt{2M\varepsilon}, \quad \|y_{t,0} - J_\lambda x\| \leq 2\sqrt{2M\varepsilon} \quad (0 < t \leq \delta(\varepsilon), \ n \geq 1).$$
$$(6.27)$$

Since $x = (I - \lambda A_{h,n})y_{h,n}$, from the definition of $A_{h,n}$ we have

$$y_{h,n} = \frac{h}{h+\lambda}x + \frac{\lambda}{h+\lambda}T_n(h)y_{h,n} \quad (h > 0, \ n = 0, 1, 2, \ldots).$$

Hence

$$\|y_{h,n} - y_{h,0}\| = \frac{\lambda}{h+\lambda}\|T_n(h)y_{h,n} - T(h)y_{h,0}\|$$
$$\leq \frac{\lambda}{h+\lambda}[\|y_{h,n} - y_{h,0}\| + \|T_n(h)y_{h,0} - T(h)y_{h,0}\|].$$

Therefore, we obtain

$$\|y_{h,n} - y_{h,0}\| \leq (\lambda/h)\|T_n(h)y_{h,0} - T(h)y_{h,0}\| \quad (h > 0, \ n \geq 1).$$

From this and (6.27) we have

$$\|J_\lambda^{(n)} x - J_\lambda x\| \leq \|J_\lambda^{(n)} x - y_{h,n}\| + \|y_{h,n} - y_{h,0}\| + \|y_{h,0} - J_\lambda x\|$$
$$\leq 4\sqrt{2M\varepsilon} + (\lambda/h)\|T_n(h)y_{h,0} - T(h)y_{h,0}\| \quad (0 < h \leq \delta(\varepsilon), \ n \geq 1).$$

Here if we let $n \to \infty$ with h fixed, then

$$\limsup_{n\to\infty} \|J_\lambda^{(n)} x - J_\lambda x\| \leq 4\sqrt{2M\varepsilon}.$$

Therefore we obtain $\lim_{n\to\infty} J_\lambda^{(n)} x = J_\lambda x$. □

REMARK. When A_n and A are dissipative operators of type ω, we also obtain the same results as those in Corollaries 6.9 through 6.11.

Finally, from Theorem 6.8 and Lemma 3.15, we obtain the following theorem on the approximation of semigroups.

THEOREM 6.12. *Let* X_n $(n = 1, 2, \ldots)$ *be closed convex subsets of* X *and let* $T_n: X_n \to X_n$ *be operators satisfying*

$$\|T_n x - T_n y\| \leq M_n \|x - y\| \qquad (x, y \in X_n), \tag{6.28}$$

where M_n *are constants, independent of* $x, y \in X_n$, *and* $M_n \geq 1$. *Also let* $\{h_n\}$ *be a sequence of positive numbers that converges to* 0, *and let* A *be a (not necessarily single-valued) operator satisfying condition* (R). *We assume* (a_1) *and* (a_2):

(a_1) *There is a constant* $\alpha \geq 0$ *such that* $M_n - 1 \leq \alpha h_n$ $(n = 1, 2, \ldots)$,

(a_2) $A \subset \mathrm{Lim}_{n \to \infty} h_n^{-1}(T_n - I)$.

Then (i) *and* (ii) *hold.*

(i) A *is a dissipative operator of type* α *and* A *generates a semigroup of type* α. *Denote this semigroup by* $\{T(t): t \geq 0\}$. *Then*

(ii) *For every* $x \in \overline{D(A)}$, *there exist* $x_n \in X_n$ *such that* $\lim_{n \to \infty} x_n = x$ *and*

$$\lim_{n \to \infty} T_n^{[t/h_n]} x_n = T(t)x \qquad (t \geq 0). \tag{6.29}$$

Moreover, (6.29) *holds for every* $x_n \in X_n$ *and* $x \in \overline{D(A)}$ *with* $\lim_{n \to \infty} x_n = x$ *and the convergence is uniform on bounded subintervals of* $[0, \infty)$.

PROOF. We set $A_n = h_n^{-1}(T_n - I)$ and $\omega_n = h_n^{-1}(M_n - 1)$ $(n = 1, 2, \ldots)$.

(i) From (6.28), A_n are dissipative operators of type ω_n. Since $\omega_n \leq \alpha$ $(n = 1, 2, \ldots)$ and $A \subset \mathrm{Lim}_{n \to \infty} A_n$, from (i) of Theorem 6.8, it follows that A is a dissipative operator of type α and generates a semigroup $\{T(t): t \geq 0\}$ of type α.

(ii) For each natural number n, because A_n is a dissipative operator of type ω_n and

$$R(I - \lambda A_n) \supset X_n = D(A_n) \; (= \overline{D(A_n)}) \qquad (0 < \lambda < \omega_n^{-1})$$

holds (we carry out the same proof as in Example 2.4), A_n generates a semigroup $\{T_n(t): t \geq 0\}$. We note that since A_n is Lipschitz continuous, from Theorem 5.14, it follows that for each $z \in X_n$, $T_n(t)z$ is a unique solution of the Cauchy problem $(CP; z)_\infty$ for A_n.

Then from Theorem 6.8, for every $x \in \overline{D(A)}$, there exist $x_n \in X_n$ such that $\lim_{n \to \infty} x_n = x$, and

$$\lim_{n \to \infty} T_n(t)x_n = T(t)x \qquad (t \geq 0). \tag{6.30}$$

Furthermore, (6.30) holds for every $x_n \in X_n$ and $x \in \overline{D(A)}$ that satisfy $\lim_{n \to \infty} x_n = x$, and the convergence is uniform on bounded subintervals of $[0, \infty)$.

On the other hand, from Lemma 3.15 (see also Lemma 3.14), for each natural number n, there exists a semigroup $\{T(t; T_n - I): t \geq 0\}$ of type $M_n - 1$ on X_n which has $T_n - I$ as its infinitesimal generator, and for an every $z \in X_n$, $T(t: T_n - I)z$ is strongly continuously differentiable on $[0, \infty)$ and satisfies

$$(d/dt)T(t; T_n - I)z = (T_n - I)T(t; T_n - I)z \qquad (t \geq 0)$$

and for every $z \in X_n$, nonnegative integer k, and $t \geq 0$,

$$\|T(t; T_n - I)z - T_n^k z\| \leq M_n^k e^{(M_n - 1)t}[(k - M_n t)^2 + M_n t]^{1/2}\|T_n z - z\|.$$

Now, if we set $T(t; A_n)z = T(t/h_n; T_n - I)z$ $(z \in X_n, \ t \geq 0)$, then $T(t; A_n)z$ $(z \in X_n)$ is strongly continuously differentiable on $[0, \infty)$ and satisfies

$$(d/dt)T(t; A_n)z = A_n T(t; A_n)z \quad (t \geq 0), \qquad T(0; A_n)z = z,$$

and hence it is a solution of the Cauchy problem $(CP; x)_\infty$ for A_n. Therefore, $T(t; A_n)z = T_n(t)z$ $(z \in X_n, \ t \geq 0)$. And from the estimation above, for every $z \in X_n$ and $t \geq 0$, we obtain

$$\|T_n(t)z - T_n^{[t/h_n]}z\| \leq e^{2\alpha t}[\alpha^2 t^2 h_n + 3\alpha t h_n + 3t]^{1/2}\sqrt{h_n}\|A_n z\|, \qquad (6.31)$$

where we used $\omega_n \leq \alpha$ and $M_n^{[t/h_n]} \leq \exp[t h_n^{-1} \log(1 + \alpha h_n)] \leq e^{\alpha t}$ $(t \geq 0)$.

Now, suppose that $x_n \in X_n$ and $x \in \overline{D(A)}$ satisfy $\lim_{n \to \infty} x_n = x$. Then, if we can show

$$\begin{cases} \lim_{n \to \infty} \|T_n(t)x_n - T_n^{[t/h_n]}x_n\| = 0 \text{ uniformly on} \\ \text{arbitrary bounded subintervals of } [0, \infty), \end{cases} \qquad (6.32)$$

then from this and (6.30), we obtain the required conclusion. So, we now prove (6.32). Choose $u \in D(A)$ and $v \in Au$ arbitrarily. Then from the assumption (a_2), there exist $u_n \in X_n$ that satisfy $\|u_n - u\| + \|A_n u_n - v\| \to 0$ as $n \to \infty$. If we put $z = u_n$ in (6.31) and let $n \to \infty$, then for any every $T > 0$ we obtain

$$\sup_{0 \leq t \leq T} \|T_n(t)u_n - T_n^{[t/h_n]}u_n\| \to 0 \qquad (n \to \infty).$$

Now

$$\begin{aligned} \|T_n(t)x_n - T_n^{[t/h_n]}x_n\| &\leq \|T_n(t)x_n - T_n(t)u_n\| + \|T_n(t)u_n - T_n^{[t/h_n]}u_n\| \\ &\quad + \|T_n^{[t/h_n]}u_n - T_n^{[t/h_n]}x_n\| \\ &\leq (e^{\omega_n t} + M_n^{[t/h_n]})\|x_n - u_n\| + \|T_n(t)u_n - T_n^{[t/h_n]}u_n\| \\ &\leq 2e^{\alpha t}\|x_n - u_n\| + \|T_n(t)u_n - T_n^{[t/h_n]}u_n\|. \end{aligned}$$

Hence for every $T > 0$

$$\limsup_{n \to \infty}\left[\sup_{0 \leq t \leq T} \|T_n(t)x_n - T_n^{[t/h_n]}x_n\|\right] \leq 2e^{\alpha T}\|x - u\|.$$

Here since $x \in \overline{D(A)}$ and $u \in D(A)$ are arbitrary, the right-hand side of the above inequality can approach 0 as closely as desired and hence (6.32) is proved. $\square$

From the above theorem and Theorem 6.5 we obtain the following corollary:

COROLLARY 6.13. *Let* X_n $(n = 1, 2, \ldots)$ *be closed convex subsets of* X, *and* $T_n \in \mathrm{Cont}(X_n)$ *(i.e.,* T_n *are contraction operators from* X_n *into* X_n *). Let* $\{h_n\}$ *be a sequence of positive numbers which converges to* 0, *and let* A *be a dissipative operator satisfying condition* (R'_1) *and let* $\{T(t) : t \geq 0\}$ *be a (contraction) semigroup generated by* A. *We set* $J_\lambda = (I - \lambda A)^{-1}$ $(\lambda > 0)$. *Then the following hold*:

(i) *Assume that* $\overline{D(A)} \subset \mathrm{Lim}_{n \to \infty} X_n$ *and* "*if* $x_n \in X_n$, $x \in \overline{D(A)}$, *and* $\lim_{n \to \infty} x_n = x$, *then* $\lim_{n \to \infty} (I - \lambda A_n)^{-1} x_n = J_\lambda x$ $(\lambda > 0)$," *where* $A_n = h_n^{-1}(T_n - I)$. *Then* (ii) *of Theorem* 6.12 *holds*.

(ii) *If* $X_n \supset \overline{D(A)}$ $(n = 1, 2, \ldots)$ *and* $\lim_{n \to \infty} (I - \lambda A_n)^{-1} x = J_\lambda x$ $(x \in \overline{D(A)}, \lambda > 0)$, *then for every* $x \in \overline{D(A)}$

$$\lim_{n \to \infty} T_n^{[t/h_n]} x = T(t)x \qquad (t \geq 0). \tag{6.33}$$

Furthermore, the convergence in (6.33) *is uniform on bounded subintervals of* $[0, \infty)$.

PROOF. (i) We define an operator B in the same way as in Theorem 6.5. Then from Theorem 6.5, our assumption is equivalent to $B \subset \mathrm{Lim}_{n \to \infty} A_n = \mathrm{Lim}_{n \to \infty} h_n^{-1}(T_n - I)$. Also, as we have shown in the proof of (i) of Corollary 6.9, B is a dissipative operator satisfying the condition (R'_1) and generates $\{T(t) : t \geq 0\}$ (the semigroup generated by A). Hence we can apply Theorem 6.12 to prove this.

(ii) is obvious from Theorem 6.5 and (i). $\square$

Furthermore, for the case where T_n satisfy (6.28), under assumption (a_1) of Theorem 6.12, we can also obtain the same results as in the above corollary.

§3. Perturbation of semigroups

We investigate the problem of whether $A + B$ generates a semigroup or not in the case where A is a dissipative operator of type ω that generates a semigroup and B is a continuous operator. If we can show that $A + B$ is a dissipative operator of type ω and $A + B$ satisfies condition (R), then from the generation theorem, we see that $A + B$ generates a semigroup. Thus, we examine what kind of continuous operator B, when added to A, causes $A + B$ (as a dissipative operator of type ω) to satisfy the condition (R). In the following, we consider the case where A and $A + B$ are dissipative operators. (The method of argument is the same when they are dissipative operators of type ω.)

LEMMA 6.14. *Let* A *be a dissipative operator and* $x \in \overline{D(A)}$. *If* $\{\lambda_n\}$ *is a sequence of positive numbers that converges to* 0, *and if* $x_n \in R(I - \lambda_n A)$ $(n = 1, 2, \ldots)$ *and* $\lim_{n \to \infty} x_n = x$, *then* $\lim_{n \to \infty} (I - \lambda_n A)^{-1} x_n = x$.

PROOF. We choose $u_k \in D(A)$ $(k = 1, 2, \ldots)$ such that $\lim_{k \to \infty} u_k = x$ and let $v_k \in A u_k$. Since $\lambda_n^{-1}[(I - \lambda_n A)^{-1} x_n - x_n] \in A(I - \lambda_n A)^{-1} x_n$ and A

is a dissipative operator, we have

$$\|(I - \lambda_n A)^{-1} x_n - u_k\|$$
$$\leq \|(I - \lambda_n A)^{-1} x_n - u_k - \lambda_n[\lambda_n^{-1}((I - \lambda_n A)^{-1} x_n - x_n) - v_k]\|$$
$$= \|x_n - u_k + \lambda_n v_k\|.$$

Hence

$$\|(I - \lambda_n A)^{-1} x_n - x\| \leq \|(I - \lambda_n A)^{-1} x_n - u_k\| + \|u_k - x\|$$
$$\leq \|x_n - u_k + \lambda_n v_k\| + \|u_k - x\|.$$

Here if we let $n \to \infty$ with k fixed, then

$$\limsup_{n \to \infty} \|(I - \lambda_n A)^{-1} x_n - x\| \leq 2\|u_k - x\|.$$

Since the above holds for every k, we let $k \to \infty$ to complete the proof. $\square$

THEOREM 6.15. *Let A be a dissipative operator, and let $B: D(B)$ $(\subset X) \to X$ satisfy $\overline{D(A)} \subset D(B)$ and be a continuous operator on $\overline{D(A)}$. We assume that $A + B$ is a dissipative operator. Then, a necessary and sufficient condition for $A + B$ to satisfy condition* (R) *(i.e., for all $x \in \overline{D(A + B)}$, $\liminf_{\lambda \to 0+} \lambda^{-1} d(R(I - \lambda(A + B)), x) = 0$) is that*

$$\liminf_{\lambda \to 0+} \lambda^{-1} d(R(I - \lambda A), x + \lambda Bx) = 0 \qquad for\ all\ x \in \overline{D(A)}. \tag{6.34}$$

PROOF. (Necessity) Let $x \in \overline{D(A)}$. Since $\overline{D(A)} = \overline{D(A + B)}$, we have $x \in \overline{D(A + B)}$. Then from $\liminf_{\lambda \to 0+} \lambda^{-1} d(R(I - \lambda(A + B)), x) = 0$, there exists a sequence $\{\lambda_n\}$ of positive numbers that converges to 0, such that $\lim_{n \to \infty} \lambda_n^{-1} d(R(I - \lambda_n(A + B)), x) = 0$. Now for each n, we choose $x_n \in D(A)$ and $y_n \in Ax_n$ that satisfy $\|x_n - \lambda_n(y_n + Bx_n) - x\| < d(R(I - \lambda_n(A + B)), x) + \lambda_n^2$, and set $z_n = \lambda_n^{-1}(x_n - \lambda_n y_n - \lambda_n Bx_n - x)$. Then we have $z_n \to 0$ as $n \to \infty$ and

$$d(R(I - \lambda_n A), x + \lambda_n Bx) \leq \lambda_n \|z_n\| + \lambda_n \|Bx - Bx_n\|, \tag{6.35}$$

Since $x + \lambda_n z_n = x_n - \lambda_n y_n - \lambda_n Bx_n \in (I - \lambda_n(A + B))x_n$ and $x + \lambda_n z_n \to x$ as $n \to \infty$, by Lemma 6.14 we have $x_n = (I - \lambda_n(A + B))^{-1}(x + \lambda_n z_n) \to x$ as $n \to \infty$. Then since B is continuous on $\overline{D(A)}$, $\|Bx_n - Bx\| \to 0$ as $n \to \infty$. Hence from (6.35), we obtain $\lim_{n \to \infty} \lambda_n^{-1} d(R(I - \lambda_n A), x + \lambda_n Bx) = 0$ and thus $\liminf_{\lambda \to 0+} \lambda^{-1} d(R(I - \lambda A), x + \lambda Bx) = 0$.

(Sufficiency) Let $x \in \overline{D(A + B)}$ $(= \overline{D(A)})$. Then from (6.34), there exists a sequence $\{\lambda_n\}$ of positive numbers that converges to 0, such that $\lim_{n \to \infty} \lambda_n^{-1} d(R(I - \lambda_n A), x + \lambda_n Bx) = 0$. Now for each n, we choose $x_n \in D(A)$ and $y_n \in Ax_n$ that satisfy $\|x + \lambda_n Bx - (x_n - \lambda_n y_n)\| < d(x + \lambda_n Bx, R(I - \lambda_n A)) + \lambda_n^2$, and set $z_n = \lambda_n^{-1}(x + \lambda_n Bx - x_n + \lambda_n y_n)$. Then $z_n \to 0$ as $n \to \infty$ and

$$d(R(I - \lambda_n(A + B)), x) \leq \lambda_n \|z_n\| + \lambda_n \|Bx_n - Bx\|.$$

Since

$$x - \lambda_n z_n + \lambda_n Bx = x_n - \lambda_n y_n \in (I - \lambda_n A)x_n \quad \text{and} \quad \lim_{n \to \infty}(x - \lambda_n z_n + \lambda_n Bx) = x,$$

by Lemma 6.14 we have $x_n = (I - \lambda_n A)^{-1}(x - \lambda_n z_n + \lambda_n Bx) \to x$ as $n \to \infty$. Then since B is continuous on $\overline{D(A)}$, $\|Bx_n - Bx\| \to 0$. Hence from the above inequality, we obtain $\lim_{n \to \infty} \lambda_n^{-1} d(R(I - \lambda_n(A + B)), x) = 0$ and thus $\liminf_{\lambda \to 0+} \lambda^{-1} d(R(I - \lambda(A + B)), x) = 0$. $\square$

COROLLARY 6.16. *Let* X_0 $(\subset X)$ *be a closed set, and let* A *be a dissipative operator with* $\overline{D(A)} \subset X_0$ *with*

$$R(I - \lambda A) \supset X_0 \quad \text{for all } \lambda > 0. \tag{6.36}$$

Also, let B *be an operator with* $\overline{D(A)} \subset D(B)$ $(\subset X)$ *and with values in* X. *Then* (i) *and* (ii) *hold:*

(i) *If* $A + B$ *is a dissipative operator,* B *is continuous on* $\overline{D(A)}$, *and* B *satisfies*

$$\liminf_{\lambda \to 0+} \lambda^{-1} d(X_0, x + \lambda Bx) = 0 \qquad (x \in \overline{D(A)}), \tag{6.37}$$

then $A + B$ *satisfies condition* (R).

(ii) *If* $D(B)$ *is a closed set with* $D(B) \subset X_0$, *and* B *is a continuous dissipative operator on* $D(B)$ *satisfying condition* (R), *then* $A + B$ *is a dissipative operator satisfying condition* (R).

PROOF. (i) From (6.36), $d(R(I - \lambda A), x + \lambda Bx) \le d(X_0, x + \lambda Bx)$ $(\lambda > 0; x \in D(B))$. Hence (6.34) holds from (6.37). From Theorem 6.15, $A + B$ satisfies condition (R).

(ii) From Theorem 5.15, since B becomes the infinitesimal generator of a semigroup of contractions on $D(B)$, B is a strictly dissipative operator. Hence $A + B$ is a dissipative operator. From Theorem 5.15 again, B satisfies condition (R) if and only if

$$\liminf_{\lambda \to 0+} \lambda^{-1} d(D(B), x + \lambda Bx) = 0 \qquad (x \in D(B)).$$

From this and $D(B) \subset X_0$, (6.37) holds. Therefore from (i), $A + B$ satisfies condition (R). $\square$

THEOREM 6.17. *Let* X_0 $(\subset X)$ *be a closed convex set, and let* A *be a dissipative operator satisfying* (6.36) *with* $\overline{D(A)} \subset X_0$. *Also, let* $B: D(B)$ $(\subset X) \to X$ *be a continuous operator on* $\overline{D(A)}$ *with* $\overline{D(A)} \subset D(B)$. *Then* $A + B$ *is a dissipative operator, and if* (6.37) *holds, then* $\overline{A + B}$ *is a dissipative operator satisfying the following* (6.38).

$$R(I - \lambda(\overline{A} + B)) \supset X_0 \quad \text{for all } \lambda > 0. \tag{6.38}$$

PROOF. Let $y \in X_0$ and $\alpha > 0$, and put $Cx = \alpha Ax - x + y$ $(x \in D(A))$. Since A is a dissipative operator, C is a dissipative operator of type -1, and

$$R(I - \lambda C) \supset X_0 \quad \text{for all } \lambda > 0. \tag{6.39}$$

In fact, let $z \in X_0$ be given arbitrarily. Then since X_0 is a convex set, for every $\lambda > 0$, $(1 + \lambda)^{-1}z + \lambda(1 + \lambda)^{-1}y \in X_0$. Also from (6.36), $X_0 \subset R(I - \lambda\alpha(1 + \lambda)^{-1}A)$ $(\lambda > 0)$. Hence

$$(1 + \lambda)^{-1}z + \lambda(1 + \lambda)^{-1}y \in R(I - \lambda\alpha(1 + \lambda)^{-1}A) \qquad (\lambda > 0).$$

From this, for each $\lambda > 0$, there exist $x_\lambda \in D(A)$ and $y_\lambda \in Ax_\lambda$ such that

$$\frac{1}{1 + \lambda}z + \frac{\lambda}{1 + \lambda}y = x_\lambda - \frac{\lambda\alpha}{1 + \lambda}y_\lambda.$$

From this, $z = x_\lambda - \lambda(\alpha y_\lambda - x_\lambda + y) \in R(I - \lambda C)$. Hence we obtain (6.39).

Now we consider $C + \alpha B = \alpha(A + B) - I + y$. From the assumption that $A + B$ is a dissipative operator, $C + \alpha B$ is a dissipative operator of type -1 (from which $C + \alpha B$ is also a dissipative operator). Then, we can apply part (i) of Corollary 6.16 to C and αB, and thus $C + \alpha B$ satisfies the condition (R). Therefore, by the generation theorem (Theorem 5.12), $C + \alpha B$ generates a semigroup of type -1 on $\overline{D(C + \alpha B)} = \overline{D(A)}$. Let this semigroup be $\{T(t): t \geq 0\}$. Then, there exists $z \in \overline{D(A)}$ which satisfies (6.40):

$$T(t)z = z \quad \text{for all } t \geq 0. \tag{6.40}$$

(In fact, since $\|T(t)x - T(t)y\| \leq e^{-t}\|x - y\|$ $(t \geq 0; x, y \in \overline{D(A)})$, by the fixed point theorem, for every $t > 0$, there exists a unique $x_t \in \overline{D(A)}$ such that $T(t)x_t = x_t$. However, from $T(s)x_t = T(s)T(t)x_t = T(t)T(s)x_t$, $T(s)x_t = x_t$. Hence $x_s = x_t$ $(t \neq s)$, and thus there is a common fixed point for all $T(t)$.)

Since $T(t)z$ $(\equiv z)$ is an integral solution of type -1 of the Cauchy problem $(CP; z)_\infty$ for $C + \alpha B = \alpha(A + B) - I + y$ (see Theorem 5.12), for every $u \in D(A)$, $v \in Au$, and $t > 0$

$$(e^{2t} - 1)\|z - u\|^2 \leq 2\int_0^t e^{2\xi}\langle \alpha v + \alpha Bu - u + y, z - u\rangle_s \, d\xi$$

$$= 2\langle \alpha v + \alpha Bu - u + y, z - u\rangle_s \int_0^t e^{2\xi} \, d\xi.$$

Therefore,

for every $u \in D(A)$ and $v \in Au$, $\langle \alpha v + \alpha Bu - u + y, z - u\rangle_s \geq 0$. (6.41)

Since $C + \alpha B = \alpha(A + B) - I + y$ satisfies condition (R),

$$\liminf_{\lambda \to 0+} \lambda^{-1} d(R(I - \lambda(\alpha(A + B) - I + y)), z) = 0.$$

From this, we can choose $x_n \in D(A)$, $y_n \in Ax_n$, and a sequence $\{\lambda_n\}$ of positive numbers that converges to 0, such that

$\lambda_n^{-1}[x_n - \lambda_n(\alpha y_n + \alpha B x_n - x_n + y) - z]$ (we denote this by z_n) $\to 0$ as $n \to \infty$.

By the definition of z_n,

$$\lambda_n(\alpha y_n + \alpha B x_n - x_n + y) = x_n - z - \lambda_n z_n. \tag{6.42}$$

Since $x_n \in D(A)$ and $y_n \in Ax_n$, from (6.41)

$$\langle \alpha y_n + \alpha Bx_n - x_n + y, z - x_n \rangle_s \geq 0.$$

From this and (6.42) we have

$$-\|z - x_n\|^2 + \langle -\lambda_n z_n, z - z_n \rangle_s = \langle x_n - z - \lambda_n z_n, z - x_n \rangle_s \geq 0.$$

From this we obtain $\|z - x_n\| \leq \lambda_n \|z_n\|$, and since $z_n \to 0$ as $n \to \infty$,

$$\lambda_n^{-1}(z - x_n) \to 0 \quad \text{as } n \to \infty.$$

Hence $\lim_{n\to\infty} x_n = z$, and from the above, (6.42), and the property that B is continuous on $\overline{D(A)}$, we have

$$y_n = \alpha^{-1}[x_n - y + \lambda_n^{-1}(x_n - z) - z_n] - Bx_n$$
$$\to \alpha^{-1}(z - y) - Bz \quad \text{as } n \to \infty.$$

Therefore, $z \in D(\overline{A})$ and $\alpha^{-1}(z - y) - Bz \in \overline{A}z$, and thus $y \in (I - \alpha(\overline{A} + B))z \subset R(I - \alpha(\overline{A} + B))$. Thus, (6.38) is proved.

Finally, since B is continuous on $\overline{D(A)}$, $\overline{A} + B = \overline{A + B}$ can easily be shown. Also, since $A + B$ is a dissipative operator, so is $\overline{A + B}$. Therefore, $\overline{A} + B$ is a dissipative operator. $\square$

COROLLARY 6.18. *Let* X_0 *($\subset X$) be a closed convex set, and let* A *be a dissipative operator with* $\overline{D(A)} \subset X_0$ *satisfying* (6.36). *Also, let* B *be a continuous dissipative operator on* $D(B)$, *satisfying condition* (R), *and let* $\overline{D(A)} \subset D(B) \subset X_0$ *and* $D(B)$ *be closed. Then* $\overline{A} + B$ *is a dissipative operator and satisfies* (6.38).

PROOF. As in the proof of part (ii) of Corollary 6.16, we show that $A + B$ is a dissipative operator and thus (6.37) holds. Hence we can apply Theorem 6.17 to complete the proof. $\square$

COROLLARY 6.19. *Let* A *be an* m-*dissipative operator and let* B *be an operator with* $\overline{D(A)} \subset D(B)$ *and with values in* X.

(i) *If* B *is continuous on* $\overline{D(A)}$ *and* $A + B$ *is a dissipative operator, then* $A + B$ *is an* m-*dissipative operator.*

(ii) *If* $D(B)$ *is a closed set and* B *is a continuous operator on* $D(B)$ *satisfying condition* (R) *then* $A + B$ *is an* m-*dissipative operator.*

(iii) *If* $D(B) = X$ *and* B *is a continuous dissipative operator on* X, *then* $A + B$ *is an* m-*dissipative operator.*

PROOF. We note that an m-dissipative operator is a closed operator. If we let $X_0 = X$ and apply Theorem 6.17 and Corollary 6.18, then we obtain (i) and (ii), respectively.

(iii) By Corollary 5.16, B is an m-dissipative operator. Hence B satisfies condition (R), and thus we obtain (iii) from (ii). $\square$

EXAMPLE 6.1. Let $\beta(x): R^1 \to R^1$ be a continuous monotone decreasing function. Then β is an m-dissipative operator (on the space R^1) (see Example 2.1). For a bounded continuous real-valued function f on R^1, we define $\|f\| = \sup\{|f(x)|: x \in R^1\}$. X, the set of "all the bounded continuous real-valued functions on R^1," forms a Banach space with the above norm $\|\cdot\|$. Now we define an operator $B: X \to X$ by

$$(Bf)(x) = \beta(f(x)) \qquad (f \in X; x \in R^1);$$

then B is a continuous operator on X. Moreover B is a dissipative operator. In fact, since β is a dissipative operator on R^1, we have

$$|f(x) - g(x)| \le |f(x) - g(x) - \lambda(\beta(f(x)) - \beta(g(x)))| \qquad (f, g \in X; \lambda > 0).$$

From this we obtain

$$\|f - g\| \le \|f - g - \lambda(Bf - Bg)\| \qquad (f, g \in X; \lambda > 0)$$

and B is a dissipative operator.

Now, we let A be an m-dissipative operator on X. (For example, the differential operator d^2/dx^2 is an m-dissipative operator on X.) Then from part (iii) of Corollary 6.19, $A + B$ is also an m-dissipative operator. Hence $A + B$ generates a semigroup of contractions. When X is the set of "bounded continuous real-valued functions on R^n," we also define the operator B in the same way as above and can show that the same results hold.

EXAMPLE 6.2. Here we consider Carleman's equations of hyperbolic type (4.67) that we examined in Example 4.10 (see also Example 2.3), using the perturbation theorems described above. We suppose that X, X_0, and $D_M(A)$ $(M > 0)$ are the same as in Example 4.10 and set $X_M = \{[u, v] \in X: 0 \le u, v \le M\}$ $(M > 0)$. Then X_0 and X_M are closed convex subsets of X and $X_0 = \overline{\bigcup_{M>0} X_M}$.

(i) We put $D(A) = (\bigcup_{M>0} D_M(A)) \cap \{[u, v] \in X: [u_x, v_y] \in X\}$ and define

$$A[u, v] = [-u_x, -v_y] \qquad ([u, v] \in D(A)).$$

Then the operator $A: D(A) \to X$ is a dissipative operator (see Example 2.3).

Next for each $M > 0$, we set $A_M = A|_{X_M \cap D(A)}$; that is, A_M is the restriction of A to $X_M \cap D(A)$. Given $\lambda > 0$ and $[f, g] \in X_M \cap D(A)$, we set

$$u(x, y) = \lambda^{-1} e^{-x/\lambda} \int_0^x e^{\xi/\lambda} f(\xi, y) \, d\xi,$$

$$v(x, y) = \lambda^{-1} e^{-y/\lambda} \int_0^y e^{\eta/\lambda} f(x, \eta) \, d\eta \qquad (x, y \ge 0).$$

Then $[u, v] \in X_M \cap D(A)$ and $(I - \lambda A_M)[u, v] = [f, g]$. Hence, A_M is a dissipative operator and satisfies $R(I - \lambda A_M) \supset X_M \cap D(A)$ $(\lambda > 0)$. From this, $\overline{A}_M$ is also a dissipative operator, and since $\overline{X_M \cap D(A)} = X_M$ and $R(I - \lambda \overline{A}_M)$ are closed sets, we obtain

$$R(I - \lambda \overline{A}_M) \supset X_M \qquad (\lambda > 0, \ M > 0). \qquad (6.43)$$

(ii) The set $\{(\xi, \eta): \xi, \eta \in R^1\}$ forms a Banach space with norm $\|(\xi, \eta)\|_1 = |\xi| + |\eta|$. We denote this space by $l^1(2)$. Now we define $\beta(\xi, \eta): (R^+)^2 (\subset l^1(2)) \to L^1(2)$ by

$$\beta(\xi, \eta) = (\eta^2 - \xi^2, \xi^2 - \eta^2) \qquad ((\xi, \eta) \in (R^+)^2).$$

Then β is a dissipative operator (on $l^2(2)$). (In fact, if we let $(\xi_i, \eta_i) \in (R^+)^2$ $(i = 1, 2,)$ then since $\xi_1 - \xi_2$ and $\xi_1^2 - \xi_2^2$ are of the same sign, for every $\lambda > 0$

$$|\xi_1 - \lambda(\eta_1^2 - \xi_1^2) - [\xi_2 - \lambda(\eta_2^2 - \xi_2^2)]| \geq |\xi_1 - \xi_2| + \lambda|\xi_1^2 - \xi_2^2| - \lambda|\eta_1^2 - \eta_2^2|.$$

Similarly,

$$|\eta_1 - \lambda(\xi_1^2 - \eta_1^2) - [\eta_2 - \lambda(\xi_2^2 - \eta_2^2)]| \geq |\eta_1 - \eta_2| + \lambda|\eta_1^2 - \eta_2^2| - \lambda|\xi_1^2 - \xi_2^2| \qquad (\lambda > 0).$$

When we add the respective sides of these two, we obtain

$$\|(\xi_1, \eta_1) - \lambda\beta(\xi_1, \eta_1) - [(\xi_2, \eta_2) - \lambda\beta(\xi_2, \eta_2)]\|_1$$
$$\geq \|(\xi_1, \eta_1) - (\xi_2, \eta_2)\|_1 \qquad (\lambda > 0).$$

Hence β is a dissipative operator.)

Now, for $[u, v] \in \bigcup_{M>0} X_M$ we define

$$B[u, v] = \beta(u, v) \ (= [v^2 - u^2, u^2 - v^2]).$$

Then, since β is a dissipative operator (in $l^1(2)$), we see easily $B: \bigcup_{M>0} X_M \to X$ is a dissipative operator. For every $M > 0$, we set $B_M = B|_{X_M}$. Then $B_M: X_M \to X$ is also a dissipative operator. Also, B_M is continuous on X_M. Furthermore,

$$R(I - \lambda B_M) \supset X_M \qquad (\lambda > 0). \qquad (6.44)$$

In fact, for $\lambda > 0$, $f, g \in X_M$ we set

$$u(x, y) = [(1 + c_\lambda(x, y))f(x, y) + (1 - c_\lambda(x, y))g(x, y)]/2,$$
$$v(x, y) = [(1 - c_\lambda(x, y))f(x, y) + (1 + c_\lambda(x, y))g(x, y)]/2 \qquad (x, y \geq 0),$$

where $c_\lambda(x, y) = (1 + 2\lambda(f(x, y) + g(x, y)))^{-1}$ $(x, y \geq 0)$. Then $[u, v] \in X_M$ and $(I - \lambda B_M)[u, v] = [f, g]$. Therefore, (6.44) holds.

(iii) From (i) and (ii) above, for every $M > 0$, we see that X_M, $\overline{A}_M$, and B_M satisfy the assumption of Corollary 6.18. Hence for every $M > 0$, $\overline{A}_M + B_M$ $(= \overline{A_M + B_M})$ is a dissipative operator and satisfies

$$R(I - \lambda(\overline{A_M + B_M})) \supset X_M \qquad (\lambda > 0).$$

Then since $A_M + B_M$ $(M > 0)$ is a dissipative operator, we see easily that $A + B$ is also a dissipative operator. Therefore, $\overline{A + B}$ is also a dissipative operator. Also, from $A + B \supset A_M + B_M$, we have $R(I - \lambda(\overline{A + B})) \supset \bigcup_{M>0} X_M$ $(\lambda > 0)$. Hence we obtain

$$R(I - \lambda(\overline{A + B})) \supset X_0 \qquad (\lambda > 0).$$

This is (4.68) shown in Example 4.10. (As we see from the definitions of A and B, $A + B$ is, in fact, the same as the operator A defined in Example 4.10.)

§4. Applications of perturbation theorems;
on the evolution equation $(d/dt)u(t) = A(t)u(t)$

By the same reasoning as used to prove Corollary 6.18 in the previous section, we can prove the following:

COROLLARY 6.18′. *Let* X_0 $(\subset X)$ *be a closed convex set and let* A *be a dissipative operator with* $\overline{D(A)} \subset X_0$ *satisfying* (6.36). *Also, let* B *be a continuous dissipative operator of type* ω *on* $D(B)$ *satisfying condition* (R), *and let* $\overline{D(A)} \subset D(B) \subset X_0$ *and* $D(B)$ *be closed. Then* $\overline{A} + B$ *is a dissipative operator of type* ω *and satisfies the following:*

For all $\lambda > 0$ *such that* $\lambda\omega < 1$, $R(I - \lambda(\overline{A} + B)) \supset X_0$. $\qquad$ (6.45)

Applying this perturbation theorem we consider a differential equation $(d/dt)u(t) = A(t)u(t)$ with operators $A(t) : D(A(t))(\subset X) \to X$ that depends on time t. First, we examine the Cauchy problem:

$$\begin{cases} (d/dt)u(t) = A(t)u(t) & (0 \leq t \leq T), \\ u(0) = x \ (\in X). \end{cases} \qquad (6.46)$$

THEOREM 6.20. *Let* $A(t)$ $(0 \leq t \leq T)$ *be an operator satisfying* (a_1) *through* (a_3):

(a_1) *There exists a bounded measurable function* $\omega(t) : [0, T] \to R^1$ *such that* $A(t)$ *is a dissipative operator of type* $\omega(t)$ *for every* $t \in [0, T]$. $D(A(t))$ *is a closed convex set, independent of* t $(\in [0, T])$. *(Hereafter, we set* $D = D(A(t))$ $(0 \leq t \leq T)$. *Hence* D *is a closed convex subset of* X.)

(a_2) $A(t)x$ *is continuous on* $[0, T] \times D$ *as a mapping of two variables* t *and* x.

(a_3) *For all* $t \in [0, T]$, *and all* $\lambda > 0$ *such that* $\lambda\omega(t) < 1$

$$R(I - \lambda A(t)) \supset D.$$

Then for every $x \in D$, *the Cauchy problem* (6.46) *has a unique solution* $u(\cdot\,; x) \in C^1([0, T]; D)$. *Moreover,*

$$\|u(t\,; x) - u(t\,; y)\| \leq \|x - y\| \exp\left(\int_0^t \omega(s)\,ds\right) \qquad (x, y \in D, \ t \in [0, T]).$$
$$(6.47)$$

Here $C^1([0, T]; D)$ denotes the set of all strongly continuously (once) differentiable functions $u(t): [0, T] \to D$.

Next, in order to prove this, we set $\omega = \max\{0, \sup_{0 \le t \le T} \omega(t)\}$. Then from (a_1) and (a_3), $A(t)$ is a dissipative operator of type ω for all $t \in [0, T]$ and (a_4) holds:

(a_4) *For all $\lambda > 0$ such that $\lambda\omega < 1$, $R(I - \lambda A(t)) \supset D$ $(0 \le t \le T)$.*

LEMMA 6.21. *We assume (a_1) through (a_3) of the above theorem and set $J_\lambda(t) = (I - \lambda A(t))^{-1}$ $(0 \le t \le T)$ for $\lambda > 0$ such that $\lambda\omega < 1$. Then $J_\lambda(t)x$ is continuous on $[0, T] \times D$ and has values in D.*

PROOF. First, we show

$$\lim_{s \to t} \|J_\lambda(s)x - J_\lambda(t)x\| = 0 \qquad (x \in D, \ 0 \le t \le T). \qquad (6.48)$$

In fact we give $x \in D$ and $0 \le t \le T$ arbitrarily. From $J_\lambda(s)x = \lambda A(s)J_\lambda(s)x + x$ $(0 \le s \le T)$, we have

$$J_\lambda(s)x - J_\lambda(t)x = \lambda A(s)J_\lambda(s)x - \lambda A(s)J_\lambda(t)x + \lambda A(s)J_\lambda(t)x + x - J_\lambda(t)x.$$

From this and the fact that $A(s)$ is a dissipative operator of type ω,

$$\|J_\lambda(s)x - J_\lambda(t)x\|$$
$$\le (1 - \lambda\omega)^{-1}\|J_\lambda(s)x - J_\lambda(t)x - \lambda[A(s)J_\lambda(s)x - A(s)J_\lambda(t)x]\|$$
$$\le (1 - \lambda\omega)^{-1}\|\lambda(A(s)J_\lambda(t)x + x - J_\lambda(t)x\|.$$

From (a_2) the right-hand side of the above inequality becomes

$$(1 - \lambda\omega)^{-1}\|\lambda A(t)J_\lambda(t)x + x - J_\lambda(t)x\| = 0$$

as $s \to t$. Hence we obtain (6.48). Now let $t_n, t \in [0, T]$, $x_n, x \in D$, and let $t_n \to t$ and $\|x_n - x\| \to 0$ as $n \to \infty$. Then from $\|J_\lambda(t_n)x_n - J_\lambda(t_n)x\| \le (1 - \lambda\omega)^{-1}\|x_n - x\|$ and (6.48) we have

$$\|J_\lambda(t_n)x_n - J_\lambda(t)x\| \le (1-\lambda\omega)^{-1}\|x_n - x\| + \|J_\lambda(t_n)x - J_\lambda(t)x\| \to 0 \qquad (n \to \infty).$$

□

PROOF OF THEOREM 6.20. $C([0, T]; X)$, the set of all continuous functions defined on $[0, T]$ with values in X, forms a Banach space with norm

$$\|u(\cdot)\| = \max_{0 \le t \le T} \|u(t)\| \qquad (u(\cdot) \in C([0, T]; X)).$$

(By considering this Banach space as the space X in Corollary 6.18', we intend to apply the corollary.)

(I) We define an operator B as follows: Let $e(t) = e^{(1+\omega)t}$ $(0 \le t \le T)$, and

$$D(B) = \{u(\cdot): e(\cdot)u(\cdot) \in C([0, T]; D)\},$$
$$(Bu)(t) = e^{-(1+\omega)t}A(t)(e^{(1+\omega)t}u(t)) \qquad (u(\cdot) \in D(B), \ 0 \le t \le T),$$

where $C([0, T]; D) = \{u(\cdot) \in C([0, T]; X) : u(t) \in D \ (0 \le t \le T)\}$. Then (i) and (ii) hold:

(i) $D(B)$ is a closed convex subset of $C([0, T]; X)$ and $B: D(B) \to C([0, T]; X)$ is a continuous dissipative operator of type ω.

(ii) For all $\lambda > 0$ such that $\lambda\omega < 1$, $R(I - \lambda B) \supset D(B)$.

(We denote the identity operator on $C([0, T]; X)$ by the same notation as the identity operator I on X.)

PROOF. (i) Since $C([0, T]; D)$ is a closed convex subset of $C([0, T]; X)$, $D(B)$ is also a closed convex subset of $C([0, T]; X)$. We obtain from (a_2) that $B: D(B) \to C([0, T]; X)$ is continuous. Next, let $u_i(\cdot) \in D(B)$ ($i = 1, 2$). Since $A(t) - \omega I$ $(0 \le t \le T)$ is a dissipative operator, for every $\lambda > 0$ we have

$$\|e^{(1+\omega)t}u_1(t) - e^{(1+\omega)t}u_2(t)\|$$
$$\le \|e^{(1+\omega)t}u_1(t) - e^{(1+\omega)t}u_2(t)$$
$$- \lambda[(A(t) - \omega I)(e^{(1+\omega)t}u_1(t)) - (A(t) - \omega I)(e^{(1+\omega)t}u_2(t))]\|,$$

that is,

$$\|u_1(t) - u_2(t)\| \le \|u_1(t) - u_2(t) - \lambda[((B - \omega I)u_1)(t) - ((B - \omega I)u_2)(t)]\|$$
$$(0 \le t \le T).$$

Hence for every $\lambda > 0$

$$\|u_1(\cdot) - u_2(\cdot)\| \le \|u_1(\cdot) - u_2(\cdot) - \lambda[(B - \omega I)u_1(\cdot) - (B - \omega I)u_2(\cdot)]\|$$

holds, and $B - \omega I$ is a dissipative operator.

(ii) Let $\lambda > 0$ be such that $\lambda\omega < 1$. Choose $u(\cdot) \in D(B)$ arbitrarily and set $v(t) = e^{-(1+\omega)t}J_\lambda(t)(e^{(1+\omega)t}u(t))$ $(0 \le t \le T)$.

Since $J_\lambda(t)(e^{(1+\omega)t}u(t)): [0, T] \to D$ is a continuous function by Lemma 6.21, we have that $v(\cdot) \in D(B)$ and

$$((I - \lambda B)v)(t) = v(t) - \lambda e^{-(1+\omega)t}A(t)(e^{(1+\omega)t}v(t))$$
$$= e^{-(1+\omega)t}[J_\lambda(t)(e^{(1+\omega)t}u(t)) - \lambda A(t)(J_\lambda(t)(e^{(1+\omega)t}u(t)))]$$
$$= e^{-(1+\omega)t}[(I - \lambda A(t))J_\lambda(t)(e^{(1+\omega)t}u(t))] = u(t)$$
$$(0 \le t \le T).$$

Hence we obtain $(I - \lambda B)v(\cdot) = u(\cdot)$, and thus $R(I - \lambda B) \supset D(B)$.

(II) Let $x \in D$ and define an operator $\tilde{A}$ as follows:

$$D(\tilde{A}) = \{u(\cdot) \in C^1([0, T]; X) : u(0) = x\},$$
$$(\tilde{A}u)(t) = -(d/dt)u(t) \qquad (u(\cdot) \in D(\tilde{A}), \ 0 \le t \le T).$$

$\tilde{A}: D(\tilde{A}) \to C([0, T]; X)$ is an m-dissipative operator. In fact, for every $f(\cdot) \in C([0, T]; X)$ and $\lambda > 0$, we set

$$(J_\lambda f)(t) = e^{-t/\lambda}x + \lambda^{-1}\int_0^t e^{(s-t)/\lambda}f(s)\, ds \qquad (0 \le t \le T);$$

then we see that $v(\cdot) = J_\lambda f(\cdot)$ is a unique solution of $(I - \lambda \widetilde{A})v(\cdot) = f(\cdot)$. Hence $R(I - \lambda \widetilde{A}) = C([0, T]; X)$. Moreover, for every $f(\cdot)$, $g(\cdot) \in C([0, T]; X)$ and $\lambda > 0$

$$\|(I - \lambda \widetilde{A})^{-1}f(\cdot) - (I - \lambda \widetilde{A})^{-1}g(\cdot)\| = \|J_\lambda f(\cdot) - J_\lambda g(\cdot)\|$$
$$\leq \|f(\cdot) - g(\cdot)\|.$$

Therefore $\widetilde{A}$ is an m-dissipative operator.

Now, we define an operator A as follows:

$$D(A) = D(B) \cap D(\widetilde{A}) \quad \text{and} \quad A = (\widetilde{A} - (1 + \omega)I)|_{D(A)}.$$

Then, (iii) and (iv) hold:

(iii) A is a dissipative operator of type $-(1 + \omega)$ and a closed operator.

(iv) For all $\lambda > 0$, $R(I - \lambda A) \supset D(B)$.

PROOF. Since $A + (1 + \omega)I \subset \widetilde{A}$ and $\widetilde{A}$ is a dissipative operator, $A + (1 + \omega)I$ is also a dissipative operator. Because every m-dissipative operator is a closed operator, $\widetilde{A}$ is a closed operator. Hence $\widetilde{A} - (1 + \omega)I$ is also a closed operator. From this and the fact that $D(B)$ is a closed set, we see that A is a closed operator.

Next, we show (iv). Let $\lambda > 0$. We take $u(\cdot) \in D(B)$ arbitrarily and set $v(t) = e^{(1+\omega)t}u(t)$. From the definition of $D(B)$, $v(t) \in D$ $(0 \leq t \leq T)$. Then D is a closed convex set and

$$(J_\lambda v)(t) = e^{-t/\lambda}x + \lambda^{-1}\int_0^t e^{(s-t)/\lambda}v(s)\,ds$$
$$= \left(1 - \lambda^{-1}\int_0^t e^{(s-t)/\lambda}\,ds\right)x$$
$$+ \lambda^{-1}\int_0^t e^{(s-t)/\lambda}\,ds \cdot \frac{\lambda^{-1}\int_0^t e^{(s-t)/\lambda}v(s)\,ds}{\lambda^{-1}\int_0^t e^{(s-t)/\lambda}\,ds} \in D \quad (0 \leq t \leq T).$$

Hence $J_\lambda v(\cdot) \in D(\widetilde{A}) \cap C([0, T]; D)$. Now we set

$$w_\lambda(t) = e^{-(1+\omega)t}(J_\lambda v)(t) \qquad (0 \leq t \leq T).$$

Then $w_\lambda(\cdot) \in D(B) \cap D(\widetilde{A}) = D(A)$. Next, from $(d/dt)(J_\lambda v)(t) = -\lambda^{-1}(J_\lambda v)(t) + \lambda^{-1}v(t)$

$$-(\widetilde{A}w_\lambda)(t) = (d/dt)w_\lambda(t) = -(1 + \omega)w_\lambda(t) + e^{-(1+\omega)t}(d/dt)(J_\lambda v)(t)$$
$$= -(1 + \omega)w_\lambda(t) - \lambda^{-1}w_\lambda(t) + \lambda^{-1}u(t) \quad (0 \leq t \leq T).$$

Hence $w_\lambda(\cdot) - \lambda(\widetilde{A} - (1 + \omega)I)w_\lambda(\cdot) = u(\cdot)$. From this and $w_\lambda(\cdot) \in D(A)$, we obtain $w_\lambda(\cdot) - \lambda A w_\lambda(\cdot) = u(\cdot)$ and thus $R(I - \lambda A) \supset D(B)$.

(III) We consider $C([0, T]; X)$ and $D(B)$ as X and X_0 in Corollary 6.18′, respectively. Since A is a dissipative operator of type $-(1 + \omega)$ with $\omega \geq 0$, A is also a dissipative operator. Then from those we examined in (I) and (II), the operators A and B satisfy the assumptions of Corollary $6.18'$. Moreover, noting that A is a closed operator, from this corollary we obtain

$$R(I - \lambda(A + B)) \supset D(B) \quad \text{for all } \lambda > 0 \text{ such that } \lambda\omega < 1. \tag{6.49}$$

Now, from (i) and (ii), B is the infinitesimal generator of a semigroup of type ω on $D(B)$ (see Theorem 5.15). Hence $B - \omega I$ is a strictly dissipative operator. Also, as we have shown in (iii), $A + (1 + \omega)I$ is a dissipative operator. Therefore, the sum of these, $A + B + I$, is a dissipative operator. In other words, $A + B$ is a dissipative operator of type -1. Hence

$$\|(I - \lambda(A + B))^{-1}u(\cdot) - (I - \lambda(A + B))^{-1}v(\cdot)\| \tag{6.50}$$
$$\leq (1 + \lambda)^{-1}\|u(\cdot) - v(\cdot)\| \quad (u(\cdot), v(\cdot) \in R(I - \lambda(A + B)), \ \lambda > 0).$$

Now we choose $\lambda_0 > 0$ such that $\lambda_0\omega < 1$. Then from (6.49) and (6.50) we obtain

$$\|(I - \lambda_0(A + B))^{-1}u(\cdot) - (I - \lambda_0(A + B))^{-1}v(\cdot)\|$$
$$\leq (1 + \lambda_0)^{-1}\|u(\cdot) - v(\cdot)\| \quad (u(\cdot), v(\cdot) \in D(B)).$$

Hence $(I - \lambda_0(A + B))^{-1}: D(B) \to D(A) \ (\subset D(B))$ is a strict contraction operator and has a unique fixed point $v(\cdot)$; that is, $(I - \lambda_0(A + B))^{-1}v(\cdot) = v(\cdot)$. Therefore we obtain $(A + B)v(\cdot) = 0$; that is, $-\widetilde{A}v(\cdot) + (1 + \omega)v(\cdot) = Bv(\cdot)$.

From the definitions of $\widetilde{A}$ and B

$$(d/dt)v(t) + (1 + \omega)v(t) = e^{-(1+\omega)t}A(t)(e^{(1+\omega)t}v(t)) \quad (0 \leq t \leq T),$$

and $v(\cdot) \in D(B) \cap C^1([0, T]; X)$ and $v(0) = x$. Now if we put $u(t) = e^{(1+\omega)t}v(t) \ (0 \leq t \leq T)$, then $u(\cdot) \in C^1([0, T]; D)$ and $u(\cdot)$ satisfies

$$\begin{cases} (d/dt)u(t) = A(t)u(t) & (0 \leq t \leq T), \\ u(0) = x. \end{cases}$$

Since $x \in D$ is arbitrary, we have shown that for every $x \in D$ the Cauchy problem (6.46) has a solution $u(\cdot; x) \in C^1([0, T]; D)$.

Finally, let $x, y \in D$ and let $u(\cdot)$ and $v(\cdot) \in C^1([0, T]; D)$ be the solutions of (6.46) with its initial values x and y, respectively. Then from Lemma 2.8

$$(d/dt)\|u(t) - v(t)\|^2 = 2\langle A(t)u(t) - A(t)v(t), u(t) - v(t)\rangle_i$$
$$\leq 2\omega(t)\|u(t) - v(t)\|^2 \quad (\text{a.e. } t \in [0, T]).$$

Hence

$$(d/dt)\left[\|u(t) - v(t)\|^2 \exp\left(-2\int_0^t \omega(s)\,ds\right)\right] \leq 0 \quad (\text{a.e. } t \in [0, T]).$$

If we integrate this over $[0, t]$ we have

$$\|u(t) - v(t)\| \leq \|x - y\| \exp\left(\int_0^t \omega(s)\,ds\right) \quad (0 \leq t \leq T).$$

From this, we obtain the uniqueness of the solution of the Cauchy problem (6.46) and have shown (6.47) at the same time. □

REMARK. Under the assumptions (a_1) and (a_2), (a_3) is equivalent to (a'_3) or (a''_3) (see Theorem 5.15):

(a'_3) $\liminf_{\lambda \to 0+} \lambda^{-1} d(D, x + \lambda A(t)x) = 0$ $(x \in D, \; 0 \le t \le T)$.

(a''_3) $\liminf_{\lambda \to 0+} \lambda^{-1} d(R(I - \lambda A(t)), x) = 0$ $(x \in D, \; 0 \le t \le T)$.

It is known that the above theorem holds even when D is a closed set in (a_1), together with the assumptions (a_2) and (a'_3) (then (a'_3) is equivalent to (a''_3)) (see [58] and also [82] in the bibliography).

COROLLARY 6.22. *Let $A(t)$ $(0 \le t \le T)$ be an operator satisfying the following (b_1) and (b_2):*

(b_1) *For all $t \in [0, T]$, $A(t)$ is a dissipative operator defined on the whole space X.*

(b_2) *$A(t)x$ is continuous on $[0, T] \times X$ as a mapping of two variables t and x.*

Then, for every $x \in X$, the Cauchy problem (6.46) has a unique solution $u(\cdot \, ; x) \in C^1([0, T]; X)$.

PROOF. Since a continuous dissipative operator defined on all of X is an m-dissipative operator, all of the $A(t)$ are m-dissipative operators. Hence $R(I - \lambda A(t)) = X$ $(\lambda > 0, \; 0 \le t \le T)$. Then if we let $\omega(t) = 0$ $(0 \le t \le T)$ and let $D = X$, then (a_1) through (a_3) of Theorem 6.20 are satisfied and hence we obtain the required conclusion. □

Next, we examine the following:

$$\begin{cases} (d/dt)u(t) = A(t)u(t) & (0 \le t \le T), \\ u(0) = u(T). \end{cases} \tag{6.51}$$

THEOREM 6.23. *Let $A(t)$ $(0 \le t \le T)$ be an operator satisfying (c_1) through (c_3):*

(c_1) *Let $\omega_0 < 0$. For every $t \in [0, T]$, $A(t)$ is a dissipative operator of type ω_0 and $D(A(t))$ is a closed convex set that does not depend on t. In the following, we set $D(A(t)) = D$.*

(c_2) *$A(t)x$ is continuous on $[0, T] \times D$ as a mapping of two variables t and x.*

(c_3) *For all $t \in [0, T]$ and $\lambda > 0$, $R(I - \lambda A(t)) \supset D$.*

Then (6.51) has a unique solution $u(\cdot) \in C^1([0, T]; D)$. In particular, (beside (c_1) through (c_3)) if we assume

(c_4) *$A(0)x = A(T)x$ $(x \in D)$,*

then the solution of (6.51), $u(\cdot)$ $(\in C^1([0, T]; D))$, satisfies $D^+ u(0) = D^- u(T)$. From this it follows that if we define $\widetilde{A}(t)$ $(-\infty < t < \infty)$ as $\widetilde{A}(t) = A(t)$ $(0 \le t \le T)$ with a period T (i.e., $\widetilde{A}(t + T) = \widetilde{A}(t)$), then the differential equation

$$(d/dt)u(t) = \widetilde{A}(t)u(t) \qquad (-\infty < t < \infty) \tag{6.52}$$

has a unique periodic solution $u(\cdot) \in C^1((-\infty, \infty); D)$ *with period* T.

PROOF. Let $\omega(t) = \omega_0$ $(0 \le t \le T)$ and apply Theorem 6.20. Then for $x \in D$, the Cauchy problem (6.46) has a unique solution $u(\cdot; x) \in C^1([0, T]; D)$ and

$$\|u(t; x) - u(t; y)\| \le e^{\omega_0 t}\|x - y\| \qquad (x, y \in D, \ t \in [0, T]).$$

Now, we define an operator $U: D \to D$ by

$$Ux = u(T; x) \qquad (x \in D).$$

Since $\|Ux - Uy\| \le e^{\omega_0 T}\|x - y\|$ $(x, y \in D)$ and $\omega_0 < 0$, $U: D \to D$ is a strict contraction operator. Hence U has a unique fixed point $x_0 \in D$, that is, $Ux_0 = x_0$. Then $u(t) = u(t; x_0)$ $(0 \le t \le T)$ satisfies $(d/dt)u(t) = A(t)u(t)$ $(0 \le t \le T)$ and $u(0) = x_0$. Also, from $Ux_0 = x_0$ we obtain $u(T) = u(T; x_0) = x_0 = u(0)$. Hence $u(\cdot) \in C^1([0, T]; D)$ is a solution of (6.51). Next, in order to show the uniqueness of the solution, we let $v(\cdot) \in C^1([0, T]; D)$ be a solution of (6.51). Then

$$(d/dt)\|u(t) - v(t)\|^2 = 2\langle A(t)u(t) - A(t)v(t), u(t) - v(t)\rangle_i$$
$$\le 2\omega_0\|u(t) - v(t)\|^2 \quad (\text{a.e. } t \in [0, T]);$$

that is,

$$(d/dt)[\|u(t) - v(t)\|^2 e^{-2\omega_0 t}] \le 0 \quad (\text{a.e. } t \in [0, T]).$$

If we integrate this on $[0, t]$, we have

$$\|u(t) - v(t)\| \le e^{\omega_0 t}\|u(0) - v(0)\| \qquad (0 \le t \le T). \qquad (6.53)$$

In particular, $\|u(T) - v(T)\| \le e^{\omega_0 T}\|u(0) - v(0)\|$. From this and $\|u(0) - v(0)\| = \|u(T) - v(T)\|$, we have $\|u(0) - v(0)\| = 0$. Therefore, from (6.53) we obtain $u(t) = v(t)$ $(0 \le t \le T)$, and the uniqueness of the solution is shown.

Next, assume (c_4) and let $u(\cdot) \in C^1([0, T]; D)$ be the solution of (6.51). Then

$$D^+ u(0) = A(0)u(0) = A(T)u(T) = D^- u(T).$$

Now, if we define $\tilde{u}(t) = u(t - nT)$ $(nT \le t < (n+1)T; n = 0, \pm 1, \pm 2, \dots)$, then $\tilde{u}(t)$ is the required solution of (6.52). (We use $D^+ u(0) = D^- u(T)$ to show that $u(t)$ is strongly differentiable at $t = nT$ $(n = 0, \pm 1, \pm 2, \dots)$.) □

From this we obtain the following corollary:

COROLLARY 6.24. *Let* $A(t)$ $(0 \le t \le T)$ *be an operator satisfying* (b_1) *and* (b_2) *of Corollary 6.22. Then for every* $\lambda > 0$,

$$\begin{cases} (d/dt)u(t) = A(t)u(t) - \lambda u(t) & (0 \le t \le T), \\ u(0) = u(T) \end{cases} \qquad (6.54)$$

has a unique solution $u(\cdot) \in C^1([0, T]; X)$.

In particular, (besides (b_1) *and* (b_2)) *if we assume*
(b_3) $A(0)x = A(T)x$ $(x \in X)$, *then for every* $\lambda > 0$,

$$(d/dt)u(t) = \tilde{A}(t)u(t) - \lambda u(t) \qquad (-\infty < t < \infty) \qquad (6.55)$$

has a unique periodic solution $u(\cdot) \in C^1((-\infty, \infty); X)$ *with period* T. *Here* $\tilde{A}(t)$ $(-\infty < t < \infty)$ *is the operator defined in the same way as in Theorem* 6.23.

PROOF. Let $\lambda > 0$ and set $B(t) = A(t) - \lambda I$ $(-\infty < t < \infty)$. Then $B(t)$ is a dissipative operator of type $-\lambda$ such that $D(B(t)) = X$, and satisfies $R(I - \mu B(t)) = X$ $(\mu > 0)$. (Since $A(t)$ is a continuous dissipative operator defined on all of X, it is an m-dissipative operator and hence $R(I - \mu A(t)) = X$ $(\mu > 0)$. From this we obtain $R(I - \mu B(t)) = X$ $(\mu > 0)$.) Moreover $B(t)x: [0, T] \times X \to X$ is continuous. From this, for $B(t)$ $(0 \le t \le T)$ we can apply Theorem 6.23 to complete the proof. $\square$

CHAPTER 7

Quasilinear Partial Differential Equations of First Order

Applying the theory of nonlinear semigroups and, in particular, the theorems on the convergence and approximation of semigroups given in the previous chapters, we consider Cauchy's problem for quasilinear partial differential equations of first order:

$$u_t + \sum_{i=1}^{d} (\phi_i(u))_{x_i} = 0 \qquad (x = (x_1, x_2, \ldots, x_d) \in R^d, \ t > 0).$$

Here, ϕ_i represent continuous, once differentiable functions defined on R^1 satisfying $\phi_i(0) = 0$, and R^d denotes the d-dimensional Euclidean space.

§1. Preparation

In this section, let X be a Banach space. We describe the theorems on the convergence and approximation of semigroups in the form that we need later.

DEFINITION 7.1. A family $\{J_\lambda : \lambda > 0\}$ of contraction operators J_λ $(\lambda > 0)$ with $D(J_\lambda) \subset X$ and $R(J_\lambda) \subset x$ is called a pseudoresolvent if (a_1) and (a_2) hold:

(a_1) $\quad R\left(\frac{\mu}{\lambda}I + \left(1 - \frac{\mu}{\lambda}\right)J_\lambda\right) \subset D(J_\mu) \qquad (\lambda, \mu > 0),$

(a_2) $\quad J_\lambda u = J_\mu\left(\frac{\mu}{\lambda}u + \left(1 - \frac{\mu}{\lambda}\right)J_\lambda u\right) \qquad (\lambda, \mu > 0, \ u \in D(J_\lambda)).$

LEMMA 7.1. (i) *Let A be a dissipative operator and put $J_\lambda = (I - \lambda A)^{-1}$ $(\lambda > 0)$. Then J_λ $(\lambda > 0)$ is a contraction operator and $\{J_\lambda : \lambda > 0\}$ is a pseudoresolvent.*

(ii) *Let a family $\{J_\lambda : \lambda > 0\}$ of contraction operators be a pseudoresolvent. Then there is a dissipative operator A such that $R(J_\lambda) = R(J_\mu)$ $(\lambda, \mu > 0)$, and $D(A) = R(J_\lambda)$ and $J_\lambda = (I - \lambda A)^{-1}$ $(\lambda > 0)$. In particular, if for some $\lambda_0 > 0$, J_{λ_0} is a one-to-one operator, then A is a single-valued operator.*

PROOF. (i) This follows from Corollary 2.10 and part (iii) of Lemma 2.11.

(ii) By (a_2), for every $\lambda, \mu > 0$ we have $R(J_\lambda) \subset R(J_\mu)$. Hence $R(J_\lambda) = R(J_\mu)$ $(\lambda, \mu > 0)$. Next, let $\lambda, \mu > 0$, and choose u and v such that $u \in R(J_\lambda)$ and $v \in \lambda^{-1}(I - J_\lambda^{-1})u$. Then from (a_2)

$$u = J_\lambda(u - \lambda v) = J_\mu\left(\frac{\mu}{\lambda}(u - \lambda v) + \frac{\lambda - \mu}{\lambda} J_\lambda(u - \lambda v)\right)$$

$$= J_\mu\left(\frac{\mu}{\lambda}(u - \lambda v) + \frac{\lambda - \mu}{\lambda} u\right) = J_\mu(u - \mu v).$$

Hence $u - \mu v \in J_\mu^{-1} u$, that is, we obtain $v \in \mu^{-1}(I - J_\mu^{-1})u$, and thus $\lambda^{-1}(I - J_\lambda^{-1})u \subset \mu^{-1}(I - J_\mu^{-1})u$. If we repeat this with λ and μ interchanged, then we obtain the inclusion relations reversed and thus have $\lambda^{-1}(I - J_\lambda^{-1})u = \mu^{-1}(I - J_\mu^{-1})u$ $(u \in R(J_\lambda) = R(J_\mu))$. That is, we obtain

$$\lambda^{-1}(I - J_\lambda^{-1}) = \mu^{-1}(I - J_\mu^{-1}) \qquad (\lambda, \mu > 0).$$

Now we define $A = \lambda^{-1}(I - J_\lambda^{-1})$ $(D(A) = R(J_\lambda))$. If we choose $u_i \in D(A)$ and $v_i \in Au_i$ $(i = 1, 2)$ arbitrarily, then $u_i = J_\lambda(u_i - \lambda v_i)$. Hence, from the contractiveness of J_λ we obtain

$$\|u_1 - u_2\| \le \|u_1 - \lambda v_1 - (u_2 - \lambda v_2)\| \qquad (\lambda > 0),$$

and A is a dissipative operator. Obviously $J_\lambda = (I - \lambda A)^{-1}$ $(\lambda > 0)$. Also, if J_{λ_0} is one-to-one, then $A = \lambda_0^{-1}(I - J_{\lambda_0}^{-1})$ is a single-valued operator. $\square$

Let X_m be a closed convex subset of X satisfying $X_m \subset X_{m+1}$ $(m = 1, 2, \dots)$ and set $X_0 = \bigcup_{m \ge 1} X_m$. Next, let $C_{m,n}$ $(m, n = 1, 2, \dots)$ be operators satisfying

$$C_{m,n} \in \text{Cont}(X_m) \qquad (m, n \ge 1); \tag{7.1}$$

also, let $h_{m,n}$ $(m, n = 1, 2, \dots)$ be positive numbers satisfying

$$\lim_{n \to \infty} h_{m,n} = 0 \qquad (m = 1, 2, \dots). \tag{7.2}$$

We define operators $A_{m,n}$ by the following:

$$A_{m,n} = h_{m,n}^{-1}(C_{m,n} - I) \qquad (m, n = 1, 2, \dots). \tag{7.3}$$

Then, each $A_{m,n}$ is a dissipative operator and satisfies $R(I - \lambda A_{m,n}) \supset X_m = D(A_{m,n})$ $(\lambda > 0)$ (see Example 2.4). Hence, by the generation theorem, each $A_{m,n}$ generates a semigroup of contractions on X_m. We denote this semigroup of contractions by $\{T_{m,n}(t) : t \ge 0\}$. Then from part (ii) of Corollary 6.9, part (ii) of Corollary 6.13, and Lemma 7.1, we obtain the following theorem:

THEOREM 7.2. *Let* X_m, X_0, $C_{m,n}$, $h_{m,n}$, $A_{m,n}$, *and* $\{T_{m,n}(t): t \geq 0\}$ $(m, n = 1, 2, \ldots)$ *be those defined above. We assume the following condition* (C):

$$
\begin{cases}
\text{There exists a pseudoresolvent } \{J_\lambda : \lambda > 0\} \text{ such that} \\
J_\lambda \in \mathrm{Cont}(X_0) \ (\lambda > 0), \text{ and} \\
J_\lambda u = \lim_{n \to \infty} (I - \lambda A_{m,n})^{-1} u \ (\lambda > 0, \ u \in X_m, \ m \geq 1).
\end{cases}
\tag{C}
$$

Then (i) *through* (iv) *hold*:

(i) *There exists a dissipative operator A that satisfies*

$$
R(I - \lambda A) = X_0 \supset D(A) \quad \text{and} \quad J_\lambda = (I - \lambda A)^{-1} \qquad (\lambda > 0).
$$

(ii) $\overline{A}$ *generates a semigroup of contractions on* $\overline{D(A)}$. *Let this semigroup be* $\{T(t): t \geq 0\}$. *Then*

$$
T(t)(X_m \cap \overline{D(A)}) \subset X_m \cap \overline{D(A)} \qquad (t \geq 0, \ m \geq 1).
$$

Hence $T(t)(X_0 \cap \overline{D(A)}) \subset X_0 \cap \overline{D(A)}$ $(t \geq 0)$ *also holds.*

(iii) $\lim_{n \to \infty} T_{m,n}(t)u = T(t)u$ $(t \geq 0, \ u \in X_m \cap \overline{d(A)}, \ m \geq 1)$, *and the convergence is uniform on bounded subintervals of* $[0, \infty)$.

(iv) $\lim_{n \to \infty} C_{m,n}^{[t/h_{m,n}]} u = T(t)u$ $(t \geq 0, \ u \in X_m \cap \overline{D(a)}, \ m \geq 1)$, *and the convergence is uniform on bounded subintervals of* $[0, \infty)$.

PROOF. Since $J_\lambda \in \mathrm{Cont}(X_0)$ $(\lambda > 0)$ and $\{J_\lambda : \lambda > 0\}$ is a pseudoresolvent, by Lemma 7.1 we obtain (i). From (i), $\overline{A}$ is a dissipative operator with $R(I - \lambda \overline{A}) \supset \overline{D(A)} = D(\overline{A})$ $(\lambda > 0)$. Hence $\overline{A}$ generates a semigroup of contractions $\{T(t): t \geq 0\}$ on $\overline{D(A)}$, and if we set $\tilde{J}_\lambda = (I - \lambda \overline{A})^{-1}$ $(\lambda > 0)$, then

$$
T(t)u = \lim_{\lambda \to 0+} \tilde{J}_\lambda^{[t/\lambda]} u \qquad (u \in \overline{D(A)}, \ t \geq 0)
\tag{7.4}
$$

(see Theorem 4.2 and Corollary 5.13). Let $u \in X_m \cap \overline{D(A)}$. Since $(I - \lambda A_{m,n})^{-1} u \in X_m$ $(\lambda > 0, \ m, n \geq 1)$ and X_m are closed sets,

$$
J_\lambda u = \lim_{n \to \infty} (I - \lambda A_{m,n})^{-1} u \in X_m \cap D(A) \subset X_m \cap \overline{D(A)} \qquad (\lambda > 0).
$$

Also $\tilde{J}_\lambda u = J_\lambda u$ $(\lambda > 0)$. Hence, from (7.4) we obtain $T(t)u \in X_m \cap \overline{D(A)}$ $(t \geq 0)$, and thus (ii) is proved.

Next, in order to show (iii) and (iv), we define $A_m = A|_{X_m \cap D(A)}$ $(m > 1)$. Then each A_m is a dissipative operator and satisfies

$$
R(I - \lambda A_m) \supset X_m \cup \overline{D(A)} = \overline{D(A_m)} \qquad (\lambda > 0),
\tag{7.5}
$$

$$
(I - \lambda A_m)^{-1} u = J_\lambda u \qquad (u \in \overline{D(A_m)}, \ \lambda > 0).
\tag{7.6}
$$

In fact, we let $u \in X_m \cap \overline{D(A)}$. We note that $\tilde{J}_\lambda v = J_\lambda v$ $(v \in R(I - \lambda A))$, $\lim_{\lambda \to 0+} \tilde{J}_\lambda v = v$ $(v \in \overline{D(A)})$ and $R(I - \lambda A) \supset X_m$, and thus we have $J_\lambda u = \tilde{J}_\lambda u \to u$ as $\lambda \to 0+$. Since $J_\lambda u \in X_m \cap D(A)$, we obtain $u \in \overline{X_m \cap D(A)}$.

From this, we have shown $X_m \cap \overline{D(A)} \subset \overline{X_m \cap D(A)}$. Since the relations with the inclusions reversed are self-evident, $X_m \cap \overline{D(A)} = \overline{X_m \cap D(A)} = \overline{D(A_m)}$ holds. Also from the definition of A_m and $J_\lambda u \in X_m \cap D(A) = D(A_m)$, we have $(I - \lambda A_m)J_\lambda u = (I - \lambda A)J_\lambda u \ni u$. From this we obtain $J_\lambda u = (I - \lambda A_m)^{-1}u$ and $R(I - \lambda A_m) \ni u$ $(\lambda > 0)$, and thus (7.5) and (7.6) hold. From (7.5), A_m generates a semigroup of contractions on $\overline{D(A_m)}$. Now, if we denote this semigroup of contractions by $\{T_m(t) : t \geq 0\}$, then we have

$$T_m(t)u = \lim_{\lambda \to 0+} (I - \lambda A_m)^{-[t/\lambda]}u \qquad (u \in \overline{D(A_m)}, \ t \geq 0).$$

Then from (7.6), (7.4), and $\tilde{J}_\lambda u = J_\lambda u$ $(u \in \overline{D(A_m)}, \ \lambda > 0)$ we obtain

$$T(t)u = T_m(t)u \qquad (u \in \overline{D(A_m)}, \ t \geq 0). \tag{7.7}$$

Now from the assumption (C) and from (7.6), we have

$$(I - \lambda A_m)^{-1}u = \lim_{n \to \infty} (I - \lambda A_{m,n})^{-1}u \qquad (u \in \overline{D(A_m)}, \ \lambda > 0).$$

Hence, we can apply part (ii) of Corollary 6.9 and part (ii) of Corollary 6.13 (we consider $A_{m,n}$, A_m, and $C_{m,n}$, as A_n, A, and T_n in these corollaries, respectively) and obtain

$$\lim_{n \to \infty} T_{m,n}(t)u = T_m(t)u, \qquad \lim_{n \to \infty} C_{m,n}^{[t/h_{m,n}]}u = T_m(t)u \qquad (u \in \overline{D(A_m)}, \ t \geq 0).$$

Here, these converge uniformly on every bounded interval of $[0, \infty)$. From this and (7.7), (iii) and (iv) are proved. $\square$

COROLLARY 7.3. *Let* X_m, X_0, $C_{m,n}$, $h_{m,n}$, $A_{m,n}$, *and* $\{T_{m,n}(t) : t \geq 0\}$ $(m, n = 1, 2, \dots)$ *be the same as those in Theorem 7.2. Let* X_0 *be a linear set. We assume condition* (C) *of Theorem 7.2 and the following condition* (C_1):

There exist a set D with $D \subset X_0$ and a single-valued
operator A_1, such that $\qquad\qquad\qquad\qquad\qquad (C_1)$
$$\lim_{n \to \infty} A_{m,n}u = A_1 u \in X_0 \ (u \in X_m \cap D, \ m \geq 1).$$

Then, in addition to (i) *through* (iv) *of Theorem 7.2, we have that* $D \subset D(A)$ *and* $A_1 u \in Au$ $(u \in D)$. *Furthermore, if* D *is dense in* X_0, *then the convergence of* (iii) *and* (iv) *of Theorem 7.2 holds for arbitrary* $u \in X_m$, *and* $T(t)u = \lim_{\lambda \to 0+}(I - \lambda A)^{-[t/\lambda]}u$ $(u \in X_0, \ t \geq 0)$, *and* $\{T(t)|_{X_0} : t \geq 0\}$ *is a semigroup of contractions on* X_0.

PROOF. Let $u \in D$ and $\lambda > 0$. (Note that $D \subset D(A_1)$ and $A_1 u \in X_0$ $(u \in D)$ are included in the above assumption (C_1).) From the assumption that X_0 is a linear set, we have $u - \lambda A_1 u \in X_0 = \bigcup_{m \geq 1} X_m$. Also, $u \in D \subset X_0$. Then from $X_m \subset X_{m+1}$, we can select an m (≥ 1) such that $u \in X_m$ and $u - \lambda A_1 u \in X_m$. If we set $v = u - \lambda A_1 u$ and $v_n = (I - \lambda A_{m,n})u$

$(n \geq 1)$, then from (C_1) $v_n \to v$ as $n \to \infty$. Since $u = (I - \lambda A_{m,n})^{-1} v_n$ and $(I - \lambda A_{m,n})^{-1}$ is a contraction operator,

$$\|u - J_\lambda v\| \leq \|v_n - v\| + \|(I - \lambda A_{m,n})^{-1} v - J_\lambda v\| \qquad (n \geq 1).$$

Here, if we let $n \to \infty$, then by the assumption (C), the right-hand side of the above inequality converges to 0. Hence $u = J_\lambda v \in D(A)$. Also, from this we obtain $u - \lambda A_1 u = v \in u - \lambda Au$, and thus $A_1 u \in Au$. Therefore, $D \subset D(A)$ and $A_1 u \in Au$ $(u \in D)$. Finally, if we suppose D is dense in X_0, then $X_0 \subset \overline{D} \subset \overline{D(A)}$. Therefore, $X_m \cap \overline{D(A)} = X_m$ $(m \geq 1)$ and the convergence of (iii) and (iv) of Theorem 7.2 holds for $u \in X_m$. Next, from (7.7), $T(t)u = T_m(t)u = \lim_{\lambda \to 0+} (I - \lambda A_m)^{-[t/\lambda]} u \in \overline{D(A_m)} = X_m$ $(u \in X_m,\ m \geq 1,\ t \geq 0)$. Hence $\{T(t)|_{X_0} : t \geq 0\}$ is a semigroup of contractions on X_0. Also, from the above equation and $(I - \lambda A_m)^{-1} u = J_\lambda u$ $(u \in X_m,\ m \geq 1,\ \lambda > 0)$ we obtain $T(t)u = \lim_{\lambda \to 0+} J_\lambda^{[t/\lambda]} u$ $(u \in X_m,\ m \geq 1,\ t \geq 0)$, and thus $T(t)u = \lim_{\lambda \to 0+} (I - \lambda A)^{-[t/\lambda]} u$ $(u \in X_0,\ t \geq 0)$. $\square$

§2. Cauchy's problem for quasilinear partial differential equations of first order and its difference approximation

We consider a Cauchy problem for the quasilinear partial differential equation of first order

$$u_t + \sum_{i=1}^{d} (\phi_i(u))_{x_i} = 0 \qquad (x = (x_1, x_2, \ldots, x_d) \in R^d,\ t > 0). \qquad \text{(DE)}$$

Here, ϕ_i are given functions that satisfy the following assumption:

ASSUMPTION. $\phi_i \in C^1(R^1)$ $(i = 1, 2, \ldots, d)$; that is, each ϕ_i is a continuous, once differentiable real-valued function defined on R^1 and $\phi_i(0) = 0$ for all i.

In this section and throughout latter sections, we suppose ϕ_i always satisfies the above assumption, and consider the Cauchy problem for the (DE) described above in the space $L^1(R^d)$ (of real-valued functions). First, we define the solution of the Cauchy problem in the weak sense as follows:

DEFINITION 7.2. Let $u_0 \in L^1(R^d) \cap L^\infty(R^d)$ be a given function. When $u(t, \cdot) : [0, \infty) \to L^1(R^d) \cap L^\infty(R^d)$ satisfies (a_1) through (a_3) below, we call $u(t, x)$ a weak solution of $(CP; u_0)(^1)$ (the Cauchy problem for (DE)):

(a_1) $u(t, \cdot)$ is continuous on $[0, \infty)$ with respect to the norm $\|\cdot\|_1$ (that is, $\lim_{s \to t} \|u(s, \cdot) - u(t, \cdot)\|_1 = 0$ $(t \geq 0)$), and $\|u(t, \cdot)\|_\infty$ is bounded on $[0, \infty)$.

$(^1)$ According to the notation used up to now, we should write $(CP; u_0)_\infty$, but since we are considering the Cauchy problem on $[0, \infty)$ only, we omit the ∞ for simplicity.

(a_2) For every $k \in R^1$ and every nonnegative function $f \in C_0^\infty((0, \infty) \times R^d)$

$$\int_0^\infty \int_{R^d} \left\{ \quad |u(t, x) - k| f_t(t, x) \right.$$

$$\left. + \operatorname{sign}(u(t, x) - k) \sum_{i=1}^d [\phi_i(u(t, x)) - \phi_i(k)] f_{x_i}(t, x) \right\} dx \, dt \geq 0.$$

(a_3) $u(0, \cdot) = u_0$.

Here, $C_0^\infty((0, \infty) \times R^d)$ is the set of all infinite differentiable (real-valued) functions defined on $(0, \infty) \times R^d$ which have compact support in $(0, \infty) \times R^d$. $\|g\|_1$ and $\|g\|_\infty$ represent the norms defined by $\|g\|_1 = \int_{R^d} |g(x)| \, dx$ ($g \in L^1(R^d)$) and $\|g\|_\infty = \operatorname{ess\,sup}_{x \in R^d} |g(x)|$ ($g \in L^\infty(R^d)$), respectively; also $\operatorname{sign}(r)$ is a function defined on R^1 by $\operatorname{sign}(r) = 1$ ($r > 0$), $\operatorname{sign}(r) = 0$ ($r = 0$), and $\operatorname{sign}(r) = -1$ ($r < 0$).

REMARK. The reason why we call $u(t, x)$ a weak solution of (CP; u_0) in the definition above is due to the following. If $u(t, x)$ is a smooth function, then we choose $k > \sup_{t \geq 0} \|u(t, \cdot)\|_\infty$ and apply the left-hand side of the inequality in (a_2) integration by parts. The result is

$$\int_{R^d} \left[\int_0^\infty (k - u(t, x)) f_t(t, x) \, dt \right] dx$$

$$- \sum_{i=1}^d \int_0^\infty \left\{ \int_{R^d} [\phi_i(u(t, x)) - \phi_i(k)] f_{x_i}(t, x) \, dx \right\} dt$$

$$= \int_0^\infty \int_{R^d} \left[u_t(t, x) + \sum_{i=1}^d (\phi_i(u(t, x)))_{x_i} \right] f(t, x) \, dx \, dt \geq 0.$$

Next, if we choose $k < -\sup_{t \geq 0} \|u(t, \cdot)\|_\infty$ and repeat the same as above, then we obtain

$$\int_0^\infty \int_{R^d} \left[u_t(t, x) + \sum_{i=1}^d (\phi_i(u(t, x)))_{x_i} \right] f(t, x) \, dx \, dt \leq 0.$$

Hence we have

$$\int_0^\infty \int_{R^d} \left[u_t(t, x) + \sum_{i=1}^d (\phi_i(u(t, x)))_{x_i} \right] f(t, x) \, dx \, dt = 0.$$

Here since $f(t, x)$ ($\in C_0^\infty((0, \infty) \times R^d)$) is arbitrary, $u(t, x)$ satisfies (DE).

The concept of weak solution introduced above is given by Kružkov, but Kružkov considers the space $L^\infty(R^d)$ instead of $L^1(R^d) \cap L^\infty(R^d)$. Since it is known that "the weak solution of (CP; u_0) in Kružkov's sense" is determined uniquely (see [75]), we can say the same thing for our case.

Now, our object is to construct the weak solution for (CP; u_0), utilizing the difference method, from the viewpoint of the theory of semigroups. We first introduce several necessary notations. Let

$$e_i = (0, \ldots, 0, \overset{i\text{th}}{1}, 0, \ldots, 0) \in R^d \qquad (i = 1, 2, \ldots, d),$$

and for every $l > 0$ and every (real-valued) function $u(x)$ defined on R^d, we define

$$[D_i^0 u](x) = (2l)^{-1}[u(x + le_i) - u(x - le_i)],$$
$$[D_i^+ u](x) = l^{-1}[u(x + le_i) - u(x)],$$
$$[D_i^- u](x) = l^{-1}[u(x) - u(x - le_i)] \qquad (x \in R^d, \ 1 \le i \le d).$$

$(D_i^0, D_i^+, \text{ and } D_i^- \text{ also depend on } l$, but since the subscript does not matter for latter discussions, we omit l.) Hence we can write $D_i^- D_i^+ u = D_i^-(D_i^+ u)$ as follows:

$$[D_i^- D_i^+ u](x) = l^{-2}[u(x + le_i) - 2u(x) + u(x - le_i)] \qquad (x \in R^d, \ 1 \le i \le d).$$

As we easily can see, the difference operators D_i^0, D_i^+, and D_i^- are bounded linear operators with $L^1(R^d) \to L^1(R^d)$, and for every $u \in L^1(R^d)$ and $f \in C_0^\infty(R^d)$

$$(D_i^0 u, f) = -(u, D_i^0 f), \qquad (D_i^- D_i^+ u, f) = (u, D_i^- D_i^+ f) \qquad (1 \le i \le d).$$

Here the notation $(\cdot, \cdot)$ is defined by the following:

$$(f, g) = \int_{R^d} f(x)g(x)\,dx \qquad (f \in L^1(R^d), \ g \in L^\infty(R^d)).$$

We now consider the following Lax-Friedrichs scheme of the difference method for $(CP; u_0)$:

$$\begin{cases} h^{-1}[u^{\nu+1}(x) - (2d)^{-1}\sum_{i=1}^d (u^\nu(x + le_i) + u^\nu(x - le_i))] \\ \quad + (2l)^{-1}\sum_{i=1}^d [\phi_i(u^\nu(x + le_i)) - \phi_i(u^\nu(x - le_i))] = 0, \qquad \text{(DS)} \\ u^0(x) = u_0(x) \qquad (h, l > 0, \ x \in R^d, \ \nu = 0, 1, 2, \ldots). \end{cases}$$

Then, (DS) can be expressed in terms of D_i^0, D_i^+, and D_i^- as follows:

$$h^{-1}[u^{\nu+1}(x) - u^\nu(x)] - \frac{l^2}{2dh}\sum_{i=1}^d [D_i^- D_i^+ u^\nu](x) + \sum_{i=1}^d [D_i^0 \phi_i(u^\nu)](x) = 0.$$

Now for every $h, l > 0$, we define an operator $\widetilde{C}_{h,l} : L^1(R^d) \to L^1(R^d)$ by

$$[\widetilde{C}_{h,l} u](x) = (2d)^{-1}\sum_{i=1}^d (u(x + le_i) + u(x - le_i)) - h\sum_{i=1}^d [D_i^0 \phi_i(u)](x) \quad (7.8)$$

$(u \in L^1(R^d))$. Then (DS) can be written as

$$u^\nu = \widetilde{C}_{h,l} u^{\nu-1} = (\widetilde{C}_{h,l})^\nu u_0 \qquad (\nu = 1, 2, \ldots). \tag{7.9}$$

Finally, in order to apply Theorem 7.2, we define X_m, $C_{m,n}$, and $h_{m,n}$, which we considered in the previous section, as follows:

$$X_m = \{u \in L^1(R^d) \cap L^\infty(R^d): \|u\|_\infty \leq m\} \quad (m = 1, 2, \ldots), \quad X_0 = \bigcup_{m \geq 1} X_m.$$

Then, obviously we have $X_0 = L^1(R^d) \cap L^\infty(R^d)$. Next, we set

$$M_m = \max_{1 \leq i \leq d} \sup_{|s| \leq m} |\phi_i'(s)| \quad (m \geq 1),$$

and select a sequence $\{\delta_m\}$ of positive numbers that satisfies $\delta_m \leq (dM_m)^{-1}$ $(m \geq 1)$. Also, we choose two (double) sequences $\{h_{m,n}\}$ and $\{l_{m,n}\}$ of positive numbers satisfying $\lim_{n\to\infty} h_{m,n} = \lim_{n\to\infty} l_{m,n} = 0$ $(m = 1, 2, \ldots)$ and

$$\delta_m \leq h_{m,n}/l_{m,n} \leq (dM_m)^{-1} \quad (m, n = 1, 2, \ldots) \qquad (7.10)$$

and define operators $C_{m,n}$ and $A_{m,n}$ for $m, n = 1, 2, \ldots$ by

$$C_{m,n} = \tilde{C}_{h_{m,n}, l_{m,n}}|_{X_m} \quad \text{and} \quad A_{m,n} = h_{m,n}^{-1}(C_{m,n} - I). \qquad (7.11)$$

We are ready now for the following theorem.

THEOREM 7.4. *Let* X_m, $C_{m,n}$, $h_{m,n}$, *and* $A_{m,n}$ $(m, n = 1, 2, \ldots)$ *be the same as those defined above. Then* (i) *through* (iv) *hold:*

(i) *There exists a single-valued dissipative operator* A *in* $L^1(R^d)$ *that satisfies* $C_0^1(R^d) \subset D(A) \subset L^1(R^d) \cap L^\infty(R^d)$ *and*

$$Au = -\sum_{i=1}^d (\phi_i(u))_{x_i} \quad (u \in D(A)).$$

Here the differential on the right-hand side of the above equation is considered in the sense of distributions. Also $C_0^1(R^d)$ *is the set of all continuous, once differentiable (real-valued) functions defined on* R^d *that have compact support.*

(ii) *For every* $\lambda > 0$ *and* $u \in X_m$, $(I - \lambda A)^{-1}u = \lim_{n\to\infty}(I - \lambda A_{m,n})^{-1}u$ *and* $(I - \lambda A)^{-1}u$ $(= v)$ *satisfies the equation*

$$v + \lambda \sum_{i=1}^d (\phi_i(v))_{x_i} = u.$$

Hence we have $R(I - \lambda A) = L^1(R^d) \cap L^\infty(R^d)$ $(\lambda > 0)$.

(iii) *There exists a semigroup of contractions* $\{T(t): t \geq 0\}$ *on the subspace* $L^1(R^d) \cap L^\infty(R^d)$ *of* $L^1(R^d)$, *that can be expressed by the following exponential formula:*

$$T(t)u = \lim_{\lambda \to 0+} (I - \lambda A)^{-[t/\lambda]}u \quad (t \geq 0, \ u \in L^1(R^d) \cap L^\infty(R^d)).$$

Moreover, for every $u \in L^1(R^d) \cap L^\infty(R^d)$, $u(t, x) = [T(t)u](x)$ *is the weak solution of* (CP; u).

(iv) *For every* $u \in X_m$

$$\lim_{n\to\infty} C_{m,n}^{[t/h_{m,n}]} u = T(t)u \qquad (t \geq 0);$$

and the convergence is uniform on bounded subintervals of $[0, \infty)$.

§3. Proof of Theorem 7.4

In order to simplify the notation, we hereafter denote $h_{m,n}$ and $l_{m,n}$ simply by h by l, respectively. To avoid any confusion, we agree that the subscripts for h and l have the same subscripts as those for the related operators $C_{m,n}$, $A_{m,n}$, and $(I - \lambda A_{m,n})^{-1}$. Hence, when we write

$$[C_{m,n}u](x) = (2d)^{-1} \sum_{i=1}^{d} (u(x + le_i) + u(x - le_i))$$

$$- (h/2l) \sum_{i=1}^{d} [\phi_i(u(x + le_i)) - \phi_i(u(x - le_i))]$$

($u \in X_m$), the subscripts for h and l are m and n, that is, $h = h_{m,n}$ and $l = l_{m,n}$. Also, if the left-hand side of the above equation is $[C_{m',n'}u](x)$, then h and l on the right-hand side represent $h_{m',n'}$ and $l_{m',n'}$, respectively.

LEMMA 7.5. *For every* $m, n \geq 1$, $C_{m,n} \in \text{Cont}(X_m)$; *that is*, $C_{m,n}$ *is an operator satisfying* $\|C_{m,n}u - C_{m,n}v\|_1 \leq \|u - v\|_1$ ($u, v \in X_m$) *and having values in* X_m. *Moreover*, $\|C_{m,n}u\|_p \leq \|u\|_p$ ($u \in X_m$; $p = 1, \infty$).

PROOF. Let $m, n \geq 1$ and $u, v \in X_m$. From the definition of $C_{m,n}$

$$\|C_{m,n}u - C_{m,n}v\|_1 = \int_{R^d} |[C_{m,n}u](x) - [C_{m,n}v](x)| \, dx$$

$$\leq \sum_{i=1}^{d} \int_{R^d} |(2d)^{-1}[u(x + le_i) - v(x + le_i)] - (h/2l)[\phi_i(u(x + le_i))$$

$$- \phi_i(v(x + le_i))]| \, dx$$

$$+ \sum_{i=1}^{d} \int_{R^d} |(2d)^{-1}[u(x - le_i) - v(x - le_i)] + (h/2l)[\phi_i(u(x - le_i))$$

$$- \phi_i(v(x - le_i))]| \, dx$$

$$= \sum_{i=1}^{d} \int_{R^d} |(2d)^{-1}[u(x) - v(x)] - (h/2l)[\phi_i(u(x)) - \phi_i(v(x))]| \, dx$$

$$+ \sum_{i=1}^{d} \int_{R^d} |(2d)^{-1}[u(x) - v(x)] + (h/2l)[\phi_i(u(x)) - \phi_i(v(x))]| \, dx.$$

By the mean value theorem, $\phi_i(u(x)) - \phi_i(v(x)) = (u(x) - v(x))\phi_i'(\theta_i(x))$, where $\theta_i(x)$ has its value between $u(x)$ and $v(x)$. Now, since $\|u\|_\infty$, $\|v\|_\infty$ $\leq m$, we have $|\theta_i(x)| \leq m$; also from (7.10), we have $d^{-1} \pm (h/l)\phi_i'(s) \geq 0$ $(|s| \leq m; \ 1 \leq i \leq d)$. Therefore we have $d^{-1} \pm (h/l)\phi_i'(\theta_i(x)) \geq 0$ and obtain (^{2})

$$\sum_{i=1}^{d} |(2d)^{-1}[u(x) - v(x)] - (h/2l)[\phi_i(u(x)) - \phi_i(v(x))]|$$

$$+ \sum_{i=1}^{d} |(2d)^{-1}[u(x) - v(x)] + (h/2l)[\phi_i(u(x)) - \phi_i(v(x))]|$$

$$= \sum_{i=1}^{d} \{[(2d)^{-1} - (h/2l)\phi_i'(\theta_i(x))]$$

$$+ [(2d)^{-1} + (h/2l)\phi_i'(\theta_i(x))]\}|u(x) - v(x)|$$

$$= |u(x) - v(x)|.$$

Hence

$$\|C_{m,n}u - C_{m,n}v\|_1 \leq \int_{R^d} |u(x) - v(x)| \, dx = \|u - v\|_1.$$

From the assumption that $\phi_i(0) = 0$ $(1 \leq i \leq d)$, we have $C_{m,n}0 = 0$. From this, if we set $v = 0$ in the above inequality, then $\|C_{m,n}u\|_1 \leq \|u\|_1$.

Next, by the mean value theorem, $\phi_i(u(x + le_i)) - \phi_i(u(x - le_i)) = \phi_i'(\theta_i(x))[u(x + le_i) - u(x - le_i)]$ $(\theta_i(x)$ has a value between $u(x - le_i)$ and $u(x + le_i))$. Hence, if we carry out the same argument as above, then we have

$$|[C_{m,n}u](x)| \leq \sum_{i=1}^{d} |(2d)^{-1}[u(x + le_i) + u(x - le_i)]$$

$$- (h/2l)\phi_i'(\theta_i(x))[u(x + le_i) - u(x - le_i)]|$$

$$\leq \sum_{i=1}^{d} \{[(2d)^{-1} - (h/2l)\phi_i'(\theta_i(x))]|u(x + le_i)|$$

$$+ [(2d)^{-1} + (h/2l)\phi_i'(\theta_i(x))]|u(x - le_i)|\}$$

$$\leq \|u\|_\infty \sum_{i=1}^{d} \{[(2d)^{-1} - (h/2l)\phi_i'(\theta_i(x))]$$

$$+ [(2d)^{-1} + (h/2l)\phi_i'(\theta_i(x))]\}$$

$$= \|u\|_\infty.$$

Therefore, $\|C_{m,n}u\|_\infty \leq \|u\|_\infty$. $\square$

(^{2}) Since $u, v \in L^\infty(R^d)$, this inequality holds for a.e. $x \in R^d$. However, this is understood in our discussion, and we will not mention it as long as there is no cause for confusion.

In the proof above, we are using the right half of the inequality (7.10), that is, $h_{m,n}/l_{m,n} \leq (dM_m)^{-1}$ $(m, n \geq 1)$ only. Hence, noting that $\{X_m\}$ and $\{M_m\}$ are monotone increasing sequences, we obtain

$$C_{p,n}|_{X_m} \in \mathrm{Cont}(X_m) \ (p > m; \ n = 1, 2, \ldots). \tag{7.12}$$

LEMMA 7.6. *Let* $u \in X_m$. *Then for every real number* k *such that* $|k| \leq m$ *and every nonnegative function* $f \in C_0^\infty(R^d)$, *the following inequality holds*:

$$(\mathrm{sign}(u - k)A_{m,n}u, f) \leq (2d)^{-1}\sum_{i=1}^{d}(|u - k|, (l^2/h)D_i^- D_i^+ f)$$

$$+ \sum_{i=1}^{d}(\mathrm{sign}(u - k)[\phi_i(u) - \phi_i(k)], D_i^0 f)$$

$$(n = 1, 2, \ldots,). \tag{7.13}$$

PROOF. Let $u \in X_m$ and let k be a real number with $|k| \leq m$. From the definition of $A_{m,n}$

$$\mathrm{sign}(u(x) - k)[A_{m,n}u](x)$$

$$= \mathrm{sign}(u(x) - k)\left\{(l^2/2dh)\sum_{i=1}^{d}[D_i^- D_i^+ u](x) - \sum_{i=1}^{d}[D_i^0\phi_i(u)](x)\right\}$$

$$= \mathrm{sign}(u(x) - k)\left\{(l^2/2dh)\sum_{i=1}^{d}[D_i^- D_i^+ (u - k)](x)\right.$$

$$\left. - \sum_{i=1}^{d}[D_i^0(\phi_i(u) - \phi_i(k))](x)\right\}.$$

We choose an arbitrary i $(1 \leq i \leq d)$. By the mean value theorem

$$\phi_i(u(x \pm le_i)) - \phi_i(k) = \phi_i'(\theta_i^\pm(x))(u(x \pm le_i) - k).$$

(Decode $\pm$ in the same order, i.e., $\theta_i^+(x)$ is a function with value between k and $u(x + le_i)$, and similarly for $\theta_i^-(x)$.) From this, and $d^{-1} \pm (h/l)\phi_i'(\theta_i^\pm(x)) \geq 0$, and

$$[\mathrm{sign}(u(x \pm le_i) - k) - \mathrm{sign}(u(x) - k)](u(x \pm le_i) - k) \geq 0,$$

we obtain

$$(2d)^{-1}[|u(x \pm le_i) - k| - \mathrm{sign}(u(x) - k)(u(x \pm le_i) - k)]$$

$$\mp (h/2l)[\mathrm{sign}(u(x \pm le_i) - k) - \mathrm{sign}(u(x) - k)][\phi_i(u(x \pm le_i)) - \phi_i(k)]$$

$$= [(2d)^{-1} \mp (h/2l)\phi_i'(\theta_i^\pm(x))][\mathrm{sign}(u(x \pm le_i) - k)$$

$$- \mathrm{sign}(u(x) - k)](u(x \pm le_i) - k) \geq 0.$$

(Decode $\pm$ in the same order.) When we add the respective sides of these two equations we obtain

$$\text{sign}(u(x) - k)(l^2/2dh)[D_i^- D_i^+ (u - k)](x)$$
$$- \text{sign}(u(x) - k)[D_i^0(\phi_i(u) - \phi_i(k))](x)$$
$$\leq (l^2/2dh)[D_i^- D_i^+ |u - k|](x) - [D_i^0(\text{sign}(u - k)(\phi_i(u) - \phi_i(k)))](x).$$

Since i $(1 \leq i \leq d)$ is arbitrary, we have

$$\text{sign}(u(x) - k)[A_{m,n}u](x) \leq (l^2/2dh) \sum_{i=1}^d [D_i^- D_i^+ |u - k|](x)$$
$$- \sum_{i=1}^d [D_i^0(\text{sign}(u - k)(\phi_i(u) - \phi_i(k)))](x).$$

Multiplying both sides by a nonnegative function $f \in C_0^\infty(R^d)$ and integrating over R^d, we obtain (7.13) by noting that $(D_i^- D_i^+ |u - k|, f) = (|u - k|, D_i^- D_i^+ f)$ and

$$(D_i^0(\text{sign}(u - k)(\phi_i(u) - \phi_i(k))), f) = -(\text{sign}(u - k)(\phi_i(u) - \phi_i(k)), D_i^0 f). \quad \square$$

REMARK. As we see form the proof above, (7.13) also holds for a nonnegative function $f \in C_0^\infty(R^d)$[3] such that f and f_{x_i} $(i = 1, 2, \ldots, d)$ are bounded on R^d. Now, let f be such a function and let $u \in X_m$. If we let $k = 0$ and apply (7.13), then we obtain

$$(\text{sign}(u)A_{m,n}u, f) \leq (l/dh) \sum_{i=1}^d (|u|, (l/2)D_i^- D_i^+ f) + \sum_{i=1}^d (\text{sign}(u)\phi_i(u), D_i^0 f)$$
$$\leq (1/d\delta_m) \sum_{i=1}^d \left(\sup_{x \in R^d} |f_{x_i}(x)| \right) \int_{R^d} |u(x)| \, dx$$
$$+ \sum_{i=1}^d M_m \left(\sup_{x \in R^d} |f_{x_i}(x)| \right) \int_{R^d} |u(x)| \, dx$$
$$\leq [(1/d\delta_m) + M_m] \left(\sum_{i=1}^d \sup_{x \in R^d} |f_{x_i}(x)| \right) \|u\|_1.$$

(7.14)

Here, we used $\phi_i(u(x)) = \phi_i(u(x)) - \phi_i(0) = \phi_i'(\theta(x))u(x)$ and (7.10).

As we see from the definition, X_m is a closed convex subset of $L^1(R^d)$ satisfying $X_m \subset X_{m+1}$ $(m \geq 1)$. On the other hand, from Lemma 7.5, $C_{m,n} \in \text{Cont}(X_m)$ $(m, n \geq 1)$. Hence each $A_{m,n}$ is a dissipative operator in $L^1(R^d)$ satisfying $R(I - \lambda A_{m,n}) \supset X_m = D(A_{m,n})$ $(\lambda > 0)$. From this,

---

[3] We denote the set of all infinitely differentiable (real-valued) functions defined on R^d by $C^\infty(R^d)$.

$(I - \lambda A_{m,n})^{-1}$ $(\lambda > 0; \; m, n \geq 1)$ is a contraction operator and each $A_{m,n}$ generates a semigroup of contractions on X_m (a closed convex subset of $L^1(R^d)$). We denote this semigroup of contractions by $\{T_{m,n}(t): t \geq 0\}$. Now, in order to apply Theorem 7.2 and Corollary 7.3, we show below that the conditions (C) and (C_1) hold.

LEMMA 7.7. *Let* $u \in X_m$, *let* $\lambda > 0$, *and set* $v_n = (I - \lambda A_{m,n})^{-1} u$ $(n = 1, 2, \dots)$. *Then* (i) *through* (iii) *hold*:

(i) $\|v_n\|_p \leq \|u\|_p$ $(n \geq 1; \; p = 1, \infty)$.

(ii) $\int_{R^d} |v_n(x+y) - v_n(x)| \, dx \leq \int_{R^d} |u(x+y) - u(x)| \, dx$ $(y \in R^d, \; n \geq 1)$.

(iii) *As* $p \to \infty$, $\int_{|x|>\rho} |v_n(x)| \, dx \to 0$ (*uniformly with respect to* n). *As a result*, $\{(I - \lambda A_{m,n})^{-1} u: n \geq 1\}$ *is a sequentially compact subset* (*of* $L^1(R^d)$) *for every* $u \in X_m$ *and* $\lambda > 0$.

PROOF. (i) Let $F: L^\infty(R^d) \to (L^\infty(R^d))^*$ be the duality mapping and choose $v_n' \in F(v_n)$ arbitrarily. Then by Lemma 7.5, $(A_{m,n} v_n, v_n') \leq 0$. Hence

$$\|u\|_\infty \|v_n\|_\infty = \|v_n - \lambda A_{m,n} v_n\|_\infty \|v_n\|_\infty \geq (v_n - \lambda A_{m,n} v_n, v_n') \geq (\|v_n\|_\infty)^2,$$

and we obtain $\|v_n\|_\infty \leq \|u\|_\infty$. Next, from $A_{m,n} 0 = 0$, we have $(I - \lambda A_{m,n})^{-1} 0 = 0$. Then since $(I - \lambda A_{m,n})^{-1}$ is a contraction operator (on $L^1(R^d)$), $\|v_n\|_1 = \|(I - \lambda A_{m,n})^{-1} u\|_1 \leq \|u\|_1$.

(ii) For arbitrary $v \in X_m$ and $y \in R^d$, we set $v_y(x) = v(x+y)$ $(x \in R^d)$. Then $(A_{m,n} v_y)(x) = (A_{m,n} v)(x+y)$. From this we obtain

$$((I - \lambda A_{m,n})^{-1} u_y)(x) = ((I - \lambda A_{m,n})^{-1} u)(x+y) = v_n(x+y).$$

Hence

$$\int_{R^d} |v_n(x+y) - v_n(x)| \, dx = \|(I - \lambda A_{m,n})^{-1} u_y - (I - \lambda A_{m,n})^{-1} u\|_1$$
$$\leq \|u_y - u\|_1$$
$$= \int_{R^d} |u(x+y) - u(x)| \, dx \qquad (y \in R^d, \; n \geq 1).$$

(iii) From (7.14), for every nonnegative function $f \in C^\infty(R^d)$ such that f and f_{x_i} $(i = 1, 2, \dots, d)$ are bounded on R^d we have

$$(\text{sign}(v_n) A_{m,n} v_n, f) \leq [(1/d\delta_m) + M_n] \left(\sum_{i=1}^{d} \sup_{x \in R^d} |f_{x_i}(x)| \right) \|v_n\|_1.$$

Since $A_{m,n}v_n = \lambda^{-1}(v_n - u)$, from the above inequality we obtain

$$
\begin{aligned}
\int_{R^d} |v_n(x)| f(x)\,dx - \int_{R^d} |u(x)| f(x)\,dx \\
\leq \lambda[(1/d\delta_m) + M_m]\left(\sum_{i=1}^d \sup_{x\in R^d} |f_{x_i}(x)|\right)\|u\|_1.
\end{aligned}
\tag{7.15}
$$

Next we consider a family of functions $\{\delta_{r,\rho}: \rho > r > 0\} \subset C^\infty(R^1)$ that satisfy the following conditions:

$$
\begin{cases}
\delta_{r,\rho}(s) = 0 \ (|s| \leq r), \ \ \delta_{r,\rho}(s) = 1 \ (|s| \geq \rho), \ \ 0 \leq \delta_{r,\rho}(s) \leq 1 \ (s \in R^1), \\
\text{and for every } r > 0, \ \sup_{s\in R^1} |\delta'_{r,\rho}(s)| \to 0 \text{ as } \rho \to \infty.
\end{cases}
$$

And we set

$$
f_{r,\rho}(x) = \sum_{i=1}^d \delta_{r,\rho}(x_i) \qquad (x = (x_1, x_2, \ldots, x_d) \in R^d, \ \rho > r > 0).
$$

Then if $|x| = (\sum_{i=1}^d |x_i|^2)^{1/2} \geq \rho\sqrt{d}$, then $f_{r,\rho}(x) \geq 1$. Also, $f_{r,\rho}(x) = 0$ $(|x| \leq r)$ and $0 \leq f_{r,\rho}(x) \leq d$ $(x \in R^d)$. Now if we let $f = f_{r,\rho}$ and apply (7.15), we obtain the following inequality:

$$
\begin{aligned}
\int_{|x|\geq\rho\sqrt{d}} |v_n(x)|\,dx &\leq \int_{R^d} |v_n(x)| f_{r,\rho}(x)\,dx \\
&\leq \int_{|x|>r} |u(x)| f_{r,\rho}(x)\,dx + \lambda[(1/d\delta_m) + M_m]d\left(\sup_{s\in R^1} |\delta'_{r,\rho}(s)|\right)\|u\|_1 \\
&\leq d\int_{|x|>r} |u(x)|\,dx + \lambda d[(1/d\delta_m) + M_m]\left(\sup_{s\in R^1} |\delta'_{r,\rho}(s)|\right)\|u\|_1.
\end{aligned}
$$

Hence

$$
\limsup_{\rho\to\infty}\left[\sup_{n\geq 1}\int_{|x|\geq\rho\sqrt{d}} |v_n(x)|\,dx\right] \leq d\int_{|x|>r} |u(x)|\,dx \to 0 \qquad (r \to \infty).
$$

Therefore (iii) is shown.

Since (i)–(iii) hold, by the M. Riesz theorem $\{(I - \lambda A_{m,n})^{-1}u: n \geq 1\}$ $(= \{v_n: n \geq 1\})$ is sequentially compact in the space $L^1(R^\infty)$. $\square$

LEMMA 7.8. *Let* $u, v \in X_m$, $\lambda > 0$, *and* $p \geq m$. *We let* $\{j_n\}$ *and* $\{j'_n\}$ *be two subsequences of* $\{n\}$ *and set* $w_n = (I - \lambda A_{m,j_n})^{-1}u$ *and* $z_n = (I - \lambda A_{p,j'_n})^{-1}v$ $(n = 1, 2, \ldots)$. *If* $\|w_n - w\|_1 \to 0$ *and* $\|z_n - z\|_1 \to 0$ *as* $n \to \infty$, *then* $\|w - z\|_1 \leq \|u - v\|_1$.

PROOF. Let $\|w_n - w\|_1 \to 0$ and $\|z_n - z\|_1 \to 0$ as $n \to \infty$. Then we can select subsequences $\{w_{n_k}\}$ of $\{w_n\}$ and $\{z_{n_k}\}$ of $\{z_n\}$ such that $\lim_{k\to\infty} w_{n_k}(x) = w(x) = \lim_{k\to\infty} z_{n_k}(x) = z(x)$ (a.e. $x \in R^d$). For simplicity, we rewrite $\{w_n\}$ and $\{z_n\}$ for $\{w_{n_k}\}$ and $\{z_{n_k}\}$, respectively. Hence $\lim_{n\to\infty} w_n(x) = w(x)$ and $\lim_{n\to\infty} z_n(x) = z(x)$ (a.e. $x \in R^d$). Also we denote h_{m,j_n} and l_{m,j_n} by h and l, respectively; and, h_{p,j'_n} and l_{p,j'_n} by h' and l', respectively.

Now, let $f(x, y) \in C_0^\infty(R^d \times R^d)$ be a nonnegative function and consider w_n, $z_n(y)$, and $f(\cdot, y)$ as u, k and f in Lemma 7.6, respectively, and apply inequality (7.13) (we note that by (7.12) $z_n \in X_m$ and hence we obtain $\|z_n\|_\infty \le m$). And, noting $A_{m,j_n} w_n = \lambda^{-1}(w_n - u)$, we integrate both sides of the inequality with respect to y over R^d and obtain the following inequality:

$$0 \le \int_{R^d \times R^d} \text{sign}(w_n(x) - z_n(y))(u(x) - w_n(x))f(x, y)\,dx\,dy$$

$$+ (\lambda/2d) \sum_{i=1}^d \int_{R^d \times R^d} |w_n(x) - z_n(y)|(l^2/h)[D_i^- D_i^+ f(\cdot, y)](x)\,dx\,dy \tag{7.16}$$

$$+ \lambda \sum_{i=1}^d \int_{R^d \times R^d} \text{sign}(w_n(x) - z_n(y))[\phi_i(w_n(x)) - \phi_i(z_n(y))]$$

$$\times [D_i^0 f(\cdot, y)](x)\,dx\,dy,$$

where $[D_i^0 f(\cdot, y)](x) = (2l)^{-1}[f(x + le_i, y) - f(x - le_i, y)]$ and the same for $[D_i^- D_i^+ f(\cdot, y)](x)$. Next, consider z_n, $w_n(x)$, and $f(x, \cdot)$ as u, k, and f in Lemma 7.6, respectively, and apply inequality (7.13). Note that $A_{p,j'_n} z_n = \lambda^{-1}(z_n - v)$ and integrate both sides of the inequality with respect to x over R^d. We obtain

$$0 \le \int_{R^d \times R^d} \text{sign}(z_n(y) - w_n(x))(v(y) - z_n(y))f(x, y)\,dy\,dx$$

$$+ (\lambda/2d) \sum_{i=1}^d \int_{R^d \times R^d} |z_n(y) - w_n(x)|(l'^2/h')[D_i^- D_i^+ f(x, \cdot)](y)\,dy\,dx$$

$$+ \lambda \sum_{i=1}^d \int_{R^d \times R^d} \text{sign}(z_n(y) - w_n(x))[\phi_i(z_n(y)) - \phi_i(w_n(x))]$$

$$\times [D_i^0 f(x, \cdot)](y)\,dy\,dx,$$

where $[D_i^0 f(x, \cdot)](y) = (2l')^{-1}[f(x, y + l'e_i) - f(x, y - l'e_i)]$ and the same for $[D_i^- D_i^+ f(x, \cdot)](y)$. Adding this to the respective sides of (7.16), we obtain the following inequality:

$$0 \le \int_{R^d \times R^d} \mathrm{sign}(w_n(x) - z_n(y))(u(x) - v(y) - w_n(x) + z_n(y))$$
$$\times f(x, y)\, dx\, dy$$

$$+ (\lambda/2d) \sum_{i=1}^{d} \int_{R^d \times R^d} |w_n(x) - z_n(y)| \{ (l^2/h)[D_i^- D_i^+ f(\cdot, y)](x) \quad (7.17)$$
$$+ (l'^2/h')[D_i^- D_i^+ f(x, \cdot)](y) \}\, dx\, dy$$

$$+ \lambda \int_{R^d \times R^d} \mathrm{sign}(w_n(x) - z_n(y)) \sum_{i=1}^{d} [\phi_i(w_n(x)) - \phi_i(z_n(y))]$$
$$\times \{ [D_i^0 f(\cdot, y)](x) + [D_i^0 f(x, \cdot)](y) \}\, dx\, dy.$$

When $n \to \infty$, since $\mathrm{sign}(w_n(x) - z_n(y)) \to \mathrm{sign}(w(x) - z(y))$ (a.e. $(x, y) \in$ $\{(x, y) \in R^d \times R^d : w(x) > z(y)\} \cup \{(x, y) \in R^d \times R^d : w(x) < z(y)\}$ and $\mathrm{sign}(w_n(x) - z_n(y)) \sum_{i=1}^{d} [\phi_i(w_n(x)) - \phi_i(z_n(y))] \to 0$ (a.e. $(x, y) \in$ $R^d \times R^d : w(x) = z(y)\}$) and

$$(l^2/h)[D_i^- D_i^+ f(\cdot, y)](x) + (l'^2/h')[D_i^- D_i^+ f(x, \cdot)](y) \to 0 \quad ((x, y) \in R^d \times R^d),$$

if we let $n \to \infty$ in (7.17), we have

$$0 \le \int_{R^d \times R^d} (|u(x) - v(y)| - |w(x) - z(y)|) f(x, y)\, dx\, dy$$

$$+ \lambda \int_{R^n \times R^d} \mathrm{sign}(w(x) - z(y)) \sum_{i=1}^{d} [\phi_i(w(x)) - \phi_i(z(y))] \quad (7.18)$$
$$\times (f_{x_i}(x, y) + f_{y_i}(x, y))\, dx\, dy.$$

Now consider a nonnegative function $\sigma \in C_0^\infty(R^1)$ with $\int_{-\infty}^{\infty} \sigma(s)\, ds = 1$ and set

$$\omega(x) = \prod_{i=1}^{d} \sigma(x_i) \qquad (x = (x_1, x_2, \dots, x_d) \in R^d),$$
$$\omega_\rho(x) = \rho^{-d} \omega(x/\rho) \quad (\rho > 0, \ x \in R^d).$$

Next, let $g \in C_0^\infty(R^d)$ be an arbitrary nonnegative function and apply (7.18) with $f(x, y) = g(\frac{x+y}{2})\omega_\rho(\frac{x+y}{2})$, and carry out the change of variables $x = \xi + \eta$ and $y = \xi - \eta$ to obtain the following inequality:

$$0 \le \int_{R^d} \left[\int_{R^d} \{ (|u(\xi + \eta) - v(\xi - \eta)| - |w(\xi + \eta) - z(\xi - \eta)|) g(\xi) \right.$$

$$+ \lambda \, \mathrm{sign}(w(\xi + \eta) - z(\xi - \eta)) \sum_{i=1}^{d} (\phi_i(w(\xi + \eta)) \quad (7.19)$$

$$\left. - \phi_i(z(\xi - \eta))) g_{\xi_i}(\xi) \} d\xi \right] \omega_\rho(\eta)\, d\eta.$$

We denote by $I_g(\eta)$ the integral on the right-hand side of (7.19) enclosed by [ ]. Then by the convergence theorem we obtain

$$0 \le \liminf_{\rho \to 0+} \int_{R^d} I_g(\eta) \omega_\rho(\eta) \, d\eta \le \limsup_{|\eta| \to 0} I_g(\eta)$$

$$\le \lim_{|\eta| \to 0} \int_{R^d} \{|u(\xi + \eta) - v(\xi - \eta)| - |w(\xi + \eta) - z(\xi - \eta)|)g(\xi)$$

$$+ \lambda \sum_{i=1}^{d} |[\phi_i(w(\xi + \eta)) - \phi_i(z(\xi - \eta))]g_{\xi_i}(\xi)|\} \, d\xi \quad (7.20)$$

$$= \int_{R^d} (|u(\xi) - v(\xi)| - |w(\xi) - z(\xi)|)g(\xi) \, d\xi$$

$$+ \lambda \sum_{i=1}^{d} \int_{R^d} |[\phi_i(w(\xi)) - \phi_i(z(\xi))]g_{\xi_i}(\xi)| \, d\xi.$$

Here we consider a nonnegative function $\kappa \in C_0^\infty(R^1)$ that satisfies $\kappa(s) = 1$ $(|s| \le 1)$, and set $g(\xi) = \kappa(|\xi|/r)$ $(\xi \in R^d, r > 0)$ and apply (7.20). If we let $r \to \infty$, then

$$0 \le \int_{R^d} (|u(\xi) - v(\xi)| - |w(\xi) - z(\xi)|) \, d\xi;$$

that is, we obtain $\|w - z\|_1 \le \|u - v\|_1$. □

From Lemmas 7.7 and 7.8 we see the following:

> For every $\lambda > 0$, $m = 1, 2, \ldots$, and for every $u \in X_m$, $\{(I - \lambda A_{m,n})^{-1} u\}_{n \ge 1}$ is a convergent sequence in $L^1(R^d)$ and the limit does not depend on m. That is, $\quad (7.21)$ $\lim_{n \to \infty} (I - \lambda A_{m,n})^{-1} u = \lim_{n \to \infty} (I - \lambda A_{p,n})^{-1} u$ $(u \in X_m, p > m)$.

In fact, let $\lambda > 0$, $m \ge 1$, and $u \in X_m$. Set $v_n = (I - \lambda A_{m,n})^{-1} u$ $(n = 1, 2, \ldots)$, and consider two arbitrary subsequences $\{v_n'\}$ and $\{v_n''\}$ of $\{v_n\}$. By Lemma 7.7, since $\{v_n : n = 1, 2, \ldots\}$ is sequentially compact (in $L^1(R^d)$), $\{v_n'\}$ and $\{v_n''\}$ have convergent subsequences $\{v_{n_k}'\}$ and $\{v_{n_k}''\}$, respectively. Set $v = \lim_{k \to \infty} v_{n_k}'$ and $z = \lim_{k \to \infty} v_{n_k}''$; that is, $\|v_{n_k}' - v\|_1 \to 0$ and $\|v_{n_k}'' - z\|_1 \to 0$ as $k \to \infty$. Then by Lemma 7.8, $\|v - z\|_1 \le \|u - u\|_1 = 0$. Therefore, $v = z$. We have shown that every subsequence of $\{v_n\}$ always contains a convergent subsequence; moreover, these subsequences all have the same limit. Hence $\{v_n\}$ itself converges in $L^1(R^d)$. Next, if $p > m$, then since $u \in X_m \subset X_p$, the limit $\lim_{n \to \infty} (I - \lambda A_{p,n})^{-1} u$ exists. Here, if we apply Lemma 7.8 again, then $\lim_{n \to \infty} (I - \lambda A_{m,n})^{-1} u = \lim_{n \to \infty} (I - \lambda A_{p,n})^{-1} u$.

LEMMA 7.9. (i) *For every* $u \in C_0^1(R^d) \cap X_m$

$$\lim_{n \to \infty} A_{m,n} u = -\sum_{i=1}^d (\phi_i(u))_{x_i} = -\sum_{i=1}^d \phi_i'(u) u_{x_i} \quad \text{uniformly on } R^d.$$

(ii) *There exists a pseudoresolvent* $\{J_\lambda : \lambda > 0\} \subset \text{Cont}(X_0)$ *satisfying the following*

$$J_\lambda u = \lim_{n \to \infty} (I - \lambda A_{m,n})^{-1} u \quad (\lambda > 0, \ u \in X_m, \ m \geq 1).$$

That is, condition (C) *holds. Moreover, for every* $\lambda > 0$ *and* $u \in X_0 = L^1(R^d) \cap L^\infty(R^d)$, $J_\lambda u$ *is a solution of the equation*

$$u = v + \lambda \sum_{i=1}^d (\phi_i(v))_{x_i},$$

where the differential is considered in the sense of distributions.

PROOF. (i) Let $u \in C_0^1(R^d) \cap X_m$. Then

$$[A_{m,n} u](x) = (l/2dh) \sum_{i=1}^d [l D_i^- D_i^+ u](x)$$

$$- \sum_{i=1}^d [D_i^0 \phi_i(u)](x) \quad (x \in R^d, \ n \geq 1).$$

We note (7.10) and let $n \to \infty$; then the first term on the right-hand side of the above equation $\to 0$ and the second term $\to -\sum_{i=1}^d \phi_i'(u(x)) u_{x_i}(x)$ $(x \in R^d)$. Here, in each case we have uniform convergence on R^d.

(ii) From (7.21), for every $\lambda > 0$ we can define an operator $J_\lambda : X_0 \to X_0$ by

$$J_\lambda u = \lim_{n \to \infty} (I - \lambda A_{m,n})^{-1} u \quad (u \in X_m, \ m \geq 1).$$

Since $(I - \lambda A_{m,n})^{-1}[X_m] \subset X_m$ and $\|(I - \lambda A_{m,n})^{-1} u - (I - \lambda A_{m,n})^{-1} v\|_1 \leq \|u - v\|_1$ $(u, v \in X_m, \ m, n \geq 1, \ \lambda > 0)$, each $J_\lambda : X_0 \to X_0$ is a contraction operator. Next, since $A_{m,n}$ is a dissipative operator, $\{(I - \lambda A_{m,n})^{-1} : \lambda > 0\}$ is a pseudoresolvent (consists of contraction operators) ((i) of Lemma 7.1). From this and the fact that $X_0 = L^1(R^d) \cap L^\infty(R^d)$ is a linear set, we see that $\{J_\lambda : \lambda > 0\}$ is also a pseudoresolvent.

Next, let $\lambda > 0$ and $u \in X_0$. Then we can select a natural number m such that $u \in X_m$, and set $v_n = (I - \lambda A_{m,n})^{-1} u$ $(n = 1, 2, \ldots)$. Then $\{A_{m,n} v_n\}_{n \geq 1}$ converges to $-\sum_{i=1}^d (\phi_i(J_\lambda u))_{x_i}$ in the sense of distributions.

In fact, for every $f \in C_0^\infty(R^d)$

$$(A_{m,n}v_n, f) = (l^2/2dh) \sum_{i=1}^{d} (D_i^- D_i^+ v_n, f) - \sum_{i=1}^{d} (D_i^0 \phi_i(v_n), f)$$

$$= (l^2/2dh) \sum_{i=1}^{d} (v_n, D_i^- D_i^+ f) + \sum_{i=1}^{d} (\phi_i(v_n), D_i^0 f).$$

Here if we let $n \to \infty$ (noting (7.10)), then we obtain

$$\lim_{n \to \infty} (A_{m,n}v_n, f) = \sum_{i=1}^{d} (\phi_i(J_\lambda u), f_{x_i}) \qquad (f \in C_0^\infty(R^d)),$$

and $\{A_{m,n}v_n\}_{n \geq 1}$ converges to $-\sum_{i=1}^{d}(\phi_i(J_\lambda u))_{x_i}$ in the sense of distributions. On the other hand, when $n \to \infty$,

$$A_{m,n}v_n = \lambda^{-1}(v_n - u) \to \lambda^{-1}(J_\lambda u - u)$$

(in the norm on $L^1(R^d)$). Hence it follows that

$$(\lambda^{-1}(J_\lambda u - u), f) = \sum_{i=1}^{d} (\phi_i(J_\lambda u), f_{x_i}) \qquad (f \in C_0^\infty(R^d)),$$

and we obtain $u = J_\lambda u + \lambda \sum_{i=1}^{d} (\phi_i(J_\lambda u))_{x_i}$. $\quad\square$

Finally

PROOF OF THEOREM 7.4. (I) In (ii) of Lemma 7.9, we showed that condition (C) holds. Also, let $D = C_0^1(R^d)$ and define $A_1 u = -\sum_{i=1}^{d}(\phi_i(u))_{x_i}$ $(= -\sum_{i=1}^{d} \phi_i'(u)u_{x_i})$ $(u \in D)$; then $D \subset X_0 = L^1(R^d) \cap L^\infty(R^d)$ and by (i) of Lemma 7.9, we obtain $\lim_{n \to \infty} \|A_{m,n}u - A_1 u\|_1 = 0$ $(u \in D \cap X_m,$ $m \geq 1)$. Hence condition (C_1) also holds and we can apply Corollary 7.3, because X_0 is a linear set. Moreover, we note that D is dense in $L^1(R^d)$. Then there exists a dissipative operator A (in $L^1(R^d)$) such that $C_0^1(R^d) \subset D(A) \subset L^1(R^d) \cap L^\infty(R^d)$ and $J_\lambda = (I - \lambda A)^{-1}$ $(\lambda > 0)$, and $A_1 u \in Au$ $(u \in C_0^1(R^d))$. Now, A is a single-valued operator. In fact, from the latter half of (ii) of Lemma 7.9, we have

$$u = J_\lambda u + \lambda \sum_{i=1}^{d} (\phi_i(J_\lambda u))_{x_i} \qquad (u \in D(J_\lambda) = X_0).$$

Hence, we obtain that if $J_\lambda u = J_\lambda v$ then $u = v$, and J_λ is a one-to-one operator. Therefore, A is a single-valued operator ((ii) of Lemma 7.1). Furthermore, from the above equation we can show $Av = -\sum_{i=1}^{d}(\phi_i(v))_{x_i}$

$(v \in D(A))$. Thus we have proved (i) and (ii) of Theorem 7.4. Also, (iv) and the first half of (iii) an be obtained immediately from Corollary 7.3.

(II) (Proof of the latter half of (iii)). For every $u \in L(R^d) \cap L^\infty(R^d)$, we show that $[T(t)u](x)$ is a weak solution of $(CP; u)$. Let $u \in L^1(R^d) \cap L^\infty(R^d)$ and set $u(t, x) = [T(t)u](x)$ $(t \geq 0, \ x \in R^d)$. Then we show that (a_1) through (a_2) of Definition 7.2 are satisfied. First, since $\{T(t): t \geq 0\}$ is a semigroup of contractions on $L^1(R^d) \cap L^\infty(R^d)$ and also $T(t)[X_m] \subset X_m$ $(t \geq 0, \ m \geq 1)$, (a_1) and (a_3) are obviously satisfied. Next, we show $u(t, x)$ satisfies (a_2). We give $k \in R^1$ and a nonnegative function $f \in C_0^\infty((0, \infty) \times R^d)$ arbitrarily. Since $X_m \subset X_{m+1}$ $(m \geq 1)$ and $L^1(R^d) \cap L^\infty(R^d) = \bigcup_{m \geq 1} X_m$, there exists a natural number m satisfying

$$u \in X_m \quad \text{and} \quad |k| \leq m.$$

Now for every $\varepsilon > 0$, we define

$$u_\varepsilon(t) = (I - \varepsilon A_{m,n})^{-[t/\varepsilon]} u(t \geq 0) \quad \text{and} \quad u_\varepsilon(t, x) = [u_\varepsilon(t)](x)$$

$(t \geq 0, \ x \in R^d)$. Since the semigroup of contractions $\{T_{m,n}(t): t \geq 0\}$ generated by $A_{m,n}$ on X_m can be written as

$$T_{m,n}(t)v = \lim_{\lambda \to 0+} (I - \lambda A_{m,n})^{-[t/\lambda]} v$$

(uniformly on bounded subintervals of $[0, \infty)$) for each $v \in X_m$, we have

$$\lim_{\varepsilon \to 0+} u_\varepsilon(t, \cdot) = \lim_{\varepsilon \to 0+} u_\varepsilon(t) = T_{m,n}(t)u \in X_m \qquad (7.22)$$

(the convergence is uniform on bounded subintervals of $[0, \infty)$). Also from the definition of $u_\varepsilon(t)$

$$\text{sign}(u_\varepsilon(t, x) - k)[A_{m,n} u_\varepsilon(t)](x)$$
$$= \varepsilon^{-1} \text{sign}(u_\varepsilon(t, x) - k)[(u_\varepsilon(t, x) - k) - (u_\varepsilon(t - \varepsilon, x) - k)]$$
$$\geq \varepsilon^{-1}(|u_\varepsilon(t, x) - k| - |u_\varepsilon(t - \varepsilon, x) - k|) \qquad (k \in R^1, \ x \in R^d).$$
$$(7.23)$$

We consider $u_\varepsilon(t, \cdot)$ and $f(t, \cdot)$ as u and f in Lemma 7.6, respectively, and apply (7.13). If we integrate both sides of (7.13) with respect to t over $[\varepsilon, \infty)$, then we obtain the following inequality by using (7.23):

$$0 \leq I_1(\varepsilon) + I_2(\varepsilon) + I_3(\varepsilon), \qquad (7.24)$$

where

$$I_1(\varepsilon) = \int_\varepsilon^\infty \int_{R^d} \varepsilon^{-1}(|u_\varepsilon(t-\varepsilon, x) - k| - |u_\varepsilon(t, x) - k|) f(t, x) \, dx \, dt$$

$$= \int_0^\infty \int_{R^d} |u_\varepsilon(t, x) - k| \varepsilon^{-1} (f(t+\varepsilon, x) - f(t, x)) \, dx \, dt$$

$$+ \int_0^\varepsilon \int_{R^d} \varepsilon^{-1} |u_\varepsilon(t, x) - k| f(t, x) \, dx \, dt,$$

$$I_2(\varepsilon) = \int_\varepsilon^\infty \int_{R^d} |u_\varepsilon(t, x) - k| (l^2/2dh) \sum_{i=1}^d [D_i^- D_i^+ f(t, \cdot)](x) \, dx \, dt,$$

$$I_3(\varepsilon) = \int_\varepsilon^\infty \int_{R^d} \operatorname{sign}(u_\varepsilon(t, x) - k) \sum_{i=1}^d [\phi_i(u_\varepsilon(t, x)) - \phi_i(k)]$$

$$\times [D_i^0 f(t, \cdot)](x) \, dx \, dt.$$

If we note that $f(t, x)$ has compact support in $(0, \infty) \times R^d$, then the second term of $I_1(\varepsilon)$ becomes 0 when we choose $\varepsilon > 0$ sufficiently small. From this and (7.22) we obtain

$$\lim_{\varepsilon \to 0+} I_1(\varepsilon) = \int_0^\infty \int_{R^d} |[T_{m,n}(t)u](x) - k| f_t(t, x) \, dx \, dt.$$

Next, from $u_\varepsilon(t, \cdot) \in X_m$ and $l/dh \le 1/d\delta_m$ (see (7.10)), we have

$$\limsup_{\varepsilon \to 0+} I_2(\varepsilon) \le K_1 l \quad \text{(where } K_1 \text{ is a constant)}.$$

Finally we estimate $I_3(\varepsilon)$. By the mean value theorem, we have

$$|\phi_i(u_\varepsilon(t, x)) - \phi_i(k)| \le 2m M_m$$

and also $|[D_i^0 f(t, \cdot)](x) - f_{x_i}(t, x)| \le K_2 l$ (K_2 is a constant). Then

$$\int_\varepsilon^\infty \int_{R^d} |\operatorname{sign}(u_\varepsilon(t, x) - k)[\phi_i(u_\varepsilon(t, x)) - \phi_i(k)]$$

$$\times ([D_i^0 f(t, \cdot)](x) - f_{x_i}(t, x))| \, dx \, dt \le K_3 l,$$

where K_3 is a constant. Next, if $t > T$, then we choose $T > 0$ such that $f(t, x) = 0$ $(x \in R^d)$. Then from (7.22)

$$\lim_{\varepsilon \to 0+} \int_0^T \int_{R^d} |u_\varepsilon(t, x) - [T_{m,n}(t)u](x)| \, dx \, dt = 0.$$

Hence there exists a sequence $\{\varepsilon_j\}$ of positive numbers with $\lim_{j \to \infty} \varepsilon_j = 0$ such that

$$\lim_{j \to \infty} u_{\varepsilon_j}(t, x) = [T_{m,n}(t)u](x) \quad \text{(a.e. } (t, x) \in (0, T) \times R^d).$$

And from this

$$\lim_{j\to\infty}\int_{\varepsilon_j}^{\infty}\int_{R^d}\operatorname{sign}(u_{\varepsilon_j}(t,x)-k)[\phi_i(u_{\varepsilon_j}(t,x)-\phi_i(k)]f_{x_i}(t,x)\,dx\,dt$$

$$=\int_0^{\infty}\int_{R^d}\operatorname{sign}([T_{m,n}(t)u](x)-k)$$

$$\times[\phi_i([T_{m,n}(t)u](x))-\phi_i(k)]f_{x_i}(t,x)\,dx\,dt$$

holds. From this and the estimation shown above

$$\limsup_{j\to\infty}I_3(\varepsilon_j)$$

$$\leq K_3 ld+\int_0^{\infty}\int_{R^d}\operatorname{sign}([T_{m,n}(t)u](x)-k)\sum_{i=1}^{d}[\phi_i([T_{m,n}(t)u](x))-\phi_i(k)]$$

$$\times f_{x_i}(t,x)\,dx\,dt.$$

As a result, we obtain the following inequality from (7.24):

$$0\leq (K_1+K_3 d)l$$

$$+\int_0^{\infty}\int_{R^d}\{|[T_{m,n}(t)u](x)-k|f_t(t,x)+\operatorname{sign}([T_{m,n}(t)u](x)-k)\}$$

$$\times\sum_{i=1}^{d}[\phi_i([T_{m,n}(t)u](x))-\phi_i(k)]f_{x_i}(t,x)\,dx\,dt.$$

$$(7.25)$$

Now from Corollary 7.3, $\lim_{n\to\infty}T_{m,n}(t)u=T(t)u$ uniformly on bounded subintervals of $[0,\infty)$. Hence if we consider the T determined above,

$$\lim_{n\to\infty}\int_0^T\int_{R^d}|[T_{m,n}(t)u](x)-[T(t)u](x)|\,dx\,dt=0.$$

Thus, we can select a subsequence $\{n_j\}$ of $\{n\}$ such that

$$\lim_{j\to\infty}[T_{m,n_j}(t)u](x)=[T(t)u](x)=u(t,x)$$

(a.e. $(t,x)\in(0,T)\times R^d$). Then if we replace n in (7.25) by n_j and let $j\to\infty$ (note that $l=l_{m,n_j}\to 0$), then we obtain

$$0\leq\int_0^{\infty}\int_{R^d}\{|u(t,x)-k|f_t(t,x)+\operatorname{sign}(u(t,x)-k)$$

$$\times\sum_{i=1}^{d}[\phi_i(u(t,x))-\phi_i(k)]f_{x_i}(t,x)\}\,dx\,dt.\quad\square$$

REMARK. Part (iii) of Theorem 7.4 was first shown by Crandall [33]. Concerning his method of proof, we refer the reader to the text [86] by K. Masuda, published in the mathematical series by the Kinokuniya Publishing Company.

Appendix. The Minimax Theorem

In this appendix both X and Y are linear topological spaces with R^1 as the field of scalars. (Hence X and Y satisfy the Hausdorff axiom of separation.)

MINIMAX THEOREM. *Let X be finite dimensional, and let both $A \subset X$ and $B \subset Y$ be compact convex sets. If $K(x, y): A \times B \to R^1$ satisfies:*

(a_1) *for every $y \in B$, $x \mapsto K(x, y)$ is a lower semicontinuous convex function;*

(a_2) *for every $x \in A$, $y \mapsto K(x, y)$ is an upper semicontinuous concave function;*

then (i) *and* (ii) *hold:*

(i) $\min_{x \in A} \max_{y \in B} K(x, y) = \max_{y \in B} \min_{x \in A} K(x, y)$.

(ii) *There exist $x_0 \in A$ and $y_0 \in B$ such that for all $x \in A$ and $y \in B$*

$$K(x_0, y) \leq K(x_0, y_0) \leq K(x, y_0).$$

To prove this, we will use the following lemma.

LEMMA. *Let $A \subset X$ and $B \subset Y$ be compact sets. If $K(x, y): A \times B \to R^1$ satisfies*

(b_1) *For every $y \in B$, $x \mapsto K(x, y)$ is lower semicontinuous.*

(b_2) *For every $x \in A$, $y \mapsto K(x, y)$ is upper semicontinuous.*

Then (i) *and* (ii) *in the statements of the Minimax Theorem are equivalent.*

PROOF. We note that both

$$\min_{x \in A} \max_{y \in B} K(x, y) \quad \text{and} \quad \max_{y \in B} \min_{x \in A} K(x, y)$$

exist. In fact, the upper semicontinuous functions and the lower semicontinuous functions defined on compact sets have maximum and minimum values, respectively. Hence from assumption (b_2) it follows that for an arbitrary $x \in A$, $\max_{y \in B} K(x, y)$ exists (we write this as $g(x)$). Now from (b_1), $g(x)$ is lower semicontinuous on A and $\min_{x \in A} g(x) = \min_{x \in A} \max_{y \in B} K(x, y)$ exists. Similarly, $\max_{y \in B} \min_{x \in A} K(x, y)$ exists.

Next we note that

$$\max_{y \in B} \min_{x \in A} K(x, y) \leq \min_{x \in A} \max_{y \in B} K(x, y) \tag{1}$$

holds.

(ii) $\Rightarrow$ (i). From the right half of the inequality in (ii), we have $K(x_0, y_0)$ $\leq \min_{x \in A} K(x, y_0) \leq \max_{y \in B} \min_{x \in A} K(x, y)$. Next, from the left half of the inequality in (ii), we have $\min_{x \in A} \max_{y \in B} K(x, y) \leq \max_{y \in B} K(x_0, y) \leq$ $K(x_0, y_0)$. Therefore, $\min_{x \in A} \max_{y \in B} K(x, y) \leq \max_{y \in B} \min_{x \in A} K(x, y)$. From this and (1), we obtain (i).

(i) $\Rightarrow$ (ii). We set $\max_{y \in B} \min_{x \in A} K(x, y) = \min_{x \in A} \max_{y \in B} K(x, y) = \alpha$. Then there exist $x_0 \in A$ and $y_0 \in B$ with $\max_{y \in B} K(x_0, y) = \alpha = \min_{x \in A} K(x, y_0)$. Hence for arbitrary $x \in A$ and $y \in B$, $K(x_0, y) \leq \alpha \leq K(x, y_0)$. In particular, $K(x_0, y_0) \leq \alpha \leq K(x_0, y_0)$; that is, $\alpha = K(x_0, y_0)$. Hence (ii) holds. $\square$

PROOF OF MINIMAX THEOREM. From the above lemma, we need only to show (i). Since X is finite dimensional, we let the dimension be d. Then X is isomorphic with the d-dimensional Euclidean space R^d as linear topological space. When x $(\in X) \leftrightarrow (\xi_1, \xi_2, \ldots, \xi_d)$ $(\in R^d)$, we define

$$\|x\| = (\xi_1^2 + \xi_2^2 + \cdots + \xi_d^2)^{1/2}.$$

Then the function $\|x\|^2 : X \to R^1$ is a strictly convex (continuous) function (that is, for arbitrary x, $x' \in X$ $(x \neq x')$ and $0 < t < 1$, $\|tx + (1-t)x'\|^2 < t\|x\|^2 + (1-t)\|x'\|^2$).

Now let $\varepsilon > 0$ and set

$$K_\varepsilon(x, y) = K(x, y) + \varepsilon\|x\|^2 \qquad (x \in A, \ y \in B).$$
$$f_\varepsilon(y) = \min_{x \in A} K_\varepsilon(x, y) \qquad (y \in B).$$

Then for each $y \in B$, since $x \mapsto K_\varepsilon(x, y)$ is a lower semicontinuous, strictly convex function, there exists a unique $x' \in A$ such that $K_\varepsilon(x', y) = \min_{x \in A} K_\varepsilon(x, y)$. Since x' depends on $y \in B$, we denote this x' by $E(y)$. Hence

$$f_\varepsilon(y) = K_\varepsilon(E(y), y) \qquad (y \in B).$$

Next for each $x \in A$, since $y \mapsto K_\varepsilon(x, y)$ is an upper semicontinuous concave function, f_ε is an upper semicontinuous concave function. From this, $f_\varepsilon(y)$ $(y \in B)$ has a maximum value. Now we choose $y^* \in B$ such that $f_\varepsilon(y^*) = \max_{y \in B} f_\varepsilon(y)$. Since $y \mapsto K_\varepsilon(x, y)$ is a concave function

$$K_\varepsilon(x, (1-t)y^* + ty) \geq (1-t)K_\varepsilon(x, y^*) + tK_\varepsilon(x, y)$$
$$\geq (1-t)f_\varepsilon(y^*) + tK_\varepsilon(x, y) \qquad (x \in A, \ y \in B, \ 0 < t < 1).$$

If we set $x = E((1-t)y^* + ty)$ in the above inequality, then

$$f_\varepsilon(y^*) \geq f_\varepsilon((1-t)y^* + ty) \geq (1-t)f_\varepsilon(y^*) + tK_\varepsilon(E((1-t)y^* + ty), y).$$

Hence we obtain

$$f_\varepsilon(y^*) \geq K_\varepsilon(E((1-t)y^* + ty), y) \qquad (y \in B, \ 0 < t < 1). \qquad (2)$$

Now

$$\text{for arbitrary } y_1, y_2 \in B, \quad E((1-t)y_1 + ty_2) \to E(y_1) \text{ as } t \to 0+. \tag{3}$$

To show this, we let $y_1, y_2 \in B$ and set $x_t = E((1-t)y_1 + ty_2)$ $(0 < t < 1)$. Then for every $x \in A$ and $0 < t < 1$

$$K_\varepsilon(x_t, (1-t)y_1 + ty_2) = f_\varepsilon((1-t)y_1 + ty_2) \le K_\varepsilon(x, (1-t)y_1 + ty_2). \tag{4}$$

If we take an arbitrary sequence $\{t_n\}$ such that $\lim_{n\to\infty} t_n = 0$ and $0 < t_n < 1$, then since $x_{t_n} \in A$ and A is compact, there is a subsequence $\{t_{n'}\}$ of $\{t_n\}$ and an element $z \in A$ such that $x_{t_{n'}} \to z$ as $n' \to \infty$. From (4) and the fact that $y \mapsto K(x, y)$ is a concave function, we have

$$(1 - t_{n'})K_\varepsilon(x_{t_{n'}}, y_1) + t_{n'}K_\varepsilon(x_{t_{n'}}, y_2) \le K_\varepsilon(x_{t_{n'}}, (1 - t_{n'})y_1 + t_{n'}y_2)$$
$$\le K_\varepsilon(x, (1 - t_{n'})y_1 + t_{n'}y_2)$$
$$(x \in A).$$

Here if we let $n' \to \infty$, then from the lower semicontinuity of $x \mapsto K_\varepsilon(x, y)$ and the upper semicontinuity of $y \mapsto K_\varepsilon(x, y)$, for an arbitrary $x \in A$ we have

$$K_\varepsilon(z, y_1) \left(\le \liminf_{n'\to\infty} K_\varepsilon(x_{t_{n'}}, y_1) \right) \le \limsup_{n'\to\infty} K_\varepsilon(x, (1 - t_{n'}y_1 + t_{n'}y_2))$$
$$\le K_\varepsilon(x, y_1).$$

From this, $K_\varepsilon(z, y_1) = \min_{x \in A} K_\varepsilon(x, y_1)$ and thus $z = E(y_1)$. Since we have shown that an arbitrary subsequence $\{x_{t_n}\}$ (of $\{x_t\}$) has a subsequence that converges to the same limit $E(y_1)$, it follows that $x_t \to E(y_1)$ as $t \to 0+$. Therefore, (3) is proved.

Now, letting $t \to 0+$ in (2). It follows from (3) that

$$f_\varepsilon(y^*) \left(\ge \liminf_{t\to 0+} K_\varepsilon(E((1 - t)y^* + ty), y) \right) \ge K_\varepsilon(E(y^*), y) \qquad (y \in B).$$

Also from the definition of f_ε, we have that $f_\varepsilon(y^*) \le K(x, y^*)$ $(x \in A)$. Hence

$$K_\varepsilon(x^*, y) \le (f_\varepsilon(y^*) =) K_\varepsilon(x^*, y^*) \le K_\varepsilon(x, y^*) \qquad (x \in A, y \in B), \tag{5}$$

where we have set $x^* = E(y^*)$. If we set

$$M = \max\{\|x\|^2 = \sum_{i=1}^{d} |\xi_i|^2 : x \in A\},$$

then from the definition of K_ε we have

$$K(x, y) - M\varepsilon \le K_\varepsilon(x, y) \le K(x, y) + M\varepsilon \qquad (x \in A, y \in B).$$

From this and (5)

$$\min_{x \in A} \max_{y \in B} K(x, y) - M\varepsilon \leq \min_{x \in A} \max_{y \in B} K_\varepsilon(x, y) \leq \max_{y \in B} K_\varepsilon(x^*, y) \leq K_\varepsilon(x^*, y^*)$$

$$\leq \min_{x \in A} K_\varepsilon(x, y^*) \leq \max_{y \in B} \min_{x \in A} K_\varepsilon(x, y)$$

$$\leq \max_{y \in B} \min_{x \in A} K(x, y) + M\varepsilon.$$

Since $\varepsilon > 0$ is arbitrary, we have

$$\min_{x \in A} \max_{y \in B} K(x, y) \leq \max_{y \in B} \min_{x \in A} K(x, y).$$

Combining this with (1), we obtain

$$\min_{x \in A} \max_{y \in B} K(x, y) = \max_{y \in B} \min_{x \in A} K(x, y). \quad \square$$

Postscript

One of the subjects that we have not touched upon at all in this book is the subdifferential evolution equations (in Hilbert spaces). For this, see Brezis [16] and Barbu [7], and also Watanabe [125] for recent results and for the references quoted by them. Also, [16] is an excellent expository text that summarizes the theory of nonlinear semigroups in Hilbert spaces and its applications. In the following we describe briefly the main references for each chapter and supplementary items.

Chapter 2. In §2.1 and §2.2 we summarized the results based mainly on Kato [50, 51]; see also Crandall-Liggett [36] and Oharu [102]. We referred to Brezis [16] for Example 2.2 and to Crandall [31] for Example 2.3. We sometimes call a strictly dissipative operator "a dissipative operator in the sense of Browder." In §2.3 we summarized the results of Kato [51], Brezis-Pazy [20], Brezis [14], Debrunner-Flor [41], Crandall-Pazy [38], and Minty [88, 89]. Theorem 2.20 is based on [20]; Theorem 2.23 and Corollary 2.24 on [14]. Theorems 2.25 and 2.26 and Corollary 2.27 are based on [41], [38] and [88, 89]. Example 2.6 is given by Crandall-Liggett [37].

Chapter 3. In §3.1 we summarized the results of Crandall-Pazy [38], Miyadera [94], and Kato [51]. Theorem 3.5 is based on [94] and Corollary 3.7 on [51] (see also [38]). Example 3.1 is given by Komura [64] and Example 3.2 by Webb [126]. The contents of §3.2 are based on the results of Crandall-Liggett [36], Komura [65], and Kato [52]; in particular, Theorem 3.13 is obtained by Komura [65]. Also, the proof given here is due to [52]. Theorem 3.11 is based on [36]; Lemma 3.15 of §3.3 on Miyadera-Oharu [95]. See also Brezis-Pazy [20]. Theorem 3.17 is based on Crandall-Pazy [38] for the case of a Hilbert space, and on Kobayashi [60] and Kobayasi [63] for the case of an extension to a Banach space. Theorems 3.18 and 3.19 are based on [63] (see also Miyadera [93]). Also, the estimation (3.44) used for the proof of Theorem 3.19 is given by Crandall-Liggett [36] when they prove Theorem 4.2 (generation theorem of semigroups). The way we apply induction in the proof is due to Rasmussen [115].

Chapter 4. In §4.1 and §4.2, we described generation theorems of semi-

groups based on Crandall-Liggett [36]. See also Miyadera [93] on Theorem
4.5 and Crandall [32] on Theorem 4.7. In §4.3 we summarized mainly those
results on generation theorems of semigroups given by Komura [64, 65], Kato
[50, 51], Crandall-Pazy [38], Dorroh [43], Martin [81], and Miyadera [94].
Theorems 4.13, 4.14, and 4.16 are based on [94]; Corollary 4.17 on Brezis
[14]; Theorem 4.18 on [81]. Theorem 4.20 is the so-called Komura theorem.
See also [38] and Komura [66]. Examples 4.5, 4.6, and 4.7 of §4.4 are based
on Crandall-Liggett [36]; Examples 4.8 and 4.9 on Crandall [34]; Example
4.10 on Crandall [31]. In this book, we have not considered the question
of compactness of the contraction semigroups $\{T(t): t \geq 0\}$ generated by
m-dissipative operators (if for each $t > 0$, $T(t)$ maps a bounded subset
of $\overline{D(A)}$ into a sequentially compact set, then this semigroup is said to be
compact), and we have not covered the generation of semigroups in Banach
lattices. See Konishi [72] and H. Brezis, *New results concerning monotone
operators and nonlinear semigroups*, Analysis of Nonlinear Problems, Res.
Inst. Math. Sci., Kyoto Univ., 1974 for the characterization of compactness
of semigroups. See Konishi [67, 69] and Calvert [25] for the generation of
semigroups in Banach lattices.

Chapter 5. Theorem 5.1 of §5.1 is based on Brezis-Pazy [20]. The concept
of the integral solution of Cauchy's problem is introduced by Benilan [9, 10].
The proof of Theorem 5.2 is quoted from Kobayashi [61]. Section 5.2 is
based mainly on Benilan [10], Takahashi [121], and Kobayashi [61]. Part (i)
of Theorem 5.7 (the convergence) is based on [121] and [61]. The proof (of
the estimation (5.11)) given in this book is shown in [61]. Also the proof
of uniqueness of the integral solution in this same theorem is due to [10].
Theorem 5.10 of §5.3 and Theorem 5.12 (generation theorem of semigroups)
are based on [61]. Theorem 5.15 is due to Martin [82] (see also [61]). Section
5.4 is an application of Theorem 5.10. Theorem 5.18 is an extension of the
result given in [10]. Also, Kenmochi-Oharu [57], Crandall [34], Crandall-
Evans [35], Kobayashi-Kobayasi [62], and Pierre [112] are closely related to
the contents of this chapter.

Chapter 6. Theorem 6.5 of §6.1, Theorems 6.8 and 6.12 of §6.2 are due
to Miyadera-Kobayashi [96] (Theorem 6.12 holds even if X_n is not convex
(see [96])). Corollaries 6.9 and 6.13 are summaries of the results obtained by
Kurtz [76], Brezis-Pazy [19], and Goldstein [47]. Corollary 6.11 is shown by
Benilan [8]. Also, its extension is given in H. Brezis, *New results concerning
monotone operators and nonlinear semigroups*, Analysis of Nonlinear Problems, Res. Inst. Math. Sci., Kyoto Univ., 1974. Theorems 6.15 and 6.17 of
§6.3 are due to Pierre [112]. Part (iii) of Corollary 6.19 is given by Barbu [6]
(for parts (i) and (ii) of the same corollary, see also Kobayashi [61]). Also,
Kato [51], Brezis [16], and Kobayashi-Kobayasi [62] are among those concerned with perturbation theorems in the form of the Corollary 6.19 (that is,
the question as to whether or not $A+B$ is an m-dissipative operator, when an

operator B is added to an m-dissipative operator A). $(d/dt)u(t) \in A(t)u(t)$ is one of the topics that we have not discussed in this book, even though it is related to the evolution equation $(d/dt)u(t) \in A(t)u(t)$ which we considered in §6.4. Here, the operator $A(t)$ is not necessarily single-valued (not even a continuous operator). Regarding this subject, see Kato [51], Crandall-Pazy [39], Evans [46], and also Masuda [86]. There is also the product formula of semigroups which is related to this chapter. For this, see Brezis-Pazy [18], Webb [127, 128], Kobayashi [59], and Miyadera-Kobayashi [96].

Chapter 7. Following Oharu-Takahashi [107], we explained a construction of the difference approximation of the solution of Cauchy's problem for quasilinear partial differential equations of first order. Furthermore, concrete applications of the theory of nonlinear semigroups to the partial differential equations have been carried out recently by many scholars. For example, Aizawa [1–3], Brezis-Strauss [21], Crandall [33], Konishi [69–71, 73, 74], Kurtz [76], and Oharu [105, 106] are among those who deal with nonreflexive function spaces (beside those quoted in §4.4).

Bibliography

1. S. Aizawa, *A semigroup treatment of the Hamilton-Jacobi equation in one space variable*, Hiroshima Math. J. **3** (1973), 367–386.
2. ___, *On the uniqueness of global generalized solutions for the equation* $F(x, u, \operatorname{grad} u) = 0$, Proc. Japan Acad. **51** (1975), 147–150.
3. ___, *A semigroup treatment of the Hamilton-Jacobi equation in several space variables*, Hiroshima Math. J. **6** (1976), 15–30.
4. H. Attouch and A. Damlamian, *On multivalued evolution equations in Hilbert spaces*, Israel J. Math. **12** (1972), 373–390.
5. V. Barbu, *Dissipative sets and nonlinear perturbed equations in Banach spaces*, Ann. Scuola Norm. Sup. Pisa **26** (1972), 365–390.
6. ___, *Continuous perturbations of nonlinear m-accretive operators in Banach spaces*, Boll. Un. Mat. Ital. **6** (1972), 270–278.
7. ___, *Nonlinear semigroups and differential equations in Banach spaces*, Noordhoff, Groningen, 1976.
8. Ph. Benilan, *Une remarque sur la convergence des semi-groupes non-linéaires*, C. R. Acad. Sci. Paris Sér. I. Math. **272** (1971), 1182–1184.
9. ___, *Solution intégrales d'équations d'évolution dans un espace de Banach*, C. R. Acad. Sci. Paris Sér. I Math. **274** (1972), 47–50.
10. ___, *Équations d'évolution dans un espace de Banach quelconque et applications*, Thèse, Orsay, 1972.
11. H. Brezis, *Monotonicity methods in Hilbert spaces and some applications to nonlinear partial differential equations*, Contributions to Nonlinear Functional Analysis (E. Zarantonello, ed.), Academic Press, San Diego, CA, 1970, pp. 101–156.
12. ___, *Semigroupes non-linéaires et applications*, Sympos. sur les Problèmes d'Évolution, Rome, (1970).
13. ___, *Propriétés régularisantes de cartains semigroupes nonlinéaires*, Israel J. Math. **9** (1971), 513–534.
14. ___, *On a problem of T. Kato*, Comm. Pure Appl. Math. **24** (1971), 1–6.
15. ___, *Équations et inéquations non linéaires dans les espaces vectoriels en dualité*, Ann. Inst. Fourier **18** (1968), 115–175.
16. ___, *Opérateurs maximaux monotones et semigroupes de contractions dans les espaces de Hilbert*, Math. Studies 5, North-Holland, Amsterdam, 1973.
17. H. Brezis, M. Crandall, and A. Pazy, *Perturbations of nonlinear maximal monotone sets*, Comm. Pure Appl. Math. **23** (1970), 123–144.
18. H. Brezis and A. Pazy, *Semigroups of nonlinear contractions on convex sets*, J. Funct. Anal. **6** (1970), 237–281.
19. ___, *Convergence and approximation of semigroups of nonlinear operators in Banach spaces*, J. Funct. Anal. **7** (1972), 63–74.
20. ___, *Accretive sets and differential equations in Banach spaces*, Israel J. Math. **8** (1970), 367–383.

21. H. Brezis and W. Strauss, *Semi-linear second-order elliptic equations in* L^1, J. Math. Soc. Japan **25** (1973), 565–590.

22. F. Browder, *Nonlinear equation of evolution*, Ann. of Math. (2) **80** (1964), 485–523.

23. ——, *Nonlinear equation of evolution and nonlinear accretive operators in Banach spaces*, Bull. Amer. Math. Soc. **73** (1967), 867–874.

24. ——, *Nonlinear operators and nonlinear equations of evolution in Banach spaces*, Proc. Sympos. Pure Math., vol. 18, part II, Amer. Math. Soc., Providence, RI, 1970.

25. B. Calvert, *Nonlinear evolution equations in Banach lattice*, Bull. Amer. Math. Soc. **76** (1970), 845–850.

26. B. Calvert and K. Gustafson, *Multiplicative perturbation of nonlinear m-accretive operators*, J. Funct. Anal. **10** (1972), 149–157.

27. J. Chambers, *Some remarks on the approximation of nonlinear semigroups*, Proc. Japan Acad. **46** (1970), 519–523.

28. J. Chambers and S. Oharu, *Semigroups of local Lipschitzian in a Banach space*, Pacific J. Math. **39** (1971), 89–112.

29. P. Chernoff, *Product formulas, nonlinear semigroups, and addition of unbounded operators*, Mem. Amer. Math. Soc., No. 140, Amer. Math. Soc., Providence, RI, 1970.

30. M. Crandall, *Differential equations on convex sets*, J. Math. Soc. Japan **22** (1970), 443–445.

31. ——, *Semigroups of nonlinear transformations in Banach spaces*, Contributions to Nonlinear Functional Analysis (E. Zarantonello, ed.), Academic Press, San Diego, CA, 1971, pp. 157–179.

32. ——, *A generalized domain for semigroup generators*, Proc. Amer. Math. Soc. **37** (1973), 434–440.

33. ——, *The semigroup approach to first order quasilinear equations in several space variables*, Israel J. Math. **12** (1972), 108–132.

34. ——, *An introduction to evolution governed by accretive operators*, Dynamical Systems— An International Symposium (L. Cesari, J. Hale and J. La Salle eds.), Academic Press, New York, 1976. pp. 131–165.

35. M. Crandall and L. Evans, *On the relation of the operator* $\partial/\partial\tau + \partial/\partial s$ *to evolution governed by accretive operators*, Israel J. Math. **21** (1975), 261–278.

36. M. Crandall and T. Liggett, *Generation of semi-groups of nonlinear transformations on general Banach spaces*, Amer. J. Math. **93** (1971), 265–293.

37. ——, *A theorem and a counterexample in the theory of semigroups of nonlinear transformations*, Trans. Amer. Math. Soc. **160** (1971), 263–278.

38. M. Crandall and A. Pazy, *Semigroups of nonlinear contractions and dissipative sets*, J. Funct. Anal. **3** (1969), 376–418.

39. ——, *Nonlinear evolution equations in Banach spaces*, Israel J. Math. **11** (1972), 57–94.

40. ——, *On accretive sets in Banach spaces*, J. Funct. Anal. **5** (1970), 204–217.

41. H. Debrunner and P. Flor, *Ein Erweiterungssatz für Monotone Mengen*, Arch. Math. **15** (1964), 445–447.

42. J. Dorroh, *Some classes of semi-groups of nonlinear transformations and their generators*, J. Math. Soc. Japan **20** (1968), 437–455.

43. ——, *A nonlinear Hille-Yosida-Phillips theorem*, J. Funct. Anal. **3** (1969), 345–353.

44. ——, *Semi-groups of nonlinear transformations*, Linear Operators and Approximation, Proc. Conf. in Oberwolfach, Internat. Ser. Numer. Math. **20** (1972), 33–53.

45. N. Dunford and J. Schwartz, *Linear operators Part*. I, Interscience, New York, 1958.

46. L. Evans, *Nonlinear evolution equations*, Dissertation, Univ. of Calif., Los Angeles, 1975.

47. J. Goldstein, *Approximation of nonlinear semigroups and evolution equations*, J. Math. Soc. Japan **24** (1972), 558–573.

48. T. Hayden and F. Massey, *Nonlinear holomorphic semigroups*, Pacific J. Math. **57** (1975), 423–439.

49. E. Hille and R. Phillips, *Functional analysis and semigroups*, Amer. Math. Soc. Colloq. Publ., vol. 31, Amer. Math. Soc., Providence, RI, 1957.

50. T. Kato, *Nonlinear semigroups and evolution equations*, J. Math. Soc. Japan **19** (1967), 508–520.

51. _____, *Accretive operators and nonlinear evolution equations in Banach spaces*, Nonlinear Functional Analysis, Proc. Sympos. Pure Math., vol. 18, part I, Amer. Math. Soc., Providence, RI, 1970, pp. 138–161.

52. _____, *Note on differentiability of nonlinear semigroups*, Global Analysis, Proc. Sympos. Pure Math., vol. 16, Amer. Math. Soc., Providence, RI, 1970, pp. 91–94.

53. N. Kenmochi, *Remarks on the m-accretiveness of nonlinear operators*, Hiroshima Math. J. **3** (1973), 61–68.

54. _____, *Nonlinear operators of monotone type in reflexive Banach spaces and nonlinear perturbations*, Hiroshima Math. J. **4** (1974), 229–263.

55. _____, *Pseudomonotone operators and nonlinear elliptic boundary value problems*, J. Math. Soc. Japan **27** (1975), 121–149.

56. _____, *Some nonlinear parabolic variational inequalities*, Israel J. Math. **22** (1975), 304–331.

57. N. Kenmochi and S. Oharu, *Difference approximation of nonlinear evolution equations*, Publ. Res. Inst. Math. Sci. **10** (1974), 147–207.

58. N. Kenmochi and T. Takahashi, *On the global existence of solutions of differential equations on closed subsets of a Banach space*, Proc. Japan Acad. **51** (1975), 520–525.

59. Y. Kobayashi, *A note on Cauchy problems of semi-linear equations and semi-groups in Banach spaces*, Proc. Japan Acad. **49** (1973), 514–519.

60. _____, *On approximation of nonlinear semigroups*, Proc. Japan Acad. **50** (1974), 729–734.

61. _____, *Difference approximation of Cauchy problems for quasi-dissipative operators and generation of nonlinear semigroups*, J. Math. Soc. Japan **27** (1975), 640–665.

62. Y. Kobayashi and K. Kobayasi, *On perturbation of nonlinear equations in Banach spaces*, Publ. Res. Inst. Math. Sci. **12** (1977), 709–725.

63. K. Kobayasi, *Note on approximation of nonlinear semi-groups*, Proc. Japan Acad. **50** (1974), 735–737.

64. Y. Kōmura, *Nonlinear semigroups in Hilbert space*, J. Math. Soc. Japan **19** (1967), 493–507.

65. _____, *Differentiability of nonlinear semi-groups*, J. Math. Soc. Japan **21** (1969), 375–402.

66. _____, *On nonlinear semigroups*, Sûgaku **25** (1973), 148–160. (Japanese)

67. Y. Konishi, *Nonlinear semigroups in Banach lattices*, Proc. Japan Acad. **47** (1971), 24–28.

68. _____, *A remark on perturbation of m-accretive operators in Banach space*, Proc. Japan Acad. **47** (1971), 452–455.

69. _____, *Some examples of nonlinear semigroups in Banach lattices*, J. Fac. Sci. Univ. Tokyo Sect. IA Math. **18** (1972), 537–543.

70. _____, *Une remarque sur la perturbation d'opérateurs m-accrétifs dans un espace de Banach*, Proc. Japan Acad. **48** (1972), 157–160.

71. _____, *On $u_t = u_{xx} - F(u_x)$ and the differentiability of the nonlinear semigroup associated with it*, Proc. Japan Acad. **48** (1972), 281–286.

72. _____, *Sur la compacité des semigroupes non linéaires dans les espaces de Hilbert*, Proc. Japan Acad. **48** (1972), 278–280.

73. _____, *Sur un systèm dégenéré dés équations paraboliques semilinéaires avec les conditions aux limites nonlinéaires*, J. Fac. Sci. Univ. Tokyo Sect. IA Math. **19** (1972), 353–361.

74. _____, *On the nonlinear semi-groups associated with $u_t = \Delta\beta(u)$ and $\varphi(u_t) = A_u$*, J. Math. Soc. Japan **25** (1973), 622–628.

75. S. Kružkov, *First order quasilinear equations in several independent variables*, Math. USSR Sb. **10** (1970), 217–243.

76 T. Kurtz, *Convergence of sequences of semigroups of nonlinear operators with an application to gas kinetics*, Trans. Amer. Math. Soc. **186** (1973), 259–272.

77. J. Lions, *Quelques méthodes de résolution des problèmes aux limites non linéaires*, Dunod and Gauthier-Villars, Paris, 1969.

78. D. Lovelady and R. Martin, Jr., *A global existence theorem for a nonautonomous differential equation in a Banach space*, Proc. Amer. Math. Soc. **35** (1972), 445–449.

79. R. Martin, Jr., *The logarithmic derivative and equations of evolution in a Banach space*, J. Math. Soc. Japan **22** (1970), 411–429.

80. _____, *A global existence theorem for autonomous differential equations in Banach space*, Proc. Amer. Math. Soc. **26** (1970), 307–314.

81. _____, *Generating an evolution system in a class of uniformly convex Banach spaces*, J. Funct. Anal. **11** (1972), 62–76.

82. _____, *Differential equations on closed subsets of a Banach space*, Trans. Amer. Math. Soc. **179** (1973), 399–414.

83. K. Maruo, *On some evolution equations of subdifferential operators*, Proc. Japan Acad. Ser. A Math. Sci. **51** (1975), 304–307.

84. F. Massey, *Analyticity of solutions of nonlinear evolution equations*, J. Differential Equations **22** (1976), 416–427.

85. _____, *Semilinear parabolic equations with L^1 initial data*, Indiana Univ. Math. J. **26** (1977), 399–412.

86. K. Masuda, *Evolution equations*, Kinokuniya, Tokyo, 1975. (Japanese)

87. J. Mermin, *An exponential limit formula for nonlinear semigroups*, Trans. Amer. Math. Soc. **150** (1970), 469–476.

88. G. Minty, *Monotone (nonlinear) operators in a Hilbert space*, Duke Math. J. **29** (1962), 341–346.

89. _____, *On a generalization of a direct method of the calculus of variations*, Bull. Amer. Math. Soc. **73** (1967), 315–321.

90. I. Miyadera, *Note on nonlinear contraction semi-groups*, Proc. Amer. Math. Soc. **21** (1969), 219–225.

91. _____, *On the convergence of nonlinear semi-groups*, Tôhoku Math. J. **21** (1969), 221–236.

92. _____, *On the convergence of nonlinear semi-groups. II*, J. Math. Soc. Japan **21** (1969), 403–412.

93. _____, *Some remarks on semi-groups of nonlinear operators*, Tôhoku Math. J. **23** (1971), 245–258.

94. _____, *Generation of semi-groups of nonlinear contractions*, J. Math. Soc. Japan **26** (1974), 389–404.

95. I. Miyadera and S. Oharu, *Approximation of semi-groups of nonlinear operators*, Tôhoku Math. J. **22** (1970), 24–47.

96. I. Miyadera and Y. Kobayashi, *Convergence and approximation of nonlinear semi-groups*, Functional Analysis and Numerical Analysis, Japan-France Seminar, Tokyo and Kyoto, 1976 (H. Fujita, ed.), Japan Society for the Promotion of Science, Tokyo, 1978, pp. 272–295.

97. I. Miyadera, *Functional analysis*, Rikogakusha, 1972. (Japanese)

98. J. Moreau, *Proximité et dualité dans un espace hilbertien*, Bull. Soc. Math. France **93** (1965), 273–299.

99. J. Neuberger, *An exponential formula for one parameter semi-groups of nonlinear transformations*, J. Math. Soc. Japan **18** (1966), 154–157.

100. _____, *Product integral formulas for nonlinear nonexpansive semi-groups and nonexpansive evolution systems*, J. Math. Mech. **19** (1969), 403–409.

101. _____, *Quasi-analyticity and semigroups*, Bull. Amer. Math. Soc. **78** (1972), 909–922.

102. S. Oharu, *Note on the representation of semi-groups of nonlinear operators*, Proc. Japan Acad. Ser. A Math. Sci. **42** (1966), 1149–1154.

103. _____, *A note on the generation of nonlinear semigroups in a locally convex space*, Proc. Japan Acad. Ser. A Math. Sci. **43** (1967), 847–851.

104. _____, *On the generation of semigroups of nonlinear contractions*, J. Math. Soc. Japan **22** (1970), 525–550.

105. _____, *On the semigroup approach to nonlinear partial differential equations of first order*. I, J. Math. Anal. Appl., to appear.

106. _____, *Nonlinear semigroups and finite-difference method*, Functional Analysis and Numerical Analysis, Japan-France Seminar, Tokyo and Kyoto, 1976 (H. Fujita ed.), Japan Society for the Promotion of Science, Tokyo, 1978, pp. 383–402.

107. S. Oharu and T. Takahashi, *A convergence theorem of nonlinear semigroups and its application to first order quasilinear equations*, J. Math. Soc. Japan **26** (1974), 124–160.

108. S. Ōuchi, *On the analyticity in time of solutions of initial boundary value problems for semilinear parabolic differential equations with monotone nonlinearity*, J. Fac. Sci. Univ. Tokyo Sect. IA Math. **20** (1974), 19–41.

109. N. Pavel, *Sur certaines équations différentielles abstraites*, Boll. Un. Mat. Ital. **6** (1972), 397–409.

110. _____, *Équations non-linéaires d'évolution*, Mathematica **14** (1972), 289–300.

111. A. Pazy, *Semigroups of nonlinear contractions in Hilbert spaces*, C.I.M.E., Varenna, 1970, Ed. Cremonese, 1971, pp. 343–430.

112. M. Pierre, *Generation et perturbations de semi-groups de contractions nonlinéaires*, Thèse de Docteur de 3 é cycle, Université de Paris VI, 1976.

113. _____, *Perturbations localement lipschitziennes et continues d'opérateurs m-accrétifs*, Proc. Amer. Math. Soc. **58** (1976), 124–128.

114. E. Poulsen, *Evolution equation, semigroups and product integrals*, Uppsala Univ., 1970.

115. S. Rasmussen, *Nonlinear semi-groups, evolution equations and product integral representation*, Various Pub. Ser., No. 2 Aarhus Univ., Aarhus, 1971/72.

116. H. Riedl and G. Webb, *Relative boundedness conditions and the perturbation of nonlinear operators*, Czech. Math. J. **24** (1974), 585–597.

117. R. Rockafeller, *On the maximal monotonicity of subdifferential mappings*, Pacific J. Math. **33** (1970), 209–216.

118. _____, *Monotone operators associated with saddle-functions and minimax problem*, Nonlinear Functional Analysis, Proc. Sympos. Pure Math., vol. 18, part I, Amer. Math. Soc., Providence RI, 1970, pp. 241–250.

119. K. Sato, *A note on nonlinear dispersive operators*, J. Fac. Sci. Univ. Tokyo Sect. IA Math. **18** (1972), 456–473.

120. I. Segal, *Nonlinear semigroups*, Ann. of Math. (2) **78** (1963), 339–364.

121. T. Takahashi, *Convergence of difference approximation of nonlinear evolution equations and generation of semigroups*, J. Math. Soc. Japan **28** (1976), 96–113.

122. J. Watanabe, *Semi-groups of nonlinear operators on closed convex sets*, Proc. Japan Acad. Ser. A Math. Sci. **45** (1969), 219–223.

123. _____, *On nonlinear semigroups generated by cyclically dissipative sets*, J. Fac. Sci. Univ. Tokyo Sect. IA Math. **18** (1971), 127–137.

124. _____, *On certain nonlinear evolution equations*, J. Math. Soc. Japan **25** (1973), 446–463.

125. _____, *Evolution equations associated with subdifferentials, recent developments in Japan*, Functional Analysis and Numerical Analysis, Japan-France Seminar, Tokyo and Kyoto, 1976 (H. Fujita ed.), Japan Society for the Promotion of Science, Tokyo, 1978, pp. 525–539.

126. G. Webb, *Representation of semigroups of nonlinear nonexpansive transformations in Banach spaces*, J. Math. Mech. **19** (1969), 159–170.

127. _____, *Continuous nonlinear perturbations of linear accretive operators in Banach spaces*, J. Funct. Anal. **10** (1972), 191–203.

128. _____, *Nonlinear perturbations of linear accretive operators in Banach spaces*, Israel J. Math. **12** (1972), 237–248.

129. _____, *Accretive operators and existence for nonlinear functional differential equations*, J. Differential Equations **14** (1973), 57–69.

130. _____, *Autonomous nonlinear functional differential equations and nonlinear semigroups*, J. Math. Anal. Appl. **46** (1974), 1–12.

131. _____, *Functional differential equations and nonlinear semigroups in L^p-spaces*, J. Differential Equations **20** (1976), 71–89.

132. K. Yosida, *Functional analysis*, 4th ed., Springer-Verlag, Berlin and New York, 1974.

Subject Index

COPYING AND REPRINTING. Individual readers of this publication, and non-profit libraries acting for them, are permitted to make fair use of the material, such as to copy a chapter for use in teaching or research. Permission is granted to quote brief passages from this publication in reviews, provided the customary acknowledgment of the source is given.

Republication, systematic copying, or multiple reproduction of any material in this publication (including abstracts) is permitted only under license from the American Mathematical Society. Requests for such permission should be addressed to the Manager of Editorial Services, American Mathematical Society, P.O. Box 6248, Providence, Rhode Island 02940.

The owner consents to copying beyond that permitted by Sections 107 or 108 of the U.S. Copyright Law, provided that a fee of $1.00 plus $.25 per page for each copy be paid directly to the Copyright Clearance Center, Inc., 27 Congress Street, Salem, Massachusetts 01970. When paying this fee please use the code 0065-9282/92 to refer to this publication. This consent does not extend to other kinds of copying, such as copying for general distribution, for advertising or promotion purposes, for creating new collective works, or for resale.

Recent Titles in This Series

(Continued from the front of this publication)

(See the AMS catalog for earlier titles)